The Imaginary of Animals

This book explores the phenomenon of animal imagination and its profound power over the human imagination. It examines the structural and ethical role that the human imagination must play to provide an interface between humans' subjectivity and the real cognitive capacities of animals.

The book offers a systematic study of the increasing importance of the metaphors, the virtual, and figures in contemporary animal studies. It explores human-animal and real-imaginary dichotomies, revealing them to be the source of oppressive cultural structures. Through an analysis of creative, playful and theatric enactments, and mimicry of animal behaviors and communication, the book establishes that human imagination is based on animal imagination. This helps redefine our traditional knowledge about animals and presents new practices and ethical concerns in regard to the animals. The book strongly contends that allowing imagination to play a role in our relation to animals will lead to the development of a more empathetic approach towards them.

Drawing on works in phenomenology, contemporary animal philosophy, as well as ethological evidence and biosemiotics, this book is the first to rethink the traditional philosophical concepts of imagination, images, the imaginary, and reality in the light of a zoocentric perspective. It will appeal to philosophers, scholars and students in the field of animal studies, as well as anyone interested in human and non-human imaginations.

Annabelle Dufourcq is Associate Professor in Philosophy at Radboud University. She researches and teaches in the areas of contemporary continental philosophy and animal studies. She is the author or editor of several books on the relation between the real and the imaginary from a phenomenological perspective.

Routledge Human–Animal Studies Series

The new *Routledge Human–Animal Studies Series* offers a much-needed forum for original, innovative and cutting-edge research and analysis to explore human–animal relations across the social sciences and humanities. Titles within the series are empirically and/or theoretically informed and explore a range of dynamic, captivating and highly relevant topics, drawing across the humanities and social sciences in an avowedly interdisciplinary perspective. This series will encourage new theoretical perspectives and highlight ground-breaking research that reflects the dynamism and vibrancy of current animal studies. The series is aimed at upper-level undergraduates, researchers and research students as well as academics and policy-makers across a wide range of social science and humanities disciplines.

Series edited by **Henry Buller**, *Professor of Geography, University of Exeter, UK*

Historical Animal Geographies
Edited by Sharon Wilcox and Stephanie Rutherford

Animals, Anthropomorphism, and Mediated Encounters
Claire Parkinson

Horse Breeds and Human Society
Purity, Identity and the Making of the Modern Horse
Edited by Kristen Guest and Monica Mattfeld

Immanence and the Animal
A Conceptual Inquiry
Krzysztof Skonieczny

The Imaginary of Animals
Annabelle Dufourcq

For more information about this series, please visit: www.routledge.com/ Routledge-Human-Animal-Studies-Series/book-series/RASS

The Imaginary of Animals

Annabelle Dufourcq

Routledge
Taylor & Francis Group

LONDON AND NEW YORK

First published 2022
by Routledge
2 Park Square, Milton Park, Abingdon, Oxon OX14 4RN

and by Routledge
605 Third Avenue, New York, NY 10158

Routledge is an imprint of the Taylor & Francis Group, an informa business

British Library Cataloguing-in-Publication Data
A catalogue record for this book is available from the British Library

Library of Congress Cataloging-in-Publication Data
A catalog record has been requested for this book

ISBN: 978-0-367-77297-0 (hbk)
ISBN: 978-0-367-77298-7 (pbk)
ISBN: 978-1-003-17070-9 (ebk)

Typeset in Times NR MT Pro
by KnowledgeWorks Global Ltd.

Contents

Figures

Acknowledgments

My first thoughts go to Veronica Vasterling, Louise Westling, Renaud Barbaras, Karel Novotný, and Ted Toadvine who gave me invaluable support during the writing of this book. My survival kit in academia was designed by following their example and advice. They are the most inspiring researchers and human beings I have met in my academic and philosophical journey

Special thanks to Louise (Molly) Westling for her careful guidance, enthusiastic encouragement, and insatiable curiosity throughout the writing process. She was a friendly yet demanding first reader and kept me working by offering thoughtful and helpful comments each step of the way. I wouldn't have written this book without her and I am lucky and proud to be her friend.

This book was also made possible by all the colleagues who, through their feedback, helpful suggestions and support have been at my side during the writing process: in particular Annemie Halsema, Karen Vintges, Cris van der Hoek, Desirée Verweij; the members of the OZSW animal studies study group; Carlos Pereira, Jacquemin Piel, the "Institut du Cheval et de l'Équitation Portugaise and the research team "horse cognition project" who kindly allowed me to observe their work and offered stimulating suggestions for my research; the members of my HDR committee whose comments and suggestions have also been a great help in the process of completing this book: Florence Burgat (pres.), Etienne Bimbenet (dir.), Renaud Barbaras, Bruce Bégout, and Ted Toadvine; my colleagues at Radboud University, especially in the department of fundamental philosophy; my colleagues at Charles University, departments of philosophy and of French and German philosophy; the professors and grad students of the department of philosophy at the University of Oregon for welcoming me and engaging in so many stimulating discussions during my stay in 2011-2012; eventually, all the researchers with whom I have discussed a wide variety of topics related to this book: Anna Barseghian, Jan Bierhanzl, Florence Cayemaex, Jakub Čapek, Leandro Cardim, Carlie Coenen, Anne Coignard, Irene Delodovici, Alex de Campos Moura, Josef Fulka, Anne Gleonec, Jean-Christophe Goddard, Maria Gyemant,

Maud Hagelstein, Urszula Idziak, Robert Karul, Alice Koubová, Stefan Kristensen, Mariana Larison, Bonnie Mann, Anton Markoš, Eva Meijer, Martin Nitsche, Delia Popa, Marcus Sacrini Ayres Ferraz, Beata Stawarska, Wojciech Starzyński, Jaroslava Vydrová.

Most crucial to the maturation of this book are also all the students at Charles University and Radboud University whose interest in animal philosophy and environmental studies have constantly revived mine. Our inspiring discussions during several seminars I have taught on the imaginary of animals offered me the opportunity to polish and refine my project.

I also gratefully acknowledge the valuable language editing assistance of Louise Westling and Annalisa Zox-Weaver during the last stages of the publication process. I would like to express my appreciation as well to the efficient and helpful staff at Routledge: Faye Leerink and Henry Buller, for supporting this project; Nonita Saha and Roshni Bhandari for their kind and expert assistance in the production process. I must also thank various institutions that helped with time and money to finish this project: Radboud University, Charles University, and, with a substantial financial support, The Czech Science Foundation: This book was written as part of the grant-funded project GACR "Life and Environment. Phenomenological Relations between Subjectivity and Natural World" (GAP15-10832S).

I wish to thank Mariana Carasco Berge for being a very old wizard in disguise and an exceptional friend. Our discussions have also been a great source of inspiration for this book. Many thanks to the friends whose high-spirited and caring presence makes life brighter: Lalia Besse, Mireille Duchastelle, Aurélie François, Valérie Founaud, Didier Besse, Xavier Brugiroux, Sébastien Devillers, George Wickes. Finally, I wish to thank my parents for their loving support and their innate and acquired capacity to kindly bear with me. Their confidence in me keeps me going. I dedicate this book to my mother and my grandmother.

Introduction
The imaginary: A human-non-human-animal interface

> And the smoke of their torment rises for ever and ever:
>> and they have no rest day nor night, who worship the beast and its
>> mage
>>> *Revelation* 14.11

§1 Animal poetry

The Book of Revelation alludes, in a thinly veiled fashion, to ancient rites where beasts and their manifold images sustain, inspire, and manifest deities. The monumental gate of Ishtar and the Processional Way in Babylon, lined with hundreds of images of wild animals and hybrid creatures, lions, aurochs, and dragons, come to mind:[1] An overabundant, vivid imagery in which the beast and its image seem to reinforce each other and become glorious through their predestined union. Beasts and their images certainly survived after the end of the era of magic rites but in a drastically contained and divisive form. They came under the law that proclaims man to be the only image of God, makes the sacrificial lamb its emblem, and attempts to discriminate between holy and evil images. The predestined association between animals and images remains as an obscure trace of archaic beliefs and leaves us with a question: What is the strange bond secretly uniting the beast and its image?

"Beasts are subject to the force of imagination,"[2] Michel de Montaigne contends in *Of the Force of Imagination*. Against the backdrop of a modern conceptuality, Montaigne's definition of imagination itself is challenging, and yet, as we will see, extremely fruitful. Imagination, in Montaigne's view, consists in a relation to the absent or the unreal, a relation that is *endured* in a fundamentally embodied way. "Horses kick and whinny in their sleep" and dogs can "die of grief for the loss of their masters."[3] Even more daringly, Montaigne refers to "the hares and partridges that the snow turns white upon the mountains,"[4] phenomena he regards as evidence of the force of imagination in animals. The bodies of hares and partridges are, as it were, amazingly affected by the look of the snow and the predators' perception. Others' perception is incorporated to the body. A remote affinity seems to link animal bodies, the way their environment look, and, as it were, the mind of observers.

Rationalism has tried to exorcise such alchemical wonders and took action to save humans from their animal nature as well as their fondness for flights of fancy, both being in fact secret accomplices, as I will contend. As the human-animal dichotomy came in for radical criticism, with the so-called crisis of modernity, the consideration of the link between imagination and animal life, in human and non-human animals,[5] made a comeback on the philosophical and scientific scene, although in a still understated and protean form.

Many thinkers have cited the connection between animals and the imaginary field but have done so rather allusively, in quite a puzzling manner. The alliances emerge in rather various contexts—literary, artistic, philosophical as well as epistemological. Adding to the difficulty of fathoming the issue at stake is the fact that the "imaginary of animals" constitutes an interesting Möbius loop—two surfaces that are one: The link between our imagination of animals and animals' imagination remains to be analyzed.

Merleau-Ponty, as well as Gaston Bachelard[6] and John Berger,[7] for instance, studies the power of animals in our imagination. An abundant body of literature considers the manner in which humans across different cultures represent and fantasize animals, and how such fantasies shape—and are shaped by—their myths, narratives, and works of art.[8] But I want to take a step further and turn the issue upside down. At stake is not only the way humans imagine animals, but how the latter inspire such an imaginative process and, indeed, take part in it. How and what do animals imagine? Donna Haraway and Merleau-Ponty, in particular, tackle the imaginary of animals on a more ontological level, developing a crucial theme that also resides at the heart of Latour's works:[9] That the being of animals is essentially in process. The question is then: To what extent do the future developments of animals depend on our imagination and on a still-to-be-corroborated, still to-be-defined non-human animal imagination? In the same vein the new trend of zoopoetics engages non-human animals as poetry-makers; "zoopoets" intend to demonstrate that non-human animals actively contribute to human poetry and that they essentially consist in the entanglement between "real animals" and "literary animals."[10] More broadly, the academic field of animal studies, such as it has developed in the early 2000s, is interestingly defined by cooperation and exchanges among biology, ethology, ecology, biosemiotics, philosophy, literature studies, critical theory (ecocriticism), and art. This cooperation is not a debate or a competition to determine who is best entitled to understand animals, but is rather seen in the field of animal studies as a necessary transversality: Non-human animals are to be *known-imagined*.

In *What Do Pictures Want?*, W.J.T. Mitchell highlights the odd recurrence—from Plato to Latour—of the idea that images are living beings.[11] They have "legs,"[12] proliferate, and "capture the life" of their model. Strangely enough—and even more radically—in ancient Greek, the word *zoographia* (animal drawing) designated pictures and the art of painting in general.[13] As such, all images are (*as if*) animals. My contention is that this esoteric idea can be turned on its head, and thus completed and even intensified, beyond a mere *as if* human way of speaking, precisely in order to uncover the condition of possibility of this *as if.*

Two issues are at stake here. First: Do non-human animals have access to an imaginary realm and what can this possibly mean? Secondly: What is the link between human imagination and the question of animal imagination?

Humans make abundant use of their imagination to refer to animals. We quite spontaneously transpose ourselves into other animals' perspectives. Anthropologists and biologists regularly examine whether it is possible to transpose human capacities, such as subjectivity, affectivity, imagination, cognition, and moral conscience, to other animals. Animals are an inexhaustible source of inspiration for myths, popular imagery, and art. The human imaginary is obsessed with animals and hybrids. But, classically, the second dimension—the importance played by human imagination in our relationship with animals—is regarded as impeding proper consideration of the former aspect: Animals' imagination. Derrida brings us face-to-face with this conundrum when he claims that his discourse about animals is chimerical to an unprecedented degree,[14] after emphasizing that the cat he is talking about is "a real cat."[15] Also, when Derrida defines animal thought as poetry,[16] to what extent is this "animal poetry" a projection stemming from human poetry? If one takes the fear of anthropomorphism seriously —and given that imagination is essentially characterized by its free creativity—it may become necessary to address both issues separately: (a) How do humans creatively imagine animals? How are their own fears, desires, and capacities expressed through such fantasies? And (b) what can we observe and know about the animals' access to the virtual?

The analysis of the imaginary of animals would be *clearer*, to be sure, if we stuck to this differentiated approach. But then a study of the knot between these two strands would be missing. The passages actually trodden by those who adventurously made their way from the human imaginary to the animals' imagination and vice versa remain to be studied. Thus, it becomes possible to assess whether we can encounter other animals, in a genuine way, through the "human" imaginary. My focus will be *the imaginary of animals* defined as the interface between two correlative dimensions: The imaginary or imagery of animals in human culture and the imaginary world of animals themselves, the animals' imaginary.

As for "the imaginary," I begin by designating three inseparable facets:

1 A "pictorial"[17] form of meaning that is ambiguous, emerges through associations, shifts, or metaphors, and cannot be perfectly fixed within clear-cut concepts.
2 A form of thought, understanding, and communication that obviously sticks to concrete material supports (colors and forms, cries, diacritical sounds, behaviors, morphology) and, therefore, includes blind spots, an irreducible opacity, or, to put it in Derridean terms, trace and dissemination. Thus, as Derrida argues in "Che cos'è la poesia?,"[18] I must learn a poem by heart, for I cannot simply aim, through its materiality, at a clearly indicated meaning. I must repeat it and *carry it with me* first as a foreign body, the spiky body of a hedgehog. I have to let its style, its structure, its musicality, and its rhythm change me and enable me to see the world with new eyes.
3 The imaginary is eventually defined by creativity. Tentative leaps into the unknown and, maybe, the unreal, are solicited by the ambiguity and the absence of a *de jure* clear conceptual ground. Understanding a poem or a picture, for instance, always implies a process of reverie as well as the more active creation of image/word associations, interpretations, and transpositions into possible worlds opened by the poem or the picture. We thus hover between the actual and virtualities. This understanding process never ends, and we constantly risk getting lost along the way.

In a similar manner, we continuously invent new ways of understanding other animals and communicating with them. These exchanges take the form of a wild multiplication of myths, transposition processes, art works, stories of talking animals, and images of human-animal hybrids. Moreover, remarkably enough, the biosemiotician Timo Maran emphasizes that ecocodes themselves are quite poetical: It is plausible to assume that codes on the ecological level are not strict regulations, but rather ambiguous and fuzzy linkages based on analogies and correspondences. (...) Ecological codes do not resemble human linguistic codes or algorithms, but are rather like archetypal imagery or patterns."[19] Thus, the oneiric dimension of animals has also begun to make its way into the realm of life sciences, in conjunction with revived debates about the fundamental paradigm that should govern the latter.

And indeed, is poetry only on the side of the human interpreter? Let me state the problem in a less abstract way by leaving the floor to poet dolphins.

Ethologists observed that, when bottlenose dolphins and Guyana dolphins, two different species, come into contact, they change their vocalizations.[20] The vocalizations of the former are usually long and have a lower frequency; with Guyana dolphins, it is the opposite. But when the two species swim side by side, bottlenose dolphins use more high-pitched, short sounds while Guyana dolphins make lower-frequency sounds. Guyana dolphins may seek to coax bottlenose dolphins—who indeed dominate them physically—but this account does not explain why bottlenose dolphins also adjust their signals.

Such interspecific communication echoes Derrida's description of genuine poetical understanding: It may be claimed, in effect, that Guyana dolphins and bottlenose dolphins attempt to find a hybrid or intermediate language through progressive adjustment to each other's vocalizations. This interaction between Guyana dolphins and bottlenose dolphins would therefore be a creative and tentative form of expression that engages the matter of existing signs into a reactivated diacritical process,[21] so that a to-be-developed but not-clearly-defined-in-advance exchange of meaning emerges.

Speaking, in the latter case, of an attempt to create a new language may, however, sound like a rash extrapolation. A second example should at least unsettle even the most skeptical minds. Dolly was a bottlenose dolphin who lived in the Port Elizabeth Oceanarium, South Africa. Tayler and Saayman reported the following imaginative behavior:

> A cloud of cigarette smoke was once deliberately released against the glass as Dolly was looking in through the viewing port. The observer was astonished when the animal immediately swam off to its mother, returned and released a mouthful of milk which engulfed her head, giving much the same effect as had the cigarette smoke.[22]

Dolly thus proved able to combine, in a startling behavior, an impressive number of imaginative skills. Her "milk smoking"[23] involves imitation, analogical reasoning, creativity, and the attempt to establish communication. This message—sent adventurously to a potential, uncertain, and quite mysterious interlocutor—also displays a touching aesthetic sense. Here, again, in the absence of any conventional common language, communication nevertheless takes place at the level of raw matter. A certain gesture, which was not a sign in the first place, is taken up and performed anew, with a mirror effect and an obvious process of analogical transposition. It thus becomes a response and a sign. Dolly's "milk smoking" refers to the man's "cigarette smoking" and has seemingly no other reason for being. Because of the transposition—or, literally, metaphor—a meaning is at once brought to the fore and blurred. Milk smoking and cigarette smoking are not identical. The analogical bridges that make their association possible do not render such a connection necessary: They must have been thrown in a quite daring move. Accordingly, they inaugurate a series of analogical associations. A playful artist may extend the metaphor, multiply the variations on the same theme, and respond with original sounds, movements, colors, and forms. This open-ended process was already germinating in the first bold de-centration[24] performed by Dolly. Indeed, because such a decentration intrinsically addresses the gap between milk-smoking and cigarette-smoking, it aims at a core of contingency or nothingness that is beyond any positive reality.

However, one may object, these are only two examples. In fact, the number of observed and meticulously documented cases presented in Kaufman and Kaufman's *Animal Creativity and Innovation* is impressive. But another question could be raised: Is the reference to animal *imagination*—or even at least to Dolly's imagination—anything more than pure speculation? Does imagination necessarily imply a mental faculty and/or self-conscious experience? What may give one the right to speak of the dolphins' intentions, attempts to communicate, or capacity to aim at the unreal? But the reductive-objective rival claim is no less extrapolation than a subjective-poetic interpretation. It is not eminently legitimate or more cautious to assert that animal behaviors are always ultimately the mere product of a—maybe now still hidden and unknown—series of blind physiological reactions and of a blind process of adaptation. Such a stance has, indeed, all the hallmarks of an *ad hoc* explanation, as pointed out by the proponents of a subjective-poetic view.[25] Let us concede that, strictly speaking, this is not a refutation. It is always possible to *presuppose* that a hidden blind mechanism is the sufficient cause of a phenomenon. The ever-renewed quest for *ad hoc* elements that will save this principle at all costs is not in itself illegitimate. What is at stake here, at the scientific level, is fundamentally a paradigm issue. Several paradigms are possible; they may bring specific benefits and present various drawbacks. I want to examine how the subjective-poetic interpretation can be legitimized at the epistemological level, as well as, more fundamentally, at the ontological level.

When we invent new intonations while trying to communicate with a non-human animal, when we attempt to imagine what it is like to be a bat, we certainly throw these interpretations into the blue, which does not necessarily mean that such interpretations are arbitrary and pointless. Indeed, that these interpretations may very well be the most accurate way of encountering animals *as such*—namely as subjects, poets, and imaginative beings—is the claim I want to flesh out and advocate. Only imagination can know imagination. In other words, the "human" imaginary is an interface that affords access to the other animals' imaginary: They must be correlative.

For introductory purposes, I will first look at the differences between the imaginary of animals and an objective/rationalist approach. Secondly, I will clarify the concepts of imagination and the imaginary to be used in the body of the discussions. Eventually, I will return to the claim at stake: Why is a revolution in the concept of knowledge of animals necessary and why should the imaginary be a key in such a revolution?

§2 Two versions of animality: Powerful myths vs mute processes and mechanisms

The starting point for this reflection is the striking contrast between, *on the one hand*, the imaginary (i.e., the imaginative tradition) of talking, metamorphic and anthropomorphic animals—as well as of zoomorphic humans—in popular imageries, myths, or art, and, *on the other hand*, a rationalist/objective representation of animals as silent black boxes locked up in determinate and predictable patterns of behavior.

In *The Animal That Therefore I Am*, Derrida has highlighted how philosophers of the Western tradition have almost unanimously advocated, or simply taken for granted, the radical difference between human beings and (all) animals—in fact quite broad and clunky categories. Although, in line with Derrida, many contemporary philosophical works question these views and show how strangely *blind* they are, the issue of *the* anthropological difference remains a burning point of contention.[26]

Also at issue here is a challenged but still dominant scientific take on life. The enduring largely predominant paradigm of objective sciences does not favor the integration of creative agency, selfhood, and subjective perspectives on the world (also called *phenomenality*) into rigorous and clear scientific analysis. The successful biochemical approach, which studies life on a molecular scale, understandably exempts the researchers from considering the relevance of such an integration of a phenomenal perspective. However, several contemporary scientists in ethology and primatology (Donald Griffin,[27] Jane Goodall,[28] Sue Savage-Rumbaugh,[29] Françoise Wemelsfelder,[30] Frans De Waal,[31] for instance) as well as in fundamental biology and genetics (Richard Lewontin,[32] Susan Oyama,[33] Jesper Hoffmeyer,[34] and Denis Walsh,[35] among others) claim that such integration is not impossible, and that the relevance of biochemistry does not make the *reduction* of life processes to chemistry and molecules necessary. But they also encounter a strong resistance,[36] and some contemplate the need for a paradigm shift.[37] The concept of the genetic *program* as well as a Neo-Darwinist conception of evolution prevails, reducing the existence of living beings to a morphology and a set of behaviors determined by a strictly one-way process of information transfer and a mechanical selection of the best genotypes.[38] An insurmountable distance is thus created between natural sciences and any approach that would seek to take into account a creative and subjective dimension in animals.

To be sure, the same creative and subjective dimension in human beings is also regarded as problematic by this objectivist paradigm. The means to gain access to human subjectivity through the hermeneutical method, for instance, are already well-developed, mainly on the basis of the humanist philosophical tradition. However, this anthropocentric origin entails that the existing methods developed to counter the objectivist paradigm predominantly resists an application to the animal realm. *Imagination is proper to humans*: Such is in effect a classical claim in philosophy, including in approaches that challenge rationalism. As a matter of fact, Martin Heidegger famously stated that animals are not opened to the world[39] and Lacan granted animals an "imagination" that is actually determined by instinct and is thus hardly worthy of the name.[40] Even Hans Jonas, whose contribution to the phenomenology of animality constitutes a landmark in contemporary philosophy, falters when he collides with the issue of animal imagination. Although Jonas consistently worked to overcome an objectivist approach to animals and to abolish a form of scrutiny that holds them at a distance, he eventually restores what he calls the "metaphysical gap" between humans and animals by stating that only humans create images.[41] It is, Jonas argues, only in the human realm that imagination, namely the "freedom (...) of distance and control,"[42] can be found.

In the absence of instinct and the openness to the virtual, the Western rationalist philosophical tradition found the alleged essential features of human beings. The myth of Prometheus already suggested this idea. The human allegedly came to the world as an indigent creature: Deprived of instinct, natural tools, and weapons; she was, on the other hand, infinitely rich in potentialities. The human is identified by her "perfectibility,"[43] her plasticity, and her ability to develop original faculties. The link between imagination and such an existential openness to the virtual is also a trope. Jean-Paul Sartre has written particularly illuminating texts on this subject. Imagination is not an organ that a living being may or may not happen to possess, in the same manner—for example—that it could possess claws, fangs, hands, or eyes. Ontologically speaking, imagination is first and foremost the being-beyond-oneself, the non-identity, the openness to what is not, which Sartre also named "for-itself" or "consciousness." As bluntly stated by Sartre in *L'imaginaire,* "imagination is not an empirical power added to consciousness, but is the whole of consciousness as it realizes its freedom."[44]

In addition, the concept of human subjectivity is primordially rooted in a first-person conscious experience *of ourselves*, while animal subjectivity, if any, must be approached in a double indirect manner: I am not in their mind, and we share a very partial common ground of similarity between our bodies and theirs, our behaviors and theirs; hence, the fear of anthropomorphism.

The representation of animals as machine-like, or black boxes or, at least, entities that may be *pragmatically considered* but not rigorously *known* as thinking subjects,[45] is commonly considered rational, realistic, and based on a stringent analysis.

Nevertheless, on the other hand, in the imaginary field —and in a remarkably consistent manner—through a tremendous number of recurring images, animals feel, think, and speak to us and to each other. They develop complex political interactions in Aesop's or La Fontaine's fables, or in Boulle's

Planet of the Apes, for instance. They are also secret-keepers and help us understand the world and the meaning of existence; the doctrine of the Eternal Recurrence is thus learned with the help of the donkey, the camel, the lion, the eagle, and the snake in Nietzsche's *Thus Spoke Zarathustra*. In totemism and in many fairy tales, non-human animals possess powers that humans do not have. They are not always friendly, far from it: The imaginary of animals also involves the recurring fantasy of beasts and devouring monsters. In the imaginary field, animals are both our closest friends and a horrifying threat: Contradictions have a place in the imaginary, which is not the least of its assets. This book considers the meaning of these contradictions, but one can easily sense that such a stifling terror must also include a dimension of unbearable proximity and even intimacy. A last trait: The imaginary field swarms with human-animal metamorphoses and hybrids, which convey ambiguous values as well. Zoo-anthropomorphic creatures are at least *unheimlich*, disquieting, often frightening, sometimes horrifying, but they are also commonly presented as linked to the realm of the sacred, as introducing a process of eye-opening and power-increasing conversion. This revelatory nature of hybrids manifests in shamanism, in the myth of the werewolf, or through the examples of Willard and rats or of little Hans and horses, studied by Deleuze and Guattari in *Mille Plateaux*.

To be sure, these fantasies are thus named because they do not simply describe the patent features or even the actual characteristics of animals. They do not seek to report facts. They extrapolate and invent beyond that which can be observed. Imagination consists in a leap toward the absent and the virtual. Correlatively, in contrast to the rational approach, the representation of animals in the imaginary field is generally regarded with suspicion and often discarded as a mere fiction—the result of an illegitimate anthropomorphic projection.

However, the consistent fascination exerted by this imaginary of animals and its status of transcultural archetype still requires explanation. It would be unfair to claim that the imagined object is a pure pretext and could be replaced with any other object so that imagination can chaotically and frantically fancy anything and everything. To imagine is a way of developing a certain dialogue *with the imagined object*, and a way of entering a quasi-experience of the latter precisely by escaping consideration of its actual appearance. We will see that this exercise is not as paradoxical as it may seem. As I will argue,[46] even exploitative forms of imagination that randomly use the reference to animals to represent totally other beings in an overbearing fashion cannot but draw on the original meaning of the "absent referent."[47] This is also why such ideological fantasies always reveal more than their authors are aware of: Metaphors draw on haunting and uncontrollable phenomena. The imaginary is never *bric-a-brac*, a random decoration, but rather a set of recurring motifs—or, in other words, melodic lines or stubborn themes, to borrow from Bachelard and Merleau-Ponty. The imaginary is not a completely elusive target, even though a specific method is required to do justice to its richness, its meaningfulness, and its ontological status. These concepts of imagination and the imaginary are developed in further detail below (§2) and in Chapter 2.

Furthermore, it is now generally accepted that the rationalist model is a construction, somehow a fiction or, to quote Derrida, a *chimera*[48] (i.e., interestingly: A fiction *and* a hybrid animal). In this regard, critical philosophical approaches are backed up by the debate about a paradigm change in contemporary research in ethology, biology, and biosemiotics. The very reference to a neutral, objective knowledge is based on a fantasy. The man of reason observes his own turbulent and violent origin from afar, with detachment and condescension, and he forgets to regard himself as that which would have not been brought into being without the support of daring narratives and striking images—from the Book of Genesis to Descartes's beast-machines—as well as of symbolic conquering actions such as animal sacrifices or the rise of meat consumption.[49]

It is therefore urgent to examine why animals have been so inspiring to the imaginary field, and what aspects of living beings could be expressed in a privileged way in fantasies while they were stifled in the rationalist approach. What exactly opens the human imaginary to non-human animals? What, in the nature of animals, makes them able to live a part of their existence and become quasi-present in an exceptionally vivid way in the human imagery? What different forms of knowledge about animality can be developed through the study of the imaginary field, and how exactly can a scientific approach cooperate with the examination of the imaginary of animals?

Obviously, we take the terms "animals" and "humans" as designating old and still existing *figures,* ossified metaphors. In this sense, these terms are not simply references to concepts or realities; they are stubborn images that we should review, challenge, and reshape.

§3 Imagination and the imaginary: What are we talking about?

Imagination is a personal faculty of creativity that aims at the non-actual and, more specifically, the unreal (in contrast with memory, which aims at another form of the non-actual: The past). Creating a new reality is not the same as imagining. Imagination makes the unreal quasi-present *as unreal.* Imagination is the capacity to move beyond the limits of actual reality and turn absence into quasi-presence. The latter occurs as a source of ubiquity and as a breach in the continuous field of objective space and time: I cannot trace a way that would start from my actual place in the world and would lead continuously to the chimera, for instance. I who imagine am *here* in the actual world and *there* in the chimera's world, as imagining is not conceiving or uttering empty words, but rather quasi-perceiving, in a concrete and sensible manner, which implies affective and bodily involvement.

Imagination could not exist and perform quasi-experiences without being rooted in **the imaginary**. The latter is not a mere figment of a subjective imagination, but rather an autonomous and transcendent field, and, eventually, a fundamental ontological structure.

The French phrase *l'imaginaire* is rather tricky to translate into English. In French, the substantive "l'imaginaire" means a field of images, fantasies, metaphors, and associations forming a huge moving system that influences individual imaginations. "The imaginary" is certainly not a clear-cut conceptual system, but, as shown in particular by Bachelard, it possesses a certain autonomy and is structured by relatively consistent melodic lines. Specific motifs recur again and again through various times and places. Some associations are particularly stubborn and crystallize into powerful, even possibly transcultural, myths, popular imageries, and persistent themes; others are empty, insignificant, and fail to convince.[50] The imaginary aura of seasons, elements, colors, body, and love, for instance, inspires us and takes hold of us. It is an anonymous flow of meaning whose power of suggestion and fascination transcends *my* imagination. In the imaginary field, in a remarkably consistent way, alcohol is liquid fire, fathers eat their children, standing waters are tears, animals turn into humans and vice versa, and dead people become ghosts. The imaginary is a source of inspiration for art, reveries, and legends; it haunts us rather than being made up according to our whim. As emphasized by Montaigne, the imaginary possesses a genuine *force* and involves metamorphoses in our body.[51] The specific interest of the notion of *the imaginary* is that it unveils a pre-personal source for our fantasies and challenges the classical theory according to which images stem from a subjective faculty of the human mind.

Even more fundamentally, as I argue elsewhere,[52] the imaginary is foremost an original dimension of reality and Being, from where our imagination derives. It is the abyssal dimension that undermines every reality. I have proposed the concept of the *imaginareal* to designate such a dimension.[53] In its image—whether a picture or a fantasy—a being appears to be nomadic. The quasi-presence always refers to a possible actual perceptive presence. If I imagine a colleague of mine in this office and engage in a heated argument with him about something that he told me yesterday, I aim at a being that was here yesterday, could be here now, and is quasi-here. The original reality—the alleged "original" of the image—made these uncanny avatars possible, since it is still the same object that appears, makes itself recognized, and is "identified" as such and as *present-as-not-present*. Therefore, the original must always already be undermined by virtuality, by the ability to be quasi-present. Emotions and body attitude emerge in this fictitious discussion as an integral dimension of the relation that my colleague and I are building and through which we discover ourselves and evolve. Through the enacted fantasy, I also put the boundaries of some of our personality traits to the test. In fact, neither perception nor cognition ever presents me with a fully identifiable, coherent, and merely actual individual. Correlatively, my *here and now* (my actual body, this actual room as well as the actual matter of a painting or a photograph) must already bear in themselves the ability to be decentered and to welcome a dimension of virtuality that cannot merge with actuality. The "original" is not a self-coincident, absolutely solid being; it neither preexists its images, nor ever splits off from them. The very possibility of images requires, in the alleged "original," an ability to become a phantom. As for fictitious beings, such as Hamlet or the chimera, they make the concept of an "original" obviously even more problematic, but their quasi-presence is also experienced as a possible presence and can take possession of an actual work of art or of the actual body of the actor. Every single situation can be sidelined through imagination, and every actual reality could be quasi-present in other circumstances. As it turns out, presence is never fully and inescapably present; absence is never radical nothingness. The imaginary is thus to be understood primarily as an ontological fragility, a ubiquity, a form of hovering, and a non-substantiality that haunt every being.

As a result, the core of the phenomenon of imagination does not reside in the *re-presentation* of an absent object through a mysterious "mental image" that would be present in my mind—a rather naïve concept, indeed. A psychological approach to imagination is superficial and insufficient. It veils the ontological essential structure that is operating in every fantasy and that challenges the alternative between real and unreal, presence and absence. To imagine is to embrace and bring to the fore an ontological ubiquity that undermines my being as well as the being of the imagined object. I undergo an ontological process of de-phasing, becoming twofold, and so does the imagined object, correlatively. I am here and there, in the chimera's world; the chimera managed to enter my world but as that which creates an irreducible breach in the objective actual space and time.

Accordingly, in the imaginary field and through the thematization of the imaginary field by a subjective imagination, a certain form of meaning appears that must be contrasted with the rational and conceptual form. In the imaginary field, x is and is not x—the chimera is present and non-present, the sovereign is and is not a lion, my grandmother is and is not my mother. The principles of identity and non-contradiction are challenged. It is somehow true that dead people become ghosts or that fathers eat their children, which, of course, implies that truth stands beyond pure actual reality, beyond the mere description of what can be actually observed here and now. We "are" our future and our virtualities as well, and Cezanne's *Sainte Victoire* is also part of the being of the mountain itself and not the mere expression of Cezanne's unique psychology.[54] Ambiguity, being-like, being-manifold: Such virtual features become ontologically relevant and can be described in a privileged way through associations, metaphors, and series of images, rather than through categorizing and positive judgements. We will explain in more detail, in the first two chapters of this book, why the study of the concept of metaphor—which plays a key role in the epistemology of life sciences—combined with a phenomenological-ontological analysis of the imaginary, entails a radical redefinition of reality, being, and meaning.

As a consequence, the issue of the imaginary of animals is not first and foremost a psychological problem. It does not primordially involve a study of conscious re-presentations. As we will see, through hesitation, play, joint attention, histrionic behavior, and mimicry, animals gain access to the imaginary field. Starting from the ontological take on imagination, we will examine questions such as: Can (some) non-human animals enter into relation with the non-actual and the unreal? Can they be at the same time here and there? Can they decenter themselves into other subjects' perspectives?

What is more, the study of the imaginary is not essentially an exploration of that which is purely fictitious; quite the contrary indeed. It essentially involves a reflection on aspects of being that are usually neglected in a rationalist approach and that challenge the opposition between the real and the unreal.

Let me rephrase once again the key concept of this work. *The imaginary of animals* consists in (1) the aspects of the being of non-human animals that prepare, make possible, and inspire the human imagery of animality, and, therefore, also manifest themselves in it; and (2) the way animals in their turn thematize the imaginary field, actively play with manifold meanings and, in this sense, imagine and have their *own* imaginary field.

We will present different levels of ubiquity.

- A "being x *and* non-x," "being here and there" (**ontological ubiquity**[55]). Through the study of animals' relation to their ambiguous world [*Umwelt*], and the instability of organic and behavioral structures, we will show that animals essentially and autonomously develop a phantom-like being.
- An intentionality that thematizes ubiquity (**enacted ubiquity**). This intentionality *actively places itself into* a state of ubiquity and holds together the *here* and the *there*. It oscillates between the actual and the virtual. It handles and explores their relation. This activity is usually designated by the verbs *to play* and *to imagine* (the former engages ubiquity in a more performative manner, the latter in a more contemplative manner). The virtual as such exists only in contrast to the actual. As a relational entity, it must be aimed at through an intentionality that thematizes the discrepancy between what appears and what is suggested beyond this appearance, between what is and what could be. Yet, we do not necessarily need to *have in mind* a re-presentation (such as an image, a symbol, or a conceptual definition) of the virtual dimension of being to aim at this dimension. When we play a role—be it a stage role or a social role—we do not necessarily mentally visualize the character into which we project ourselves or that haunts us. As we will see, enacted ubiquity in animals may for instance take the form of phenomena of

self-depiction through animal appearance. Other important examples are, I will argue, mimicry, deceit, playful behaviors, and joint attention.

- An intentionality that takes the form of a representation through symbols (**symbolic ubiquity**). At this third level of ubiquity, a virtual aspect of beings is represented by an actual reality (a word, a picture, etc.) whose function is essentially to be other than itself and to give the virtual a permanent and recognizable form of presence in the actual world. I will elucidate how, in the animal realm, at the evolutionary level—through processes of ritualization—as well as at the level of interindividual relations, some gestures, facial expressions, and sounds crystallize into conventional signs. In this manner, they become symbols, namely changeovers embedded in the actual as permanent means of switching from presence to quasi-presence, from the real to the virtual.

§4 Imagination and/or knowledge

Suppose that, as I intend to demonstrate, non-human animals imagine, or at least enter into relation with the imaginary field in their own way. Does that mean that we need to use imagination to know them? Shouldn't we rather stick to observation and meticulous description? Don't we want to unveil what they *actually* can do? How could knowledge be founded on a venturesome leap into the unreal?

To be sure, such a leap into the virtual needs not be absolutely arbitrary. Yet, imagination essentially refuses to stick to the facts. It extrapolates and somehow deliberately aims wide of its object. Imagination is not good at concepts. It tolerates and interprets contradictions, and refuses to confine any being within a clear-cut essence.

However, categorization and delimitation have precisely been the scourge of modern life sciences. Concepts, classes, order, and clarity are essential characteristics of knowledge, together with openness to the complexity of the nature of the studied object. For several reasons that I have already sketched and that we will examine more thoroughly in the next chapters, living beings resist categorization in a particularly strong way, so much that the *knowledge of animals* is intrinsically a paradox and a brain-teaser. But imagination is more pragmatic than reason and can become the pillar of a renewed form of knowledge. In this respect, Coetzee's *The Lives of Animals* offers a revolutionary approach that will help us understand more concretely what a knowledge-through-the-imaginary can be.

In Coetzee's novel, Elizabeth Costello—a fictitious, successful Australian writer who has become obsessed with the question of animal suffering—sharply responds to Thomas Nagel. Contrary to Nagel's claim, Costello argues, we obviously can imagine what it is like to be a bat: "[T]here is no limit to the extent to which we can think ourselves into the being of another. There are no bounds to the sympathetic imagination."[56] Costello's first argument ("if you want proof...")[57]) is that, as a novelist, she thought her way into the existence of Molly Bloom, a character imagined by James Joyce. Did she succeed? Most certainly, since her novel was successful enough to turn her into a sought-after speaker who "has been invited to Appleton college to speak on any subject she elects."[58] *Therefore*, she claims in a second step, to imagine what it is like to be this or that particular animal is possible as well. "If I can think my way into the existence of a being who has never existed, then I can think my way into the existence of a bat or a chimpanzee or an oyster, any being with whom I share the substrate of life."[59]. The reasoning is surrealistic and toys with logic. Just like Derrida, Costello is not convinced by the power of the dazzling Cartesian "therefore."[60]

One may object that, concerning a "being who has never existed," the issue of knowledge is irrelevant. Joyce defines Molly Bloom by a set of details, but the *not-yet defined* features remained absolutely undetermined and open not only to interpretation but also, more radically, to free creation. It is neither true nor false that Little Red Riding Hood learned to speak Italian. But, as it seems, the knowledge of real animals is a whole different matter.

Costello's point is actually not naïve at all, although she insists on presenting herself as a nonexpert who deliberately refuses to use a "cool rather than heated, philosophical rather than polemical way of speaking."[61] She was offered the choice of a topic for her talk and decided to discuss, instead of her novels, "a *hobbyhorse* of hers, animals."[62] For that matter, she strives to talk about animals *as a novelist,* "a person whose sole claim to your attention is to have written stories about made-up people."[63] And the only principle that she reluctantly accepts to phrase is "open your heart and listen to what your heart says;"[64] quite a limited, sentimentalist, and puzzling—if not appalling—doctrine for the audience. But let us not be misled by her suggestion that she lacks the skills to fully manage the philosophical method.[65] Costello also firmly rejects a form of knowledge that she calls rational, one that emphasizes discrimination, separation, identity, and clear-cut differences. This analytical knowledge is, she claims, both inaccurate and dangerous. The imaginative approach is presented as a necessary alternative to a deficient rational analysis of animality.

According to Costello, the Cartesian cogito considers the soul as "a pea imprisoned in a shell."[66] In a like manner, Nagel takes for granted a radical separation between beings when he writes: "I want to *know* what it is like for a *bat* to be a bat. Yet if I try to imagine this, I am restricted to the resources of my own mind, and those resources are inadequate to the task."[67] Likewise, the accusation of anthropomorphism always begs the question: Why should we presuppose that projecting human characters onto animals will necessarily illegitimately bring together two radically heterogeneous realms? In Nagel's approach, "my own mind" must mean *a human mind*, closed onto itself or shut within a human world, and in no way a bat-like mind: Quite a set of heavy metaphysical assumptions. Such assumptions are impugned for instance, as I will recall in Chapter 2, by a phenomenological approach.

Nagel asserts that "our own experience provides the basic material for our imagination, whose range is therefore limited."[68] I cannot imagine what it is like to be a bat because I do not have the same internal neurophysiological constitution as the bat. By the same token, I cannot know what it is like to be a person deaf and blind from birth. As Nagel keeps referring to a mysterious "us" or "human viewpoint,"[69] he is taking for granted a unique human world that is actually not a reality. These arguments show that Nagel applies to imagination a completely inadequate pattern coming from a rationalist approach and based upon the duality between identity and pure difference. From this perspective, I can only imagine what I already fully experience as mine. But this position is a denial of imagination as such, namely as creativity. Put otherwise, this is not imagination, but instead, the knowledge of what I actually already know: "Imagination" then becomes a vain word. In addition, I never share exactly *the same* neurophysiological constitution with any human being. If the human world has some sort of reality, then analogy and adventurous projection must be reckoned with as reality-builders, precisely because they do not consist in projecting onto the other what is already an experience that we have in common (quite an absurd process in fact). We can distinguish between different degrees of analogy and extrapolation, but doing so will certainly not justify a radical rejection of the imagination of the bat as impossible or irrelevant.

Furthermore, by comparing the work of a novelist and the imagining of what it is like to be a bat, Costello discards the opposition between imagining something real and imagining a fictitious character. And, by referring to the success of her work as a novelist, she also prizes narrative imagination as the capacity to give life to characters. It is precisely the ontological interrelation between reality and quasi-reality that is at stake here. The transposition into an already quasi-existing fictitious character—exactly like the transposition of an actor into a role—requires the ability to *sense* the style of this character and the main motifs sketched, as it were, between the lines, via the already existing partial descriptions provided by the novel or the script. Similarly, when I imagine what it is like to be a bat, I build on hints such as the morphology of the bat, its apparent behavior, its specific environment, possibly biochemical measures, or the description of the mode of operating of sonars. My fantasies will have to be consistent with these aspects, in the same way as the actor's performance is consistent with the play. And, in fact, an animal may be closer to a play than to a digital program,[70] such is the claim that we will advocate.

In Costello's view, through imagination, humans are in fact quite familiar with other animals. For instance, we actually know very well, via a transposition process, what it is like to be a cow led to the slaughter. The problem is not that we cannot imagine or that such a fantasy gives us insufficient or irrelevant evidence. Otherwise, would our societies spend so much energy locating slaughterhouses in remote suburban places, and carefully avoiding any reference to them and to animals as individuals in food advertisements and food packaging?[71] What is more, we experience the possibility of bracketing and defusing this compassion by reasoning ourselves or letting ourselves be convinced by the reasoning of others. As Rousseau puts it, analysis comes after the feeling and only stifles it.[72]

Imaginative transposition may thus be an accurate, although paradoxical, way of knowing animals.

Costello lays the groundwork for an approach that dovetails an affective with an imaginative approach to animals. Empathy can never be reduced to an actual mystical fusion with the other's perspective. Speaking of "my" perspective is not entirely meaningless; and the other *is not* me. Claiming otherwise would be a form of violence, which does not mean, however, that individuals are confined to their own limited perspective. We can overcome this limitation, but we are never *fully* in the other's place. The encounter with the other is a decentration, a form of uprooting, and it must take place at the ontological level of non-identity and non-coincidence with oneself, in which "*x is and is not y*" is the law. If empathy is achieved, the other—into whom I transpose myself and whose being is thus open to such a transposition and, indeed, makes it possible—cannot be self-coincident either. If empathy is taken seriously, it cannot be separated from imagination. Empathy essentially includes assumption, invention, and possibly improper projection. Such is the basis of a respectful empathy with others duly recognized as such.[73]

By accepting imagination as an integral part of the knowledge of animals, we admit that we do not know *in advance* what a bat is. We refuse to claim that *what a bat is* consists in a set of determinate and fixed characteristics that may be either fully grasped or fully missed. This approach offers a new cognitive and ethical relationship with animals. The objectivist take on animality is no less an inseparable combination of cognitive and ethical inventive strategies, but it focuses on clarity, order, and mastery. By contrast, through imagination, under certain conditions, we can know and respect animals' alterity as well as their openness to the virtual.

Such a deep reconsideration of the concept of knowledge of animals has become necessary. Animals have too often been studied in the context of a barely hidden anthropocentric agenda. As De Waal highlights, "Books and articles commonly state that one of the central issues of evolutionary cognition is to find out what sets us apart. Entire conferences have been organized around the human essence, asking 'what makes us human?'"[74] Looking for differences is not *per se* a problem: Who could deny that human beings and other animals are different in many ways? But looking for a difference between humans and *all the other animals* is a biased question and an inauspicious starting point. This dichotomous approach results in a study of animals that is exclusively structured by the will to categorize and the conviction that *the* specific difference lies somewhere and simply has to be ascribed properly; this is knowledge, science, and philosophy with a saber.

Carol Adams convincingly shows, in *The Sexual Politics of Meat*, that analytical thinking develops hand-in-hand with the butchering of animals and the fragmentation of their bodies into renamed, categorized, and consumable parts. The appetite for categorization thrives on the haunting figure of the body turned into a map of meat cuts, the "edible body divided into ribs, loin, shank, and legs."[75] In the same vein Derrida mocked the obsessive hygienism that drives the quest for the proper of man [in French: *le propre de l'homme*, to wit: "The proper" and "that which is clean"], as if only *pure* humans—a dangerous concept, indeed—were tolerable. Through this approach, we ineluctably miss life as a process of adjustment and creativity, and as a diversity not only of species but also of groups and individuals.

I want to advocate the notion of a *knowledge* that will *integrate imagination*. A major asset of the imaginative approach is its pragmatism. Imagination does not exclude reason. Indeed, the latter actually operates with a set of ossified fantasies. Imagination accepts categories, concepts, and specific differences: They are the material and the support without which its plays, metaphors, and transpositions would be absolutely chaotic. Imagination integrates categories but takes them only half seriously. To put it in Nietzschean terms: Dionysus is inseparable from Apollo. Knowledge must combine two necessary qualities: Clarity and a genuine openness to the being of the studied object. As a result, I do not think that a knowledge of animals can be worthy of the name outside of a framework in which, on the one hand, imagination and the imaginary steer the investigation, and, on the other hand, concepts, genera, and specific differences are regarded as tentative and provisional distinctions with which we must also play.

§5 Background, phenomenological approach, outline

What makes the issue of animals both difficult and fascinating is that it simultaneously involves epistemological, ethical, political, and esthetic problems. It is impossible to study animals without, at the same time, considering the stories that human beings tell about themselves when they speak of animals. In this regard, one of the key references that lie in the background of the present work, the fundamental eye-opening text for this research, is Derrida's *The Animal That Therefore I Am*. Derrida left us with a riddle: Animals are chimeras and chimeras animals, reality and fiction are to be held together in the chimerical discourse about animals, and poetry is the thought of animals: What does that mean? Where does that come from and what does that entail?

I will build on Derrida's heritage while giving a stronger phenomenological, rather than deconstructionist, turn to the study of animals as chimeras or myths. Derrida's deconstructionism is totally compatible with phenomenology, but deconstruction focuses on decomposing, unpacking, and undoing complex and sedimented cultural constructs, while phenomenology investigates the source of myths and theories in everyday experience. Although the present book does not exclusively focus on texts and theories pertaining to the phenomenological tradition, I have chosen to develop a phenomenological approach to the issue at stake.

What does it mean exactly to opt for a phenomenological approach? It should first be mentioned for the sake of methodological clarity that "Phenomenology" designates a specific set of methods and concepts, many of which were put forward by Husserl or his critical heirs. Husserl himself has amended a great many of his theories and was quite forthright as to the numerous remaining sticking points and the points of contention with which phenomenology has to grapple. This open attitude established phenomenology as a field of debate and a general methodical and conceptual toolbox rather than as a closed-ended theory. Chapter 2 spells out how this phenomenological approach can be developed for use in overcoming the human/animal and real/imaginary dichotomies. I do not intend to provide an exhaustive picture of Husserl's, Heidegger's, Jonas', Merleau-Ponty's, or Buytendijk's thoughts on animals, for the imaginary of animals was simply not one of their points of focus. But some of their analyses will serve as an inspiration for the novel insights I want to develop in this book.

Why I have chosen to develop a phenomenological approach to the imaginary of animals is threefold, as more fully explicated in the second chapter of this book.

First, phenomenology ("science of *phenomena,*" namely of the way things appear to subjects) has the considerable advantage of not reducing everything to social constructs. It calls us to reflect on our experience, which contains givens that clash with ideological constructs so that we can at least sense that something "shifty" is afoot in such constructs. By returning to our very experience, beyond a process of endless deconstruction, by engaging biases and usual patterns of interpretation, and by dialoguing with empirical sciences without taking any paradigm for granted, phenomenology wants to understand the true nature of what it studies. Back to the things themselves![76] My approach is phenomenological because it addresses the connection between a human subjective perspective and the knowledge of what animals really are. The second distinguishing feature of phenomenology is that, for ontological and ethical reasons, it foregrounds a subject-centered perspective. The study of being, according to a phenomenological perspective, entails the development of a method that enables us to acknowledge and describe that which is irreducible to objects: Subjects, their experiences, their perspectives, and their agency, which is also why focusing on the concepts of the imaginary and imagination of animals is particularly important. Such concepts carve out some space for animal subjectivity, and I will contend that, among all the concepts that undertake to describe subjectivity, imagination and the imaginary are the most pertinent and illuminating for understanding animal subjectivity. I examine in

Chapter 5 why we cannot content ourselves with object-oriented approaches and the definition of animals as processes or becomings.

Lastly and above all, "fiction [*Fiktion*] makes up the vital element of phenomenology,"[77] Husserl famously stated. Phenomenology is less interested in ossified categories than in a fine-grained description of the lineaments and possibilities of lived experiences. Phenomena are protean and multifaceted. Let us suspend (phenomenological *epoché*) the presupposition of concepts such as the *human and the animal realm*, and start from a complex experience that occasionally implies some tentative distinctions, but also displays many emotional and imaginative intermixtures between these vaguely delineated realms. In this regard, the abundant imaginary of animals has already opened the way—and can serve as a guide—for phenomenology:[78] It provides a wealth of original experiences and forms of subjectivity that philosophy has too often neglected.

The first chapter concerns itself with the problem of the metaphorical relation between human and non-human animals: Are the concepts of animal subjectivity, personality, consciousness, and language pure metaphors? Do they simply result from the imposition of concepts coming from our human experience upon the non-human realm? Or do they manifest a deep unity that undergirds the human-animal difference? I compare two takes on these metaphors that, although helping us better understand their nature, largely miss their fundamental dynamism: First, a purifying approach that unfolds, for instance, in Michael Tomasello's theory of "pure imitation," and, second, Tim Ingold's holistic approach, based on the study of the ontology of hunter-gatherers. What is at stake, as I will contend, is embracing metaphor, namely accepting it as a fundamental experience, instead of trying to reduce it to identity or difference.

A phenomenological method precisely enables me to achieve that goal: **The second chapter** explains why phenomenology is in a position to provide a methodical and a conceptual apparatus to properly embrace the human-non-human metaphor. This chapter introduces a more specific phenomenological theory of the imaginary, which constitutes a new paradigm for thinking the relation between human and non-human animals. I also discuss animal subjectivity and its essential relation to an ambiguous world, following Uexküll, Buytendijk, and recent ethological studies. This chapter eventually analyzes the relation between the transcendental subject and animals. It dwells in particular on the issue of empathy and transpositions: Can we know what non-human animals think and feel? What is the role played by imagination in empathy with non-human animals?

Once the existence of the interface between our human imaginary of animals and non-human animals' imagination has been established as entailed by an ontologico-phenomenological perspective, the different aspects of this interface and its concrete manifestations remain to be investigated.

The third chapter focuses on the ontological ubiquity of (human and non-human) animals. This chapter demonstrates that a tentative, creative relation to meaning is at work in the very flesh of every living being so that animals can *never* be essentially defined as self-coincident, bound-to-the-actual realities: They are rather, literally, phantoms. This idea is tackled starting from a close reading of Merleau-Ponty's *The Structure of Behavior.* I also show that such an hauntology of animals operates at the heart of Darwin's *Origin of Species.* The imaginary of animals thus appears as the real subject of empirical animal sciences.

In **the fourth chapter**, I turn to enacted ubiquity and symbolic ubiquity in animals. How do non-human animals refer to, thematize, and even represent the virtual? What are the forms taken by symbolism in the absence of a superstructure of institutional languages? I focus on images and symbols in animal communication and unfold the different possible meanings of a recurring comparison—in Portmann, Bateson, Merleau-Ponty among others—between human dreams and non-human animal thinking and communication. This comparison, in many cases, pushes animals' mental life back into a mythic numbness and drowsiness so that it becomes far removed from us as humans. By contrast this chapter shows that an oneiric fundamental relation to meaning unfolds both in human and non-human animals, and that human thought is rooted in animal imagination: Images and symbols in the animal realm and our rich cultural imagery of animality actually continuously dialogue with and sustain each other.

The fifth and last chapter is devoted to the role that should be given to human and non-human *individual* subjectivity and imagination in the inter-action between humans and other animals. Our imagination of animals actually involves deep body metamorphoses both in us and in other animals. Does that mean that only the conceptuality of *real,* material trans-formations should be retained as relevant? Contra Deleuze and Guattari's dismissal of the reference to the conceptuality of images, imagination, and the imaginary, I contend that a subject-centered approach and the recognition of an individual imagination, both in human and other animals, can-not be substituted with a theory of becoming-animal. A comparison among Bachelard's, Deleuze and Guattari's, and Haraway's theories serves as a ground for the analyses developed in this chapter.

The conclusion addresses three remaining sticking points for clarification purposes. *First*, I examine the oppressive forms of the imaginary of ani-mals. I discuss Berger's "Why Look at Animals?" and show that even poor and manipulative images remain in fact grafted to the meaningful imagi-nary that belongs to the very being of animals. *Second*, I tackle the tricky question of animal agency and animal consciousness. At stake is a renewed understanding of viscous agency and imaginative thinking in animals, beyond the conscious-unconscious dichotomy, which also entails significant ethical and political consequences. *Eventually*, examining the thorny issue of the origin and meaning of the symbol of "the Bull and the Goddess" in the Neolithic period, the conclusion shows that the theory of the imaginary of animals requires the introduction of a circular temporality in the history of species and human symbols.

Notes

1. See Liu, Rebekah Yi. *The Background and Meaning of the Image of the Beast in Rev 13:14, 15.* PhD diss., Andrews University, 2016. On the goddess and wild animals. See also infra §39. The Latin word *bestia* (in Greek and in Rev 14: *thêrion*) designates a wild animal that is radically foreign and hostile to humans. The term "beast" thus dramatizes the human-animal dichotomy and stands as a symptom of the tension between humans and other animals as well as of the attempt to idealize humanity and draw a clear-cut distinction between humans and animals.
2. de Montaigne, Michel. "Of the Force of Imagination." In *Essays*. Trans. Charles Cotton. Project Gutenberg ebook, 2006, p.147.
3. *Ibid.*
4. *Ibid.*
5. I sparingly use the still-anthropocentric phrase "non-human animal" for disambiguation purpose. The term "animal" without specification is used to designate human and non-human animals.
6. Bachelard, Gaston. *Lautréamont,* Paris: José Corti, 1939.
7. Berger, John. "Why Look at Animals?" In *About Looking,* London: Writers & Readers, 1980.
8. See for instance Norris, Margot. *Beasts of the Modern Imagination: Darwin, Nietzsche, Kafka, Ernst, & Lawrence.* Baltimore and London: The Johns Hopkins University Press, 1985; and Gross, Aaron and Vallely, Anne (Eds). *Animals and the Human Imagination: A Companion to Animal Studies.* New York: Columbia University Press, 2012. See also Payne, Mark. *The Animal Part: Human and Other Animals in the Poetic Imagination.* Chicago: The University of Chicago Press, 2010.
9. See for instance Latour, Bruno. *Politiques de la nature. Comment faire entrer les sciences en démocratie.* Paris: La Découverte, Armillaire, 1999. Trans. Catherine Porter. Cambridge: Harvard University Press, 2004.
10. Driscoll, Kári and Hoffmann, Eva (Eds.). *What is Zoopoetics? Texts, Bodies, Entanglement.* Basingstoke: Palgrave MacMillan, 2018, p.6. See also Aaron Moe's beautiful *Zoopoetics: Animals and the Making of Poetry.* Lanham: Lexington Books, 2014.
11. Mitchell, William J. Thomas. *What Do Pictures Want? The Lives and Loves of Images.* Chicago: University of Chicago Press, 2005, in particular: p.2, 6, 10–27, 52–5, 88–103.
12. *Ibid.,* p.31–2, p.88–9.
13. *Ibid.,* p.88. See also Derrida, Jacques. *Of Grammatology.* Trans. Gayatri Spivak, Baltimore: Johns Hopkins University Press, 1976, p.292–3.
14. Derrida, Jacques. *L'animal que donc je suis,* Paris: Galilée, 2006, p.43. Trans. David Wills. New York: Fordham University Press, 2008, p.23.
15. *Ibid.,* p.20, Trans. p.6
16. "The thought of the animal, if there is such a thing, amounts to poetry," *Ibid.,* p.23, Trans., p.7. I have modified the English translation: "Thinking concerning the animal": Derrida's phrase in French is precisely a Möbius loop that both means *thinking concerning the animal* and *thinking as performed by the animal.*

17. I am referring here to the concept of "pictorial" and "iconic" turn such as defined by William John Thomas Mitchell or Gottfried Boehm. See for instance Boehm, G. *Was ist ein Bild?* München: Fink, 1994; and Mitchell, W.J.T. "The pictorial turn", *Picture Theory. Essays on Verbal and Visual Representation*. Chicago: University of Chicago Press, 1994.
18. Derrida, Jacques. "Che cos'è la poesia?" in *Poesia*, I, 11, Nov 1988.
19. Maran, Timo. "Are Ecological Codes Archetypal Structures?" In Maran, T., Lindström, K., Magnus, R., and Tønnessen, M. (Eds.), *Semiotics in the Wild: Essays in Honour of Kalevi Kull on the Occasion of His 60th Birthday*. Tartu: University of Tartu Press, 2012, p.149–51.
20. May-Collado, Laura J. "Changes in Whistle Structure of Two Dolphin Species During Interspecific Associations", *Ethology*, Vol. 116, No. 11, pages 1065-74, November 2010.
21. A diacritical process exploits the expressive capacities that are borne by raw matter. Within such a process, meaning can be expressed in a wild and poetic manner, through analogies, resemblance, the emotional power of sounds and forms, and the structural power of contrasts and repetitions. This form of expression can be typically experienced when one contemplates a painting, even more radically an abstract painting, or when one listens to a speech in an unknown language.
22. Tayler, C., and Saayman, G. "Imitative Behaviour by Indian Ocean Bottlenose Dolphins (Tursiops aduncus) in Captivity." In *Behaviour*, 44(3–4), 1973.
23. Glăveanu, Vlad Petre. "Commentary on Chapter 4: Proto-c Creativity?", in Kaufman, Allison B. and Kaufman, James C. (Eds). *Animal Creativity and Innovation*. Elsevier, 2015, p.73.
24. To de-center something is to estrange it from its original identity. One can also become decentered or decenter oneself, for instance when putting oneself in someone else's place. Decentrations actually manifest the lability of original identities. This term will prove crucial in the framework of my definition of the imaginary as the realm of ubiquity.
25. "Like the more and more subtle play of epicycles invented by astronomers to uphold the geocentric universe, thoughtful biologists have had to invent more and more subtle arguments in order to uphold a view of organic evolution as obeying the irreproachable scheme of efficient causation as the only basic explanatory tool in the natural world," Hoffmeyer, Jesper. *Biosemiotics: An Examination into the Signs of Life and the Life of Signs*. Scranton: University of Scranton Press, 2008, p.54. See also Merleau-Ponty, *The Structure of Behavior*, Boston: Beacon Press, 1960, p.19, and infra §21
26. See for instance Tomasello, *The Cultural Origins of Human Cognition*. Cambridge, Harvard University Press, 1999 and for an overview of the debate, Bimbenet, Etienne, *L'animal que je ne suis plus*, Paris: Gallimard, 2011, and Dufourcq (Ed.) *Broken bonds? Questioning Anthropological Difference*, Vol. 11, No. 1 of *Environmental Philosophy*, Spring 2014.
27. Griffin, Donald. *Animal Thinking*. Cambridge: Harvard University Press, 1984.
28. See for instance Goodall, Jane. *The Chimpanzees of Gombe: Patterns of Behavior*. Cambridge: Harvard University Press, 1986.
29. See for instance Savage-Rumbaugh, Sue and Lewin, Roger. *Kanzi: The Ape at the Brink of the Human Mind*, New York: Wiley, 1994.
30. See for instance Wemelsfelder, Françoise. "The Scientific Validity of Subjective Concepts in Models of Animal Welfare." *Applied Animal Behaviour Science* 53, p.75–88, 1997.
31. De Waal, Frans. *Are We Smart Enough to Know How Smart Animals Are?* London: Granta Books, 2016.

32. For instance Lewontin, Richard. *The Triple Helix*, Cambridge: Harvard University Press, 2000.

33. Oyama, Susan. *The Ontogeny of Information: Developmental Systems and Evolution*. Cambridge: Cambridge University Press, 1985.

34. Hoffmeyer, Jesper. *Biosemiotics: An Examination into the Signs of Life and the Life of Signs.*

35. Walsh, D.M. *Organisms, Agency, and Evolution*, Cambridge: Cambridge University Press, 2015.

36. Hoffmeyer, *Biosemiotics: An Examination into the Signs of Life and the Life of Signs*. p.59. See also Oyama, Susan. *The Ontogeny of Information*, p.26–7.

37. See for instance Levins and Lewontin (1985), Hoffmeyer (2008), and Wheeler, Wendy *The Whole Creature: Complexity, Biosemiotics and the Evolution of Culture*, London: Lawrence And Wishart Ltd, 2006.

38. For striking versions of genetic determinism and modern synthesis. See for instance Dawkins, Richard. *The Selfish Gene*. Oxford: Oxford University Press, 1976; or Nusslein-Volhard, Christiane. *Coming to Life: How genes drive development*. Carlsbad: Kales Press, 2006, p.32: "The genetic structure is a language of four letters that can be read faultlessly in order to make working copies and eventually to replicate precisely."

39. Heidegger, Martin. *Die Grundbegriffe der Metaphysik. Welt—Endlichkeit—Einsamkeit*. Frankfurt: Klostermann, 1983, see infra §6, 15 and 18.

40. Lacan, Jacques. *Le Séminaire. Les écrits techniques de Freud* (1953–1954), tome 1, Paris: Seuil, 1975, p.159. Trans. J. Forrester, New York: W.W. Norton, 1988, p.138. See infra §32.

41. Jonas, Hans. *The Phenomenon of Life. Toward a Philosophical Biology*. New York: A Delta Book, 1966, p.175.

42. Ibid. p.171.

43. Rousseau, A *Discourse on the Origin of Inequality*, Trans. G.D.H. Cole, London: Cosmo Classic, 2005, p.35.

44. Sartre, Jean-Paul. *L'imaginaire*, Paris: Gallimard, 1940, p.358, Trans. J. Webber. London: Routledge, 2004, p.186. Indeed, behind every consciousness, including perceptual or conceptual consciousness, a fundamental form of imagination is at work. It consists in the power to project oneself towards the future and to aim at the virtual. Even non-conscious engagement in actions and more practical forms of existence suppose the imagination thus defined.

45. See for instance Davidson, Donald. "Rational Animals." In *Dialectica* 36(4), 1982; and Dennett, Daniel, "Do Animals Have Beliefs?" in Roitblat, Herbert (Ed.), *Comparative Approaches to Cognitive Sciences*, London: MIT Press, 1995.

46. See infra §§37–38.

47. The concept of *the absent referent* was put forward in particular by Margaret Homans in *Bearing the Word. Language and Female Experience in Nineteenth-Century Women's Writing* (Chicago: University of Chicago Press, 1986). In *The Sexual Politics of Meat* (New York: Bloomsbury Academic, 1990), Carol Adams applies this concept to a typically human way of using animal metaphors or pictures in the symbolic realm, while butchering non-human animals and denying their own subjectivity in covert practices. Adams thus calls for a return to a focus on real animals. I fully endorse this demand, although with a twist: real animals are imaginative beings, it is also crucial to do justice to their own imaginary.

48. Derrida, *The Animal That Therefore I am*, p.41. Through a bold stretch of the term, "chimera" has become a synonym for fantasy. What secret kinship between animality and imagination is at work here and how should we come to regard the notoriously mischievous behaviors of young goats (χίμαιραι in Ancient Greek) as an archetypal form of imagination?

49. See in particular Burgat, Florence, *L'humanité carnivore* (Paris: Seuil, 2017): The carnivorous destiny of humanity cannot be properly explained if the symbolic dimension of meat-eating is not taken into account.
50. See Bachelard's numerous examples and illuminating analyses in his works on the imaginary of elements. For instance, in *Air and dreams*, about the mediocre image of the wing in poetry or paintings (Bachelard, *L'air et les songes. Essai sur l'imagination du mouvement,* Paris: José Corti, 1943, chapter 2, "La poétique des ailes").
51. Montaigne, "Of the Force of imagination", in *Essays.*
52. I summarize here the main claims and arguments that were put forward in my books devoted to the imaginary in Husserl's and Merleau-Ponty's phenomenologies. See Dufourcq, Annabelle. *La dimension imaginaire du réel dans la philosophie de Husserl.* Dordrecht: Springer, Phaenomenologica, 2010. And *Merleau-Ponty: une ontologie de l'imaginaire.* Dordrecht: Springer, Phaenomenologica, 2012.
53. I thank Ted Toadvine for suggesting this surrealistic portmanteau word to me. See Dufourcq "The fundamental imaginary dimension of the real in Merleau-Ponty's philosophy", in *Research in phenomenology*, Brill, Volume 45, Issue 1, 2015, p.47.
54. See Maurice Merleau-Ponty, *L'Œil et l'Esprit.* Paris: Gallimard. 1961. Trans. Michael B. Smith. in *The Merleau-Ponty Aesthetics Reader*, Galen A. Johnson, Michael B. Smith (Eds.). Evanston: Northwestern University Press, 1993.
55. "Ubiquity" is here defined as "presence in different places simultaneously." I contend that this plural ubiquity is characteristic of both of imagination and animal nature, it certainly prefigures the divine serene omnipresence (a classical sense of "ubiquity"), but constitutes a more embodied and imbalanced version of it.
56. Coetzee, John Maxwell. *The Lives of Animals.* Princeton: Princeton University Press, 1999, p.133.
57. *Ibid.*
58. *Ibid.*, p.114.
59. *Ibid.*, p.133.
60. Derrida, *The Animal That Therefore I am*, p.54, p.74-8. A close reading of Descartes' *Meditations* and a quick look at the many controversies they raised show that Descartes' "therefore" is largely incantatory. When the logic of reasoning actually falters, the recurring magic word "donc" conjures up the figure of rationality, conceals the cracks, and dazes the reader.
61. Coetzee, *The Lives of Animals,* p.120.
62. *Ibid.*, 114.
63. *Ibid.*, 120.
64. *Ibid.*, 131.
65. *Ibid.*, 121.
66. *Ibid.*, 132.
67. Nagel, Thomas. "What Is It Like to Be a Bat?". *The Philosophical Review*, Vol. 83, No. 4. 1974, p.439.
68. *Ibid.*
69. *Ibid.* p.443 and 444. See also "The Real Nature of Human Experience" p.444; "Our Human Point of View" (*Ibid.*), and the recurring use of "we" and "us."
70. See Hoffmeyer, J. & Emmeche. C. "Code-Duality and the Semiotics of Nature," in Anderson, M. and Merrell, F. (Eds). *On Semiotic Modelling*, Berlin & New York: Mouton & De Gruyter. 1991. See also infra Chapters 3 and 4.
71. See for instance Burgat, Florence. *L'animal dans les pratiques de la consommation*, Paris: PUF, coll. "Que sais-je? ", 1995.

72. "A murder may with impunity be committed under his window; he has only to put his hands to his ears and argue a little with himself, to prevent nature, which is shocked within him, from identifying itself with the unfortunate sufferer" (Jean-Jacques Rousseau, A *Discourse on the Origin of Inequality,* p.53).

73. For a further analysis of empathetic imagination see infra §18.

74. De Waal, *Are We Smart Enough to Know How Smart Animals Are?,* p.158.

75. For instance Adams, *The Sexual Politics of Meat,* p.73–74.

76. "Wir wollen auf die" Sachen selbst "zurückgehen", Husserl, Edmund. *Logische Untersuchungen, Zweiter Teil, Untersuchungen zur Phänomenologie und Theorie der Erkenntnis,* Tübingen: Max Niemeyer, 1901, p.7.

77. Husserl, Edmund. *Ideen zu einer reinen Phänomenologie und phänomenologischen Philosophie, Erstes Buch: Allgemeine Einführung in die reine Phänomenologie.* The Hague: M. Nijhoff, 1950. Trans. F. Kersten. The Hague, Nijhoff, 1982, §70, p.160.

78. Artists and phenomenologists are akin, Husserl emphasized in his letter to Hofmannsthal (January 12 1907, in *Husserliana. Briefwechsel,* Vol. 3.7, *Wissenschaftlerkorrespondenz.* Dordrecht: Kluwer, 1994), with the difference that artists primarily play with phenomena and their possibilities, while phenomenologists aim at formulating the meaning that unfolds in phenomena and, to this end, seek to develop fine philosophical conceptual patterns.

1 Human-animal metaphors
Identity and similarity at issue

§6 Similar but not the same: An ontological issue

One of the most stubborn, in fact ever-recurring, issues in the field of animal studies is whether it is legitimate or not to apply words that classically describe human beings to non-human animals: Subjectivity, cognition, language, feelings, imagination, morals, agency, will, decision, or culture, for instance. Even when scientific studies show that certain animal behaviors display characteristics that can be, point-to-point, likened to a cognitive or moral behavior such as it can be observed in human existence, the objection remains unshaken: Is animal "cognition" *genuine* cognition? Is animal moral sense *genuinely* "moral"? De Waal, who regularly came up against this objection, consistently responds by reasserting the importance of using anthropomorphism in ethology: "The curious situation in which scientists who work with these fascinating animals find themselves is that they cannot help but interpret many of their actions in human terms, which then automatically provokes the wrath of philosophers and other scientists."[1] By contrast, De Waal argues, those who strive to make the meaning of concepts narrower and narrower so that only human behaviors can fit them[2] are simply secretly driven by the same foolish human pride that made so many people try to outperform the "chimpion" Ayumu, the chimpanzee who can recall a series of number appearing randomly, for a fraction of second, on a screen.[3] But the issue also arises at an apparently more rational level: Differences matter as much as—if not more than—common features or similarities, especially if we want to give a clear and analytical scientific account of beings. Surely, a non-human and a human behavior can never be *exactly the same*. The issue at stake is perfectly summarized by the sentence spelled out at the end of a YouTube video posted on the channel of the Primate Research Institute of the University of Kyoto, about Ayumu's incredible skills: "Chimpanzees are so similar to us... But not the same."[4] The problem is whether and how *similarity* may gain an ontological status and find a place in the framework of the search for a relevant scientific method for life studies.

In a famous passage of *The Fundamental Concepts of Metaphysics*, Heidegger affords us a striking instance of this similarity problem. As the next chapter will show, Heidegger actually remained entangled in the web of this issue, but, at least, he succeeded in phrasing it without eluding its intricate and paradoxical nature. "The dog feeds [frißt] with us... no we do not feed. It eats [ißt] with us... no it does not eat. And yet, with us. A going along with... a transposedness, and yet not."[5] Indeed, in German one does not use the same verb to designate the non-human and human action of eating (*fressen-essen*). Admittedly, strictly speaking, there are differences between animal and human "eating," the most fundamental of which being the huge symbolic dimension that has deeply permeated the human food rituals and ceremonials. But how exactly should we account for the not-less stubborn feeling of similarity? Heidegger's description of the "transposedness and yet not" does not help much: It is hardly conceptual, rather wavering, and almost stammering. The logically unacceptable "X is and is not X" form shows the *tension* between the phenomenon of similarity and classical logics.

At least those who proclaim the prohibition of anthropomorphism cut to the chase. A non-human behavior or mode of existence is either the same as a human one, or different. And, in the latter case, it is simply rigorous to avoid using the same word to designate them. After all, this is a regular application of the principle of the excluded third.

Similarity stands beyond this dichotomy, but can it be turned into a fundamental concept? Would an ontology of resemblance be compatible with a scientific study of life? Is such a shift worth the cost that we would necessarily have to pay, namely a logical and methodological revolution?

Before examining, in the next chapter of this book, how an ontology inspired by the phenomenology of the imaginary can shed new light on animal studies, I will present two antithetical and equally problematic contemporary approaches that grapple with this problem of similarity—in the field of anthropology and cognitive ethology. The ontological issue at stake will thus be defined more concretely as a thorny question at the heart of animal studies that engages the attention of researchers beyond philosophical studies. The study of animal life is doomed to face ontological dilemmas and the connection between such ontological consideration and the question of the proper method for a scientific study of animals will also appear more clearly.

We will first focus on the misadventures of the theory of true imitation at the end of the twentieth century in cognitive ethology (§**7-8**). Such misadventures demonstrate that refusing to use the same words to denote similar behaviors in human and non-human animals—allegedly for the sake of scientific meticulousness—is not as metaphysically neutral as it may seem, nor even scientifically fruitful. We shall then turn to a thinker who takes the opposite view, by examining Ingold's criticism of the concept of anthropomorphic metaphors (§**9-11**). Ingold unmasks the metaphysical biases of the classical analytical approaches, but I contend that he lapses into the opposite mistake. Ingold explicitly calls for a new ontology, but his approach reduces similarity to identity and thus fails to provide a satisfactory ontological framework to animal studies.

PART I. HUMANS APE BETTER

§7 Can animals imitate? The "true imitation" theory and its flaws

If a non-human animal copies the general pattern of behavior that provokes an interesting result—even a behavior that is unfamiliar to her and not part of the action repertoire of her species—this cannot be called *true imitation*, a number of eminent scientists assert.[6] The argument goes as follows: Non-human animals do not focus on the detailed topography of the gestures performed by the model, but, instead, on the result of the action, the affordances of the tools, or a rough pattern of actions that can be exploited by each individual according to its own behavioral strategies. A non-human animal can *roughly* do as the model does: She selectively reproduces *some* aspects of the model's gestures.[7] Does she not imitate? This *looks like* imitation, but, according to the "true imitation" theory, this should be called *emulation learning*[8] based, for instance, on *local* and/or *stimulus enhancement*.[9] In other words, apes do not ape:[10] Smart trick, indeed. Michael Tomasello, a prominent researcher in the field of Evolutionary Anthropology, did not dither too long before concluding that *only* humans are capable of true imitation[11] and he contended that the latter is the backbone of culture, namely a highly cumulative evolution[12], a.k.a. *true* culture.[13] Cultural evolution, the argument goes, is a process in which ever new *ways* of doing things are noticed and transmitted. Without a capacity for true imitation, such unique features would never be retained; they would be lost in a schematic reproduction of general patterns of behavior.

Several daring and debatable metaphysical presuppositions are actually at work in this approach under the guise of a fine-grained description of the lineaments of behaviors. Even a half-attentive reading can ferret out symptoms of such biases: The phrase "true imitation," for instance, can only set off some alarms, as its proponents themselves sense it.[14] Certainly, the issues raised by this platonic approach warrant a more detailed look.

The "true imitation" theory chooses to radically break with the common sense of "imitation" and loses touch with vernacular language.[15] To be sure, the latter is often loose and sometimes conflates what should be kept separate. Nonetheless, in this case, as often, everyday language acknowledges a similarity between behaviors, that is an integral part of the phenomenality of these behaviors and that a fine-grained description should integrate into its account. *Common* language possesses a fluidity and a flexibility that makes a more comprehensive description possible and, correlatively, sustains an ethics of community that is also crucial in animal studies. Indeed, in this particular theory of true imitation, is Tomasello interested in paying attention to animals or, eventually, in finding a way of reasserting human uniqueness? Nevertheless, shouldn't a scientific approach use clearly circumscribed concepts? It is significant, however, that De Waal chooses, for instance, the phrase "selective *imitation*" to designate behaviors that the proponents of the "true imitation" theory would call "emulation learning."

More specifically, several researchers have strongly challenged the true imitation theory. Andrew Whiten and Deborah Custance,[16] for instance, develop criticism based on meticulous experiments, but also on a reflection on the most accurate method that should be used to shrewdly account for such experiments. Many experiments show that animals of several species can actually *truly* imitate;[17] hence, they challenge Tomasello's theory without challenging his concept of imitation. By contrast, the general thrust of Whiten and Custance's work, such as it is presented in Heyes' and Galef's *Social Learning in Animals: The Roots of Culture,* is that drawing a clear boundary between emulation and imitation is impossible.

There is indeed a dimension of imitation in the behaviors that some have tried to describe exclusively in terms of emulation.

a   First, no imitation can be perfect, not only because perfection is only divine—which makes true imitation an "ever elusive," although still pertinent, ideal—but more fundamentally because imitation essentially implies duplication, de-centration, and, consequently, some modifications. The body of the imitator is not exactly the same as the body of the model, even more so in frequent cases of interspecific emulation learning. As pointed out by Whiten and Custance, "All imitative copies must have a schematic or program-like character, because not every muscle twitch is going to be copied."[18] Identity and imitation are antinomic. The very concept of true imitation as a slavish process that sticks to every possible detail of an action is an unviable monster. This concept additionally enters into tension with another version of the true imitation theory. Tomasello sometimes argues that proper imitation should imply understanding of the intentions that rule the acts of the imitated model.[19] But such an understanding essentially consists in generalizing and catching, at least performatively, the outline structure of a task that can be reapplied to different *similar* but never identical situations.[20] Olivier Morin also mockingly highlights that the human talent for imitation would not be truly culturally fruitful if it took the form of a systematic copy of every tiniest detail without discernment.[21] If one acknowledges that imitation is essentially partial, therefore "an act may be recognized as imitative even if the replication is only 'in outline,' or if it involves just one or two of several features which could potentially be copied."[22]

b Second, ghost experiments show that the actual presence of a model who manipulates the tools and objects of interest is critical to making the learning of a behavior happen, while the tested animals failed to learn several acts on the basis of the mere display of cause-and-effect sequences of events. In ghost experiments, indeed, the model is replaced "with some hidden ('ghostly') means of making the objects of interest do what they would if the model were moving them."[23] Whiten explicitly refers to Tomasello's claim that emulation, unlike imitation, focuses exclusively on the movements of the objects and the result of the actions, so that the behaviors that a young chimpanzee apparently learns from his mother could actually have been learned "if the wind, rather than the mother, had caused the log to roll over and expose the ants."[24] Thus, the presence of a concrete acting model, "a conspecific or other adequate animal model,"[25] matters even when the animal learns a general pattern of behaviors, which fleshes out the claim that so-called emulation learning may include a part of imitation.

c The concept of genuine imitation is also based on another rather artificial distinction: Emulation focuses on tools, manipulated objects, their affordances, and the result of the action, rather than on the body and behavior of another individual. Obviously, a behavior could not be properly recognized and understood without also considering the used tool. "Hammering" makes no sense if one does not take the hammer into view. To be sure, ghost experiments are possible, but they do not result in the same learning performances as experiments with actual models—and for good reason: Phenomenally, they are not the same. Furthermore, when a model is present, how can we draw a line between the body and the tool, the body and its environment? Consequently, between alleged pure emulation and pure imitation?[26] Again, trenchant statements are based on clear-cut forced distinctions.

d Why should behaviors necessarily function in a binary way? Why would a consideration of—or even a focus on—the general outline and the result of an action necessarily *exclude* consideration of its unique topography? This postulate always implicitly supports Tomasello's reasoning. "An animal observing the instrumental behavior of another animal may learn something about either: (1) the environment and the changes in the environment that result from the observed behavior, or (2) the observed behavior itself."[27] There is no reason *per se* to admit such premises. In addition, the experiments conducted by Whiten actually do not lend themselves to such dichotomies (see [e]). Hence, they demonstrate that a fine-grained description requires a less analytical approach.

e And lastly: In many cases, animals copy only *some* aspects of the model's action (its shape, its speed, the trajectories of this or that object and/or limb, a certain sequencing in the series of action, or the general structure of the action, for instance[28]), and nonetheless prove able to acquire and pass on *original* traits of behaviors to others. In two groups respectively exposed to two models whose *way* of using the same tool was different, "each technique spread preferentially in the group in which it was seeded."[29] Another experiment shows that an original trait of behavior can be successfully transmitted through a linear chain of individuals.[30]

§8 Analytical approach vs holistic approach, and correlative ontological positions

Interestingly, the above arguments gave rise to a new *holistic* pattern of description and understanding of imitative/emulative behaviors.

Imitation is essentially imperfect and always more or less selective. Schematization is thus constantly at work in imitation, without making the transmission of cultural features impossible. As a consequence, an analytical model should give place to a synthetic one. The former is based on the isolation of exclusive behavioral units (like emulative learning and true imitation), while the latter conceives of behaviors as complex, polyvalent, and versatile wholes.[31] According to Whiten and Custance, imitation and emulation are not absolutes but "phenomena which need to be defined relative to each other within any particular frame of reference."[32] They further explain, "We can envisage a continuum of fidelity in imitation, from the close to the inexact. The inexactitude could take different forms as suggested above: The imitator might perform just a subset of the elements of the task, or it might perform them all but in a manner only vaguely like the model."[33] Accordingly, Whiten and Custance regard behaviors as moving along a spectrum of *more or less* faithful imitations. The continuum also includes a whole range of possible focuses and emphases for the imitation: Speed, shape, and so forth. Each possible aspect may be reproduced by the imitator in a *more or less* schematic/detailed way.

The continuum paradigm is applicable to both human and non-human animals. It constitutes a fluid and flexible pattern of description and understanding. It also takes into account the fact that a very selective imitation as well as a slavish copy are *both* useful learning strategies that complement each other. These two strategies may be applied preferentially "in the appropriate circumstances,"[34] and are actually both implemented by many animals in a remarkably polyvalent tactic. It is thus possible to combine a holistic approach with a fine-grained and clearly articulated characterization of behaviors.

This debate can be formulated to bring a broad range of philosophical and epistemological implications to the fore—from an ontological reflection to the definition of a method in which several major ontological shifts can find expression. Moreover, interestingly, all these new developments are not a matter of mere speculation, they are carried out under the pressure of experiments.

Another crucial lesson is to be drawn from the true imitation debate. Beyond the particular case of imitation, a more general relationship can be observed between the analytical approach and what Derrida describes as anthropological hygienism; namely the idea that a concept of *"pure human- ity"* must be demarcated at all costs.

It would be unfair to claim that Tomasello entirely conceals his obses- sion for the anthropological difference or that his research is devoid of a rich empirical dimension, but it is nonetheless striking that he usually does not make the fundamental principles that govern the concept of "true imi- tation" explicit and that, quite the reverse, he even sometimes enunciates such principles in the form of assertive statements.[35] Since there is no reason *per se* to presuppose a binary pattern of understanding, and, furthermore, since such an analytical approach is not a prerequisite for describing and understanding experiments in a meticulous manner, this binary approach must be actually supported by what Pascal called "a thought from behind the head."

The same stubborn thought must be at work, as De Waal suspects, in the relentless definition of new discriminating characteristics (use of tools, joint attention, perspective taking, theory of mind, etc.; imitation, imitation of the topology, imitation on the basis of the understanding of the other's intention, etc.) that permits the reassertion of the radical anthropological difference when recent experiments have just undermined it.[36] The *explicit hypothesis* of the anthropological difference may be a *tentative* guide for a researcher, but the theory of true imitation reveals a much more insidious and questionable tactic. The phrase "true imitation" itself is quite telling in this regard and cannot but leave an unmistakable metaphysical aftertaste. It implicitly indicates that some entities *look like* imitation—otherwise we would not have to refer to a "true" imitation—but should not be regarded as imitation. They are fake imitations, *Ersätze* of imitation. Resemblance here is simply brutally denied. A transcendent norm is applied to the diver- sity of particular phenomena: Only conformity to a highly demanding and sharply circumscribed definition of imitation—regarded as connected to the mysterious source of an absolute *truth*—will make possible the *god- approved* classification of a behavior under the category of imitation. What is at stake is the joint establishment of separations and a hierarchy. Tomasello also describes "local enhancement, emulation learning, ontoge- netic ritualization" as "weaker forms of social learning."[37] And, of course, it is entirely coincidental that, in this approach, truth falls on the human side, which results in a quite ironic reversal—another symptom that should raise suspicion: The stupidest form of imitation, namely a blind slavish copy, is regarded as an unmistakable sign of superiority. Humans ape better.

The "thought behind" that is at work in the imitation theory is an anthro- pological hygienism, or purism, inseparable from a more general onto- logical positivism. The latter can be defined as the presupposition that all that truly is must be clearly circumscribed, self-identical, and perfectly identifiable. The systematic search for clarity, distinction, and categorization is the structural accomplice of the spectator attitude, and the objectifying relation to the world through which humans can constitute themselves as master-minds,[38] the only true subjects of science. As the example of the true imitation theories shows, such presuppositions can secretly shape and bias scientific study strategies. The analytical approach cannot be regarded as simply justified by the disinterested pursuit of truth, or even clarity for science's sake.

Nevertheless, in the debate between the proponents of the true imitation theory and the promotors of a more synthetic approach, the quest for discriminating features has spurred research and contributed to the development of an increasingly meticulous description of human and non-human animal behaviors. A differentiating component remains crucial in the study of animals, if only because language and, to an even greater extent, scientific knowledge entail distinction. But the said differentiating component must be complemented by a holistic one that will keep trenchant categorizations at bay. The method based on the definition of flexible concepts is admirable in the sense that it succeeds in accounting for both "non-human and human behaviors are similar" and "they are not the same." If the true imitation theory exemplifies a first possible pitfall—that of the reduction to difference—, another peril is the reduction to identity.

PART II. WE ARE JUST THE SAME. INGOLD AND THE ONTOLOGICAL PROBLEM OF METAPHOR

§9 Ingold's critique of social constructivism: Transition to the ontological level

Tim Ingold's analyses of metaphors and anthropomorphism in *The Perception of the Environment* present somehow the mirror image of the "true imitation" approach. As a social anthropologist, Ingold is perfectly aware that ontological biases may insidiously color even a scientific investigation. And his diagnosis is clear: Western thought, which consistently relies upon concepts such as body-mind, nature-culture, animal-human, is plagued by a set of correlative great ontological divisions.[39] With characteristic argumentative vigor, Ingold succeeds in clearing the air for a much more lucid debate; however, the ontological alternative Ingold proposes replaces radical differences by identity and, in its turn, misses the complex nature of similarity.

When the Waswanipi Cree of northeastern Canada say that the animals themselves give the people what they need to live, do they metaphorically talk about nature by using patterns coming from the domain of social relations to make sense of their environment, as Adrian Tanner claimed, for instance?[40] Do they "*reinterpret* the facts about particular animals (...) *as if* they had personal relations with the hunters"?[41] A similar analysis of anthropomorphism has been advocated by Feit ("In the culturally constructed world of the Waswanipi the animals (...) are thought of as being 'like persons' in that they act intelligently and have wills and idiosyncrasies, and understand and are understood by men"),[42] Gudeman,[43] or Bird-Davis.[44]

At stake here for Ingold is the value of a constructivism that he takes to be "a commonplace in anthropological literature."[45] According to this "commonly adopted position,"[46] humans, as intentional beings, originally perceive their environment and necessarily refer to it through a system of mental representations. The latter can only emerge by applying a certain set of categories and interpretation patterns to raw sensory data.[47] The meaning thus projected is always defined within a certain culture. And, indeed, Western thought has often been forced to recognize that its classical logic and its attachment to the *logos* were historically situated and contingent.

Ingold actually wants to show that the constructivist theory—at least in the specific form that I have just described, paraphrasing Ingold—is nothing but a new avatar of Western biases. In effect, constructivism assumes the old dichotomy between humans and animals and cannot but fail to achieve its explicit goal—namely, to overcome a naturalistic vision of the hunter-gatherer economy.[48]

If, the constructivist approach claims, there is such a thing as a meaningless, pure nature upon which meaning can be projected—as opposed to a culture defined as a system of categories, concepts, and patterns of interpretation—it is at least tempting to claim that animals belong to the realm of pure nature. From this perspective, only humans are cultural beings, insofar as they *question* their own naturality and reach the concept of nature, which necessitates a "disengagement from the world"[49] and the ability to be at the same time half in nature and half out, with access to an elevated perspective.

Moreover, constructivist theory entails that hunter-gatherers interact with animals as persons only at the level of interpretation, "in a domain of virtual reality,"[50] in the realm of the "as if." The constructivist approach "creates a separate logical space,"[51] in which a whole dimension of the experience of the hunter-gatherers is considered as being "lived" (but in no way perceived or represented) at the organic level, as a set of pure mechanical or instinct-driven interactions between material entities.[52] This constructivist stance indeed justifies the commonly held claim that hunter-gatherers "in deriving their subsistence from hunting and trapping 'wild' animals and gathering 'wild' plants, honey, shellfish and so on, *are somehow comparable* in their mode of life to non-human animals in a way that farmers, herdsmen and urban dwellers are not."[53] A likely consequence of this approach is that many researchers still designate the activities of hunting and gathering as "foraging."

Ingold argues that constructivism, along with the metaphor theory that derives from it, is absolutely not neutral.

First, hunter-gatherers themselves do not characterize their relationships with nature and animals in terms of construction, interpretation, "perspective on," or metaphor.[54] Therefore, the projection theory, in radically distinguishing between nature and culture, actually imposes a Western worldview upon other cultures while claiming not to do so.[55] In other words, proponents of constructivism claim to simply compare two cultural patterns of interpretation of the world, while their method itself is shaped by their own cultural interpretative framework.

Furthermore, such a constructivism is paradoxical. According to this theory, nature "appears on two sides: On one as the product of a constructional process, on the other as its precondition."[56] In effect, for the constructivist theory to be true, there must exist a pure nature; a raw, unshaped, and meaningless-in-itself reality, in contrast to cultural constructions. Yet, according to this same theory, "nature" cannot but denote a concept that was elaborated within the framework of a specific culture.

Ingold thus repudiates cultural constructivism and posits an alternative non-dualist ontology inspired by the perspective of hunter-gatherers. This ontology is, as a result, no longer regarded as a mere perspective among others. Ingold's bold move is implicitly based on the idea that it is possible to reach an ontology of the *real* world itself, beyond alleged cultural differences. Such a claim actually does not necessarily follow from the criticism of constructivism as a paradoxical stance: It is always possible to conclude that, indeed, relativism is relative and that everybody will simply be free to advocate a dogmatic or a relativist position. In addition, the concept of the world itself, of *the real* reality beyond perspectives remains problematic, and Ingold does not address such objections. But, as I will argue in the next chapter, a phenomenological approach will help unfold the arguments that can support Ingold's commitment to a strong ontological stance.

§10 Ingold's ontology of dwelling

Let us listen more attentively to the discourse of the hunter-gatherers instead of trying to superimpose our concepts of *interpretation, social construction*, and *metaphor* onto their experience. Ingold intends to do so—however problematic his claim to understanding hunter-gatherers from a *de facto* Western situation may be—and submits the following description, for which he coins the name "ontology of dwelling."[57]

Human and non-human animals are "immersed from the start in an active, practical and perceptual engagement with constituents of the dwelt-in world."[58] Through this dwelling, the world is "apprehended"[59] but not re-presented. Ingold systematically contrasts *dwelling in* the world and *objectivizing* the world, adopting the spectator attitude, re-presenting the world, and building a view of the world. The dwelling is intrinsically holistic: "From the perspective of dwelling, the dichotomy [between one world and many worlds] is meaningless. For the dwelt-in world is a continuous field of relationships unfolding through time."[60] Our entanglement within a set of relations comes prior to the analytical process that distinguishes and separates the different terms of the relations as well-circumscribed, potentially autonomous beings. Bodies-behaviors-tools-environment-home-fellows-threats-foreigners-horizons are primordially intertwined. This phrasing is still slightly misleading, since intertwinement precedes to the intertwined entities. The latter progressively emerge against this inclusive background, without ever becoming radically separated, otherwise they would never form a world and we would not perceive them. The concept of dialogue may be illuminating here: When I truly dialogue with another subject, it is impossible to discriminate between what rightfully and exclusively belongs to *my* contribution and what, strictly speaking, belongs to hers. "My" ideas are aroused by the very situation of dialogue, the other's expectations, attitudes, and responses. Listening to the other compels me to transpose myself into her place and introject the style, the rhythms, and the intentions of her speech. Talking to her entails striving to adapt my speech to her comprehension.[61] We are by no means one and the same and we do not agree on everything: Distinct subjects emerge through the dialogue, but clear-cut boundaries will come second and must remain tentative.

Ingold does not expand much on why this paradigm of dwelling is universally fundamental—beyond the particular perspective of hunter-gatherers—and why it applies to every aspect of our existence. I will elaborate on this point in the next two paragraphs by developing an ontological concept of dwelling that Ingold merely adumbrates; doing so will elucidate that Ingold's version of this ontology pays too little attention to the status of differences and similarities.

My body, to begin with, can only be roughly delimited by the surface of my skin. Such a delineation is not entirely absurd of course, but it is also partly artificial. For instance, speaking of *my* skin as one personal, perfectly homogenous entity is problematic. The skin floras are incredibly diverse. Resident, transient, or opportunistic entities coexist in a fragile balance and may always become pathogenic. Similarly, it does not make much sense to draw a harsh line between an organism and its milieu. At a more phenomenological/existential level, one can also argue that there is only a rough overlap between an objective delimitation of my body seen from outside [*der Körper*]—as a material object exactly comparable to this table or this chair—and the body schema, namely the experienced body, as shown, for instance, by the experiments and studies presented in Schilder's *The Image and Appearance of the Human Body*.[62] The body schema integrates phantom limbs, clothes, adornments, tools, prostheses, and familiar spaces, for instance. By referring to engagement in practices and perception, Ingold—exactly as Heidegger did—highlights that the analytical representational process that divides, conceptualizes, and separates entities must take place in the framework of a more original and holistic relationship with the world, in other words: A being-in-the-world. In everyday practices, I do not with-draw into the spectator position, I am rather thrown into the situation. I am entirely absorbed by the door and, through it, by social conventions, social meanings, bodily negotiations about who should enter first, and power rela-tions with others. I am thus multicentered. "There is no place for *me* on this level," Sartre has argued.[63] Moreover, the priority is ontological: How could essentially unrelated entities ever enter into relation? On the basis of what common ground? If fragments came first, there would never be any such thing as a world.

Likewise, Ingold contends, perception is a form of *engagement*. The whole body must be active in exploring its surroundings. It must adjust to the solici-tations and the affordances of the environments, but it also brings its own style, rhythms, accommodations, and focus. Consider the example that was developed by Merleau-Ponty in *The Structure of Behavior*,[64] on the basis of observations and experiments achieved in the field of Gestalt psychology. Let us suppose that a room is lit by the yellow light coming from an electric light bulb. We perceive the walls as white and the daylight coming from the window as blueish. We do not notice that the ambient light in the room is yellow, unless, for instance, we just entered the room, or we were, an instant earlier, staring out of the window. An interpretive process is at play here, but it is performed at the level of the body attitude and behaviors. My body, through exploratory gestures, and accommodation processes responds to hints and solicitations coming from the environment. For instance, it may spontaneously perceive the electric light as the ground light, namely as a neutral *white* light, while, correlatively, the colors of lights and objects in the room undergo a change of quality proportional to the one undergone by what has now become the ground light (for instance, the light coming from the window looks blueish). The intellectual or theoretical decision not to perceive the electric light as neutral will be fruitless. However, we can always adopt a specific bodily attitude that enables us to perceive the light coming in through the window—instead of the electric light—as the ground light. Perception is a practical dialogue, a process of adjustments and nego-tiation between the exploratory gestures and behaviors of my body and an environment that solicits them, guides them, but also lends itself to different interpretations. The dichotomies between the given and the constructed, body and mind, subject and object cease to be fundamentally relevant here, as suggested by Ingold.[65]

Within this "dwelling" approach, nurturing, care, economical exchanges, and personhood are immanent in a field of practical and perceptual rela-tionships that preexist cultural "interpretations" of the world. For instance, the structure "attending to" is fundamentally *reciprocal* in the practice of hunting. "What is certain (…) is that humans figure in the perceptual world of geese just as geese figure in that of humans. It is clearly of vital impor-tance to geese that they should be as attentive to the human presence as to the presence of any other potential predator. On the basis of past experience, they learn to pick up the relevant warning signs, and continually adjust their behavior accordingly. And human hunters, for their part, attend to the presence of geese in the knowledge that geese are attending to them."[66]

And, indeed, it certainly makes much more sense to list many common points between the existence of non-human animals and the lives of human animals now that intentionality is no longer primarily understood as an intellectual and representational state. At the level of dwelling, intentionality first takes the form of an engagement in practices and perception. In fact, as the next chapter indicates, a similar concept of intentionality allows Uexküll to contend that non-human animals shape their own world [*Umwelt*] through a certain selectivity implemented by their body. Non-human animals indeed focus on and give meaning to certain constituents of their environment through their way of using them, through their perceptual senses (instruments of focus, accommodation, selection), their interests/desires/attention, and their feelings of pleasure and pain.

Care or personhood is accordingly not secondarily projected onto nature: They are directly apprehended, within the original engagement of hunter-gatherers, as characteristics of animals and the environment. They do not exclusively belong to the realm of cultural existence, since the latter is only secondarily and tentatively separated from the realm of nature.[67] Structures of personality, social existence, and meaning-giving are consequently not *incidentally re-presented* as features of non-human animals. They are all the less so that, in order to metaphorically project certain characteristics onto an entity, the latter must prepare, allow for, and lend itself to such an interpretation. A metaphor would boil down to a meaningless and arbitrary connection between two words if absolutely no link existed between the two realities that it convenes. This argument—a key in a phenomenological approach to the imaginary field—underpins Ingold's position: If we start from an ontological dualism, the Cree's description of animals as persons appears to be a completely artificial figment of the cultural imagination,[68] a pure construction, the *building of a view* rather than the *taking up of a view in* the world.[69] To take up a view is to enhance, develop, but also—and first of all—resume a process of meaning-giving that actually starts in the world, in the perceived and not only in the perceiver. Thus, in the dialogue between my body and the different light sources in a room, such as described above, my perception is first *guided* by a prevailing ground light and by the relations between lights and colors in the room. More generally, perception would be a pure fantasy, it would lose its consistency, if it were produced by the application of a new pattern of interpretation to a meaningless set of raw data.[70]

§11 Are non-human and human animals *just the same*? Beyond metaphor or at the heart of metaphor?

Although the foregoing represents a major step forward in the acknowledgement of *similarity* as a key to human-animal relations, Ingold discards too quickly the concept of metaphor, and taints the subtle complexity of the ontology of dwelling by sticking to a more robust ontology of identity.

Ingold, as well as Jackson[71]—to whom he refers—and Nadasdy,[72] who claims that Ingold did not go far enough in overcoming the metaphor theory, persist in thinking within the framework of an exclusive alternative between "metaphorical" and "real." Ingold contends that, for hunter-gatherers, "animals are not like persons, they are persons,"[73] and he quotes Jackson: "Metaphor reveals, not the thisness of a that but rather that *this is that*."[74] Therefore, Ingold concludes, "at root the constitutive quality of intimate relations with non-human and human components of the environment is one and the same."[75] Even the idea of a consciousness of non-human animals becomes simply unproblematic according to Ingold.[76]

But these assertions actually do not follow from the ontology of dwelling. The criticism of dualisms and clear-cut divisions does not entail that every being should be reduced to one and the same ontological plane,[77] on which differences will be flattened out.

And differences do immensely matter. The concept of *one and the same dwelt-in world* remains troublesome. Assuredly, beings are interlaced, existences overlap, and even if I want to claim that I have actually no clear idea of the way an oyster "attends" its world and is aware of it, I still have to acknowledge that the oyster is part of my world. We must share a minimal common ground, even in order to observe that we are a complete mystery to each other. Nevertheless, "the world," as a minimal common ground or a horizon, is not exclusive of the persistent formation of world*s*. The latter can be obscure to each other, foreign to each other, even in conflict or at war: Many features that are crucial in the realm of animal studies. Gaps, blind spots, muddling effects, and daring, tentative approaches are essential aspects of communication and especially of interspecific communication. Hunter-gatherers themselves may occasionally emphasize such discrepancies: Ojibwe, for instance, claim that non-human animals can be encountered in a privileged way through dreams.[78]

Life is indeed a source of accentuated differentiation, as suggested by the concept of personhood that Ingold puts forward. The living is unable to exist without the world, or other living beings, but it subsists only in the formation of differences of potentials and energy buildup in the service of a certain structure, as if life was building dams to articulate and differentiate a massive flow of energy. Living beings develop morphological differentiation—yet a relative one and a more or less advanced one depending on the considered species—between the interior and the exterior of the organism. As Uexküll argues, they organize their world *around* them (*Umwelt*). They shape *their Umwelt* on the basis of endogenous norms. They see routes, territories, aspects (certain sensory qualities or functions, for instance), and entities (the dog's omnipotent master, or the imaginary flies chased by the starling, for instance[79]) that remain *invisible* to others.

Uexküll's approach in *A Foray into the Worlds of Animals and Humans* is especially interesting in comparison to Ingold's tendency to reduce non-human and human animal worlds to one and the same. Uexküll identifies the *Umwelten* as *"closed"* soap bubbles, explaining: "The bubble represents each animal's environment and contains all the features accessible to the subject. As soon as we enter into one such bubble, the previous surroundings of the subject are completely reconfigured. Many qualities of the colorful meadow vanish completely, others lose their coherence with one another, and new connections are created. A new world arises in each bubble."[80]

Uexküll's theory is at the very least confusing for he wants to emphasize both the differences and the kinship between animals. On the one hand, the image of the soap bubbles suggests, maybe unintentionally, that each *Umwelt* is not entirely opaque or closed to us. In effect, Uexküll undertakes strolls (*Streifzüge*) through the *Umwelten* of human and non-human animals. He has even included in his book drawings that supposedly depict the world such as it appears to this or that animal. Yet, Uexküll specifies that animals are irremediably "enclosed" in bubbles "closed on all sides."[81] To be sure, the image of a soap bubble may convey the idea of a certain transparency, but, in Uexküll's view, it means, first and foremost, that *Umwelten* are strange invisible entities that do not subsist as objective, real, positive things. Each *Umwelt* is *for* a particular animal. The only way of reconstituting the main features of *Umwelten* is to observe the behaviors of the said individual as well as, to a certain extent, of conspecifics. We cannot, Uexküll is keen to emphasize, fully share their world. A subject is defined by a first-person perspective, which is essentially a critical blind spot for other subjects. Uexküll thus distrusts an approach that would be based on empathy: Animal worlds "will not be revealed to our body's eyes but to our mind's eyes."[82] Similarly, Uexküll does not believe that human beings, and particularly scientists, have access to *the* world of all of these worlds: We are also living in our own bubble. Getting a glimpse of certain animal worlds should not be identified with "being able to fully overcome the biases of our human perspective." We are not stuck in one unique blind perspective, but each new discovery may still be biased and remains deeply human in its own way.

Differences, although they are not chasms and would not be properly described through dichotomies, must not be overlooked, as they yield concrete and crucial difficulties in the relationships between animals, and, especially, between human and non-human animals. Interspecific communication does exist and can be further developed, but it may also fail, leave room for huge misunderstandings, even make way for severe conflicts. In Herzog's *Grizzly Man*, Timothy Treadwell's relationship to bears was certainly not a complete failure. After all, he managed to live in their territories for months. But his immoderate recourse to empathy and his feeling of being connected by ties of deep friendship with them led to a tragic ending. And several times, in the documentary, grizzly bears only respond with silence and apparent indifference to Treadwell's babbling. It remains doubtful that Treadwell ever succeeded in being integrated, rather than tolerated, within their territory. It would be excessive to claim that he built a *common world* with them.

Further, haven't humans been at war with animals for thousands of years? Something in the world of dwelling made possible the emergence of the Cartesian project.[83] Even if the idea that human beings are not animals any longer is a myth, such a fantasy is not a mere nothing. This anthropocentric myth is associated with an *actual* incredibly violent exploitation of animals. As put by Derrida: The destruction and annihilation of animals by humans is not a mere fiction.[84] Asserting that animals and humans do not belong to the same world proved to possess an imperious performative efficiency. Again, we have to understand an entity—the effective myth of the divorce between the human realm as the animal realm—that lies below the dualism between real and imaginary.

Ingold does not entirely deny differences. He points out that "humans may of course be unique in their capacity to *narrate* such encounters [with animals],"[85] or that "It is in their capacity to *construct imagined worlds* that humans surely differ from other animal species."[86] But he also consistently considers differences to be anecdotal:[87] "Non-human animals can constitute their environments, *just as humans can*, through the very fact of their dwelling in the world. There is no *fundamental* difference here. It is in their capacity to *construct imagined worlds* that humans surely differ from other animal species, and in this both language and culture are directly implicated. (...) let me remind you that such imagining is not a necessary prelude to our contact with reality, but rather an epilogue, and *an optional one at that.*"[88]

Ingold develops an ontology that cannot account for metaphors in a better way than Western ontology did. Western ontology, Ingold contends, regards only differences as ontologically relevant and, therefore, could not account for the existence of relationships, connections, and hybrids. By contrast, Ingold regards only common points as ontologically relevant. He thus fails to acknowledge the distance that must emerge between different subjects and different *Umwelten*. Correlatively, he strangely reduces the human *playful interpretation* of the world to a vain and superficial imagination, ascribing the latter to the realm of non-being or at least deficient being.[89] But even in dwelling, at the level of perception, an important dimension of creative interpretation is at work, as was already suggested by the example of the ground light in a room: The way we practically explore a place and enter into dialogue with the structure of our environment can evoke new perspectives and reveal new qualities. As Uexküll suggests, there are good reasons to contend that non-human animals do the same. Animal individuals can create new ways of using some tools or even invent new fashion trends.[90] They can lure their fellows or predators (see the example of the broken-wing display used by many birds to lure predators away from the nests[91]) or, like Dolly,[92] invent bold metaphors. This dimension of dynamic transposition (meta-phora in Greek) and creativity also makes each perspective on the world developed by a living being contingent, which is a source of possible divergences, misunderstandings, and gaps between various worlds.

Discarding one dimension of our experience on the basis of a discriminatory ontology amounts to shirking the essential and most difficult task of ontology, such as it was defined, for instance, by Aristotle. Ontology must, first, acknowledge the different modes of being in their specificity (metaphors *are*, in their way) and, second, give a definition of Being in general that can account for the emergence of *all* these different modes of beings, not only a limited selection of them.

Notes

1. De Waal, Frans, *Primates and Philosophers: How Morality Evolved.* Princeton: Princeton University Press, 2006, p.60.
2. For instance: Understanding the others' intention cannot simply consist in considering their perspective, but must, allegedly, take the form of a theory of mind. See, for instance, De Waal, Frans, *Are We Smart Enough to Know How Smart Animals Are?*, p.130–2.
3. *Ibid.* p.119–20, 128.
4. "Symbolic representation and working memory in chimpanzees" YouTube video uploaded by TheFriendsAndAi, https://www.youtube.com/watch?v=DqoImw2ZWmI
5. Heidegger, Martin. *Die Grundbegriffe der Metaphysik: Welt—Endlichkeit—Einsamkeit*, §50, p.308. Trans. William McNeill and Nicholas Walker, Bloomington: Indiana University Press, 1995, p.210.
6. See Thorpe, William Homan. *Learning and Instinct in Animals.* London: Methuen. 1956; Tomasello, Michael, Ann C. Kruger, and Ratner, Hilary H. "Cultural Learning." In *Behavioral and Brain Sciences.* 495–552, 16.3: 1993, and many others.
7. Whiten, Andrew and Custance, Deborah. "Studies of Imitation in Chimpanzees and Children." In *Social Learning in Animals: The Roots of Culture* (Heyes, C.M., & Galef, B.G., Jr. Eds.). San Diego: Academic Press, 1996.
8. In *emulation* the subject endeavors to equal or excel others. The stress is laid on the results of the actions and the animal is simply encouraged to achieve a task, in a very general sense, by seeing others achieving it. The intention of mimicking every movement of the other is absent. "The individual observing and learning some affordances of the behavior of another animal, and then using what it has learned in devising its own behavioral strategies, is what I have called emulation learning." Tomasello, Michael. "Do Apes Ape?" In *Social Learning in Animals: The Roots of Culture* (Galef, B.G. Jr. Ed.). San Diego: Academic Press, 1996, p.321.
9. Spence, K. "Experimental Studies of Learning and Higher Mental Processes in Infrahuman Primates." *Psychological Bulletin,* 34, 1937. See also Thorpe 1956 and Tomasello et al. 1993.
10. See Tomasello, "Do Apes Ape?"
11. *Ibid.*
12. Tomasello, Michael. *The Cultural Origins of Human Cognition*, Cambridge: Harvard University Press, 1999, p.39.

13. Tomasello's works on imitation are explicitly presented as a response to the studies showing that new techniques may be invented and transmitted within an animal society (see Tomasello, Michael, Savage-Rumbaugh, Sue, and Kruger, Ann Cale. "Imitative Learning of Actions on Objects by Children, Chimpanzees, and Enculturated Chimpanzees." In *Child Development*, Vol 64 No. 6, 1993, p.1689. The authors refer to Jane Goodall, *The Chimpanzees of Gombe: Patterns of Behavior*. Cambridge: Harvard University Press, 1986 and Boesch, Christophe, Boesch, Hedwige. "Tool Use and Tool Making in Wild Chimpanzees" *Folia Primatologica*, 54, 1990). This may look like non-human animal culture. And many researchers claimed it was. Tomasello, however, stuck to the cut-and-dried statement: "Only humans live in culture" (Tomasello, Kruger, and Ratner, "Cultural Learning." p.495). The reasoning goes as follows. a) Such new techniques *may* have been learnt by chimpanzees through emulation or through imitation. b) Animals cannot imitate. c) Therefore, these techniques have been learnt through emulation. d) Therefore, such techniques cannot be regarded as a form of culture. The passage from a speculative hypothesis to a trenchant statement and the reassertion of a radical anthropological difference are based on the actually very questionable assertion "animals cannot imitate."

14. Tomasello thus refers to the "ever-elusive 'true' imitation." Nevertheless, he does not comment upon his use of quotation marks and clearly asserts that the concept of *imitative learning* is his version of this ever-elusive "true" imitation (Tomasello, "Do Apes Ape?", p.324).

15. Whiten, Andrew, et al. "Emulation, Imitation, Over-Imitation and the Scope of Culture for Child and Chimpanzee," In *Philosophical Transactions of the Royal Society B: Biological Sciences,* 364 (1528), 2009, p.2418.

16. *Ibid.*, see also Whiten & Custance, "Studies of Imitation in Chimpanzees and Children," 1996.

17. See Moore, Bruce. R. "Avian Movement Imitation and a New Form of Mimicry: Tracing the Evolution of a Complex Form of Learning." *Behaviour* 122, 1992. See also Stoinsky, Tara S., Wrate, Joanna L., Ure, Nicky, and Whiten, Andrew. "Imitative Learning by Captive Western Lowland Gorillas (Gorilla Gorilla) in a Simulated Food-Processing Task." *Journal of Comparative Psychology* 115, 2001. Voelkl, Bernhard & Huber, Ludwig. "True Imitation in Marmosets." In *Animal Behaviour*, 60, 2000. Voelkl, Bernhard & Huber, Ludwig. "Imitation as Faithful Copying of a Novel Technique in Marmoset Monkeys." PLoS One, 2(7). 2007. Ferrari, Pier F., Visalberghi, Elisabetta, Paukner, Annika, Fogassi, L., and Ruggiero, A. "Neonatal Imitation in Rhesus Macaques." *PLoS Biology* 4 (9): 2006.

18. Whiten & Custance, "Studies of Imitation in Chimpanzees and Children," p.309.

19. Tomasello, "Do Apes Ape?", p.323, 324, 329, 331.

20. Against the idea that animals cannot imitate because they cannot understand the intentionality in the behavior of another individual, it has been argued that many animals are actually able to take the others' perspective into account (see for instance De Waal, *Are We Smart Enough to Know How Smart Animals Are?* p.130–2, 146–8. Thierry, Bernard. "Social Transmission, Tradition and Culture in Primates: From the Epiphenomenon to the Phenomenon." In *Techniques & Cultures* 23 (24). 1994. And Morin, Olivier. *How Traditions Live and Die*. Oxford: Oxford University Press, 2016, p.60).

21. Morin, Olivier, *How Traditions Live and Die*, p.59. "Once I had as a neighbor a man who kept every single wrapping, box, tin can, old newspaper he ever had; the police eventually forced him to empty his basement (which had become a fire hazard). (...). True imitation, as defined so far, is such a hoarder. Of all the social learning mechanisms that have been described, it is one of the stupidest."

22. Whiten & Custance, "Studies of Imitation in Chimpanzees and Children," p.305.
23. Whiten et al. "Emulation, Imitation, Over-Imitation and the Scope of Culture for Child and Chimpanzee," p.2422.
24. *Ibid.* p.2422. See Tomasello, Michael, *The Cultural Origins of Human Cognition*, p.29.
25. Whiten et al. "Emulation, Imitation, Over-Imitation and the Scope of Culture for Child and Chimpanzee," p.2422.
26. Whiten & Custance, "Studies of Imitation in Chimpanzees and Children," p.305-6.
27. Tomasello, "Do Apes Ape?", p.320. See also p.321: "In local or stimulus enhancement *nothing* is actually learned from the behavior of others" (my emphasis) and p.322: "Despite the power of local enhancement, stimulus enhancement, observational learning, and emulation learning to help individuals benefit from the skills of others, all of these processes of social learning operate without the individual organism *paying any attention whatsoever* to the actual behavior of other organisms" (my emphasis).
28. See Whiten and Custance, "Studies of Imitation in Chimpanzees and Children," p.310-1. Byrne, R.W. and Byrne, J.M.E. "Complex Leaf-Gathering Skills of Mountain Gorillas (Gorilla g. beringei): Variability and Standardization." In *American Journal of Primatology*, 31. 1993. Byrne, R.W. "The Evolution of Intelligence". In *Behaviour and Evolution* (Slater, P.J.B. & Halliday, T.R. Eds.). Cambridge: Cambridge University Press. 1994. Huber, L., Range, F., Voelkl, B., Szucsich, A., Viranyi, Z. & Miklosi, A. "The Evolution of Imitation: What do the Capacities of Non-Human Animals Tell Us About the Mechanisms of Imitation?". In *Philosophical Transactions of the Royal Society of London. B. Biological Sciences*, 364. 2009.
29. Whiten et al. "Emulation, Imitation, Over-Imitation and the Scope of Culture for Child and Chimpanzee," p.2420.
30. *Ibid.*
31. Zentall, Thomas R. "An Analysis of Imitative Learning in animals." In *Social Learning in Animals: The Roots of Culture* (Heyes, C.M., & Galef, B.G. Jr, Eds). San Diego: Academic Press, 1996, p.235-6. See also Whiten & Custance, "Studies of Imitation in Chimpanzees and Children," p.306-8, and Whiten, A., and Ham, R. "On the Nature and Evolution of Imitation in the Animal Kingdom: Reappraisal of a Century of Research." In *Advances in the Study of Behavior* (Vol. 21) (Slater, P.J.B., Rosenblatt, J.S., Beer, C., & Milinski., M. Eds.). New York: Academic Press, 1992.
32. Whiten & Custance, "Studies of Imitation in Chimpanzees and Children," p.308.
33. *Ibid.*, p.306.
34. *Ibid.*
35. "Only humans live in culture" (Tomasello, Kruger, and Ratner, "Cultural Learning," p.495).
36. De Waal, Frans, *Are We Smart Enough to Know How Smart Animals Are?*, p.122-126.
37. Tomasello, Michael. *The Cultural Origins of Human Cognition*, p.39.
38. I thank my colleague Veronica Vasterling for suggesting this phrase to me.
39. Ingold, Tim. (Ed.). *Key Debates in Anthropology*, London & New York: Routledge, 1996, p.93-4. See also Ingold *The Perception of the Environment: Essays in Livelihood, Dwelling, and Skill*. London & New York: Routledge 2000, p.40, 44, and 50-1.
40. Ingold, *The Perception of the Environment*, p.48. See Tanner, A. *Bringing Home Animals: Religious Ideology and Mode of Production of the Mistassini Cree Hunters*. New York: St Martin's Press, 1979, p.136.

41. Tanner, *Bringing Home Animals*, p.136 (my emphasis).
42. Feit, Harvey. "The Ethnoecology of the Waswanipi Cree: or How Hunters can Manage their Resources." In *Cultural Ecology: Readings on the Canadian Indians and Eskimos*, B. Cox. (Ed.). Toronto: McClelland and Stewart. 1973. p.116.
43. Gudeman, Stephen. *Economics as Culture: Models and Metaphors of Livelihood*. London: Routledge & Kegan Paul, 1986, p.43–4.
44. Bird-Davis, Nurit. "Beyond 'the original affluent society': A Culturalist Reformulation." In *Curent Anthropology* 33, 1992, p.31.
45. Ingold, *The Perception of the Environment*, p.41.
46. *Ibid.*, p.40.
47. Ingold, *Key Debates in Anthropology*, p.94.
48. Ingold, *The Perception of the Environment*, p.43.
49. Ingold, *Key Debates in Anthropology*, p.93.
50. Ingold, *The Perception of the Environment*, p.52.
51. *Ibid.*, p.59.
52. *Ibid.*, p.57, 59.
53. *Ibid.*, p.58, my emphasis.
54. *Ibid.*, p.44.
55. *Ibid.*
56. *Ibid.*, p.41.
57. *Ibid.*, p.42.
58. *Ibid.*
59. *Ibid.*
60. Ingold, *Key Debates in Anthropology*, p.96–7. See also "They see themselves as involved in an intimate relationship of interdependence with the plants, animals and hala' (including the deities) that inhabit their world." (Ingold 2000, p.43, he refers to Endicott, Kirk. *Batek Negrito religion*. Oxford: Clarendon Press, 1979, p.82). See also Ingold, 2000, p.51.
61. Merleau-Ponty, Maurice. *Psychologie et pédagogie de l'enfant, Cours de Sorbonne 1949–1952*. Lagrasse: Verdier, 2001, p.58–9.
62. Schilder, Paul. *The Image and Appearance of the Human Body: Studies in the Constructive Energies of the Psyche*. London: Routledge, 1950, p., tr. fr. par F. Gantheret et P.Truffert, L'image du corps. *Etude des forces constructives de la Psyché*, Paris, Gallimard, 1968, p.85.
63. Sartre, Jean-Paul. *The Transcendence of the Ego. An Existentialist Theory of Consciousness*. Trans. Forrest Williams and Robert Kirkpatrick. New York: Hill and Wang, 1957, p.49.
64. Merleau-Ponty, Maurice. *La structure du comportement*. Paris: PUF, 1942, Quadrige, 1990. Trans. Alden Fisher, Boston: Beacon Press, 1963, p.82.
65. Ingold, *Key Debates in Anthropology*, p.95.
66. Ingold, *The Perception of the Environment*, p.51.
67. Ingold, *Key Debates in Anthropology*, p.96 "he who would rebuild the world in his imagination must already dwell in it."
68. *Ibid.*, p.95. And p.94: "As Geertz would have it, culture consists in 'the imposition of an arbitrary framework of symbolic meaning upon reality.'" See also Ingold's critique of the distinction between a given reality and a constructed reality (Ingold, 1996, p.94 and 96).
69. Ingold, *The Perception of the Environment*, p.42.
70. Merleau-Ponty, *Phenomenology of Perception*, Trans. D. Landes. New York: Routledge, 2012, p.XXIV and 22.
71. Jackson, Michael. "Thinking through the body: An essay on understanding Metaphor". In *Social Analysis*, 14, 1983

72. Nadasdy, Paul. "The gift in the animal: The ontology of hunting and human–animal sociality". In *American Ethnologist*, Volume 34, Issue 1, February 2007.
73. Ingold, *The Perception of the Environment*, p.51.
74. *Ibid.*, p.50, Jackson, 1983, p.132. see also Jackson, p.132: "In my view metaphor reveals unities," and Nadasdy, 2007, p.27.
75. Ingold, *The Perception of the Environment*, p.47. See also p.50: "When Cree hunters claim that a goose is in some sense like a man, far from drawing a figurative parallel across two fundamentally separate domains, they are rather pointing to the real unity that underwrites their differentiation," and Ingold, *Key Debates in Anthropology*, p.97"
76. *The Perception of the Environment*, p.51.
77. *Ibid.*, p.59: "In the lower diagram, representing the hunter-gatherer ontology, there is but one plane, in which humans engage, as whole organism-persons, with components of the environment, in the activities of procurement.
78. Ingold, *The Perception of Environment*, p.94. See infra, §33.
79. Uexküll, Jakob von. *A Foray into the Worlds of Animals and Humans. With A Theory of Meaning*, Minneapolis, London: University of Minnesota Press, 2010, p.120–1.
80. *Ibid.*, p.43.
81. "We must therefore imagine all the animals that animate Nature around us (...) as having a soap bubble around them, closed on all sides, which closes off their visual space and in which everything visible for the subject is likewise enclosed." (*Ibid.*, p.69).
82. *Ibid.*, p.42.
83. Another problematic aspect of Ingold's approach lies in the clear-cut distinction that he makes between the hunter-gatherers' ontology of dwelling and the Western ontology. The ontology of dwelling has also been developed by many Western thinkers, as, for example, the concept of the lifeworld [*Lebenswelt*] in Husserl's phenomenology demonstrates. Westerners are no more the pure proponents of ontological dichotomies (see also Bruno Latour's *We have never been modern*, trans. by Catherine Porter, Harvard University Press, 1993) than the hunter-gatherers are immersed in a fusional world, which actually validates Ingold's general refusal of dichotomies, but shows that he has to pay closer attention to the ambiguity of both these ontologies. We also have to understand how the dream of the Great Divide between humans and animals could come into being within the framework of the dwelling and its holistic relationship to the world, in such a way that such a schismatic dream was able to *actually* threaten the unity of the latter.
84. Derrida, *The Animal That Therefore I am*, p.80.
85. Ingold, *The Perception of the Environment*, p.52.
86. Ingold, *Key Debates in Anthropology*, p.97.
87. In this regard, the persistence in Ingold's analyses of the platonic concepts of "what truly/really is," in contrast to "superficial illusory non-beings," is quite telling. "*At root*, the constitutive quality of intimate relations with non-human and human components of the environment is one and the same." (Ingold, *The Perception of the Environment*, p.47 my emphasis). See also p.50 and 52, and *Key Debates in Anthropology*, p.93: "I shall proceed to put forward an alternative view which restores people to *where they belong*, in an active practical engagement with constituents of the real world".
88. Ingold, *Key Debates in Anthropology*, p.97.
89. Likewise, taking a theoretical stance is, according to Ingold, a way of "rebuilding the world in one's imagination" (Ingold, *Key Debates in Anthropology*, p.96).

90. See Huffman, Michael A. "Acquisition of Innovative Cultural Behaviors in Nonhuman Primates: A Case Study of Stone Handling, a Socially Transmitted Behavior in Japanese Macaques." In *Social Learning in Animals. The Roots of Culture* (Heyes and Galef. Eds.). San Diego: Academic Press, 1996.
91. See for instance Caro, Timothy M. and Girling, Sheila. *Antipredator Defenses in Birds and Mammals*, University of Chicago Press, 2005, p.343–7.
92. cf. supra, p.5.

2 Phenomenology of the animal imaginary

Non-human subjects, ambiguous worlds, empathy

> "The minnows have come out and are swimming so leisurely," said Master Chuang, "this is the joy of fishes."
> "You're not a fish," said Master Hui." How do you know what the joy of fishes is?"
> "You're not me," said Master Chuang, "so how do you know that I don't know what the joy of fishes is?"
>
> Zhuangzi, quoted by Buytendijk[1]

§12 Why a phenomenological approach to non-human animals?

In this chapter, I focus on the imaginary of animals understood in the framework of a phenomenological approach and I contend that phenomenology provides a fruitful methodological and conceptual apparatus to solve the ontological problems highlighted in the previous chapter. Before delving into a detailed investigation of this issue, let us consider more broadly what contribution a phenomenological approach can make to animal studies.[2]

A pluralist approach: Objectivation, reflection, empathy, through everyday experiences, science, and fantasies

First of all, the strength of the phenomenological approach is that it is **intrinsically pluralist**. It aims first and foremost at *describing* the manifoldness of that which appears (*phenomena*), in other words, experience. Phenomenology thus seeks to investigate the phenomenality of non-human animals, the way they manifest themselves. Its scope extends beyond the objective study of nature and brings together, on an equal footing, all facets of our relation to animals as well as of animals' own relations to their world. Phenomenology therefore examines how animals and animality become both meaningful and mysterious to us through perceptions, emotions, fantasies, ever-recurring metaphors, transpositions, as well as in the scientific-naturalist attempt to turn them into measurable and predictable objects of knowledge. A phenomenological perspective enables us to avoid the pitfall of what I have described as reductive approaches.

From this pluralist perspective, the second specific asset of phenomenology is that it **confers the most crucial role to subjectivity** and it forges a method and concepts that enable us to study subjectivity *as such,* without reducing it to a radically hidden, private phenomenon.

Nagel's famous article describes how phenomenology looms on the horizon of an objectivist approach, arguing that we urgently need some form of phenomenology. Nagel shows the significance of a phenomenological approach that *would* concern itself with what it is like to be a bat, but he fails to overcome the issue of the alleged inaccessibility of such a subjectivity. In his view, I can only know what it would be like for *me* to be a bat, not what it is like for a *bat* to be a bat. The absence of reference to the school of phenomenology, founded by Husserl, in this article attends the conservation of the objectivist take as the unquestioned "realist" way toward knowledge. Nagel does not examine the source of knowledge that enables him to talk about *what it is like to be a bat* in an intelligible manner. He discards the richness of imagination and reduces it to a pure empty fiction, while using imagination to declare perspectives (of Martians or bats, for instance) inaccessible that he nonetheless includes in the scope of his descriptions. Moreover, Nagel does not see that the *project* of objectivity—the project of reducing everything to measurable entities and allegedly universally understandable formula—is the result of an extraordinary bold whim of creative imagination, a project that a Martian, and even many non-Western thinkers as well as countless Western philosophers, could regard as fascinating, relatively fruitful, but, above all, wildly baroque.[3]

This story Nagel tells us—about inaccessible subjects that are nonetheless somehow accessible—makes more sense in the framework of the phenomenological conversion, misleadingly called "**reduction.**" A close-up on the eye of a character—human or non-human animal—followed by a point-of-view shot that unveils the world such as it is perceived by this character is a recurring motif in films and confronts us with the paradoxical coexistence of two perspectives. On the one hand, the subject is presented as that which is hidden "behind" this opaque sense organ, behind a contingent bit of matter somewhere in the universe; on the other hand, the same subject is also that which engages the world through its perspective and imbues everything with a certain viewpoint, moods, dynamic intentions, and projects. Furthermore, this other subject then appears as a perspective into which I can transpose myself, namely as grafted upon an unlimited and fluid intersubjective world. Phenomenology performs the same strange switch through reduction. Phenomenology makes these two aspects coexist, but, less ambiguously than in a movie shot, it also highlights that the idea of a subject contained in a body is itself already an idea *for* a subject. The latter actually encompasses the eye, the subject, the world-for-this-subject, the world-for-other-subjects, and the ideal of the world-in-itself. The subject is therefore in no way confinable to any intracranial space: It is all other the place, here, there, and beyond. The objective world is not nothing, but it emerges from subjectivity. Husserl's phenomenology unveils that **we always start from subjectivity**, while Nagel was wondering how to reach subjectivity. Subjective experience is not reducible to a private, solipsist, and exclusively or primarily self-focused stream of consciousness. It is rather given as a world that is for me, but also tantalizes me, a world with various perspectives, through which I happen to circulate more or less fluidly, via exploratory practices, perceptions, anticipations, memories, and imaginations. In this world, I perceive and imagine my friend's anger, this dog's joy, and what it is like to be a bat. The "I" mentioned in these sentences is obviously the operator of such fluid transpositions: It is indeed, in this case hopefully, easily taken up by readers as theirs.

The **phenomenological epoché**—whereby I bracket all assertions regarding the reality of anything—does not filter anything out. Through the phenomenological **reduction**, one learns to stop focusing on objects and to reflect on one's own perceptive, affective, and intellectual relation to them. When performing the epoché, I still believe that the world exists "outside of me." I still see myself as a part of the world and cast the others as belonging over there. I may even still endeavor to measure time and circumscribe things and persons through an objectivizing perspective. But I simultaneously observe myself achieving this natural attitude as well as the epoché, which makes me realize that these are particular ways of engaging the world, among many others. Defining phenomenological correlationism as an anti-realism, as some authors in speculative realism do,[4] is simply wrong. The stubborn opacity of some things, blind spots, the transcendence of the world, its irreducibility to my thoughts, to human thoughts, or to any thought, are essential objects for phenomenology. The latter does not intend to *reduce* such objects *to* the subject's representation.[5] Phenomenology stands beyond the idealist/realist duality. Thought, subjectivity, and meaning are an irreducible dimension of the real, *together with* transcendence, resistance, and opacity. Hence, as this chapter will explain, we must deeply redefine subjectivity, along with reality. This is why it can be said that a phenomenological approach must be pluralist, which entails major consequences for animal studies.

Françoise Wemelsfelder stresses that scientists should not content themselves with the two-faced approach consisting in proclaiming a mechanistic vocabulary as the only officially legitimate and rigorous language, while the same scientists remain in fact dependent upon agency-related terminology when describing animals' behavior.[6] Similarly, as Richie Nimmo notes, "in their everyday working practices field primatologists tend intuitively to locate themselves in a common existential and relational space with their primate subjects, to whom they routinely attribute many of the features of 'culture' understood as lived intersubjectivity, individuality, embodied selfhood, and self-consciousness," but they avoid incorporating this approach to research publications.[7] As I have emphasized in the first chapter, Whiten is likewise annoyed that the concept of "true imitation" led Tomasello to break with the vernacular language and with a spontaneous perception of imitative behaviors in human and non-human animals. Scientists live, think, and experiment within what Husserl calls the *lifeworld* [*Lebenswelt*]—the world as experienced in everyday life and as extending far beyond the mathematized, objective world posited by modern science. Scientists may decide to focus exclusively on measurable properties, but the material they deal with in the first place is much broader and wilder: The manifold lifeworld regularly catches up with them. In the case of animal studies, a reductive approach ossifies the question of animal subjectivity into a "hard problem" that appears insoluble, simply clashing with many everyday experiences, and that, moreover, poses serious ethical issues. Phenomenology places objective analyses in a broader context: It **adds a dimension of reflection upon one's methodical choices to the scientific work**.[8] Furthermore, a phenomenological approach grants as much importance to **everyday experiences, metaphors, and artistic representation of non-human animals, as to objective descriptions** and mechanistic explanations.

As a consequence, the project of naturalizing phenomenology—in other words, the project of integrating every phenomenological stance "into an explanatory framework where every acceptable property is made continuous with the properties admitted by the natural sciences"[9]—is limited. It can prove particularly fruitful in a phenomenology of animal life, as Thompson's *Mind in Life*[10] demonstrates, for instance. Moreover, Husserl himself and many phenomenologists after him regard thought as essentially linked to this twofold entity that is my body, as *Leib*—the body that *I am*, the subject of my *I can, I see, I feel ...*—and the objective and measurable *Körper*. Phenomenology does not make the search for *correspondences*[11] between, on the one hand, measurable states of the body (and more particularly, but not exclusively, the brain), and, on the other hand, experiences in the first person irrelevant. Similarly, it is through animal behaviors that Uexküll, Buytendijk, or Plessner, for instance, using a phenomenological approach, find the basis for recognizing animal subjectivity. However, "naturalizing phenomenology" is too strong: A reductionist approach misses out on the pluralism of a phenomenological approach. Phenomenology does not boil down to a purely theoretical quest for clear knowledge. There are good reasons why Husserl's phenomenology *also* developed into existentialist phenomenology. The phenomenological method is irreducible to the search for means of prediction and explicative mathematical formulas. **The phenomenological reduction originally involves an ethical and practical dimension**. A respectful relation to subjects necessitates the preservation of a first- and second-person approach. It requires a practice of empathy, dialogue, and play that also carves space for the initiative of an unpredictable subject over there.[12] Such a respectful relation goes way beyond the scientific project of circumscribing, studying, knowing the properties of an object, and casting them into the rigid body of a systematic and explanatory account. Equally relevant objects of interest for phenomenology are the journey from naïve practices, everyday metaphors, and artistic metaphors to emerging new paradigms and ossified theories. Phenomenologizing naturalism is at least as important as naturalizing phenomenology.[13]

Animal subjectivity

Phenomenology enables us to study the subjectivity of animals through a **conceptual and methodological apparatus whose purpose is to describe subjectivity without reducing it**.

A major conceptual change, when one moves from reductive-objective approach to the phenomenological study of subjectivity, is that **"motives" must be substituted for "causes."** Accordingly, the **method becomes descriptive** rather than explanatory/causal. As §13 will indicate, a phenomenological approach emphasizes the role played by the interpretative adjustments and exploratory agency of subjects in the way this world appears to them: Things, events, and signs are not causes triggering reactions but motives to be interpreted.[14] Moreover, the phenomenal field is a relational whole. Subjects and objects are correlates. In a holistic approach, it does not make sense to isolate a circumscribable original cause, or to try to *reductively* account for complex experiences on the basis of a so-called lower level of explanation—for instance, to account for behaviors on the basis of neurobiological explanations.

Phenomenologists, starting with Husserl, coined a set of concepts that adequately describe the **dynamic and holistic nature of subjectivity**. "Intentionality," for instance, begins by designating the relation between consciousness and its objects. Intentionality becomes more broadly, in Heidegger, the characteristic of everyday practices, and even, for instance in Merleau-Ponty's approach, the fundamental structure of life, conscious or unconscious behaviors, even alleged reflexes.[15] Intentionality does not necessarily take the form of deliberated, self-conscious projects, although it does make them possible. "Intentionality" is the process in which one projects oneself beyond oneself toward new relations. It is also, to a certain extent, a way of losing oneself (self-consciousness, self-identity) by being absorbed into always new objects, new situations, and new perspectives. "Every consciousness is consciousness of something" means, according to Husserl, that consciousness is essentially a manner of *relating to* and of turning oneself *toward*. Intentionality has modes: Objects or situations can be apprehended in different ways, *as* present or *as* that symbol of something absent, for instance, *as* real or *as* pure appearances. Correlatively, things lend themselves to the contingent orientation of intentionality, to an agency that maneuvers this orientation and infuses the world with a certain perspective, particular goals, and interpretations. Nonetheless, things and situations do not coincide with the subject that apprehends them; they precisely appear *as* a dimension of transcendence and resistance in my experience. Consequently, the phenomenological concept of intentionality also entails study of the phenomenal world, the world as that which appears, imbued with meaning and emotional features. The categories of interiority/exteriority and identity/difference are challenged by this concept of intentionality: We are beyond ourselves, thrown into the world, and the world has, intrinsically, myriad appearances. Overcoming the dichotomies of interior/exterior and identity/difference is needed to account for the world as it is and as it can be experienced. Mental pictures in cranial boxes and fixed in-themselves realities remain side by side, juxtaposed: They never *appear*, they do not think, they never *make a world*. Phenomenology thus essentially compels us to depart from an egocentric and logocentric conception of intentionality; it studies subjectivity also through the way the world appears, through structures of the apprehended world, and through non-representational forms of apprehension: This quality will prove crucial in the description of animal subjectivity.

But do actual phenomenological works involve a study of animal subjectivity? Many thinkers in the phenomenological and phenomenological-existentialist tradition address this question. Because phenomenology must start from the broadest description of experience, phenomenologists are tasked to reckon with what I have called the metaphoricity of our "naïve" relation with non-human animals.

Moreover, the question of animal subjectivity is made particularly relevant by the eidetic approach that is essential to Husserl's project.

The eidetic dimension of phenomenology is crucial to constitute it as a genuine phenomeno-*logy* [science of phenomena] beyond a personal description of moods, impressions, and feelings. When Husserl describes what happens exactly when "I perceive," "I imagine," "I perform the epoché," he claims to describe the essence of subjectivity, the structures and processes that constitute every possible subjectivity.[16] This eidetic aspect of phenomenology is directly linked to the power Husserl grants to imagination. Husserl, for instance, audaciously claims that every subject—even a perfect God—would perceive things in a perspectival fashion. How can we come to know God's subjectivity? We are, Husserl claims, able to imagine that we could perceive a particular thing under a different light, from a different location in the room, with a totally different body, and so forth. We can try to imagine that we perceive it without any remainder, without the persistence of hidden sides. But this eidetic variation fails and thus unveils a stubborn essential characteristic of perception and the sensible world, namely that things *must* be perceived through "adumbrations [*Abschattungen*],"[17] a *de jure* endless series of facets. As based on concrete imaginative variations, the intuition of essences is both made possible by our experience and, nevertheless, always adventurous and provisional. We cannot be sure that the variation we have just achieved was not in fact limited by some hidden biases; it is therefore crucial to always return to new examples that stretch our imagination to unprecedented distances. Examples of animal subjectivity precisely play such a challenging role. Indeed, the question of the essence of subjectivity inevitably leads us to the issue of the baffling nature of animal subjectivity. Phenomenology cannot but confront itself with the question: Do non-human animals actively take part, as subjects, in the fluid intersubjective field unveiled through the reduction? When I imagine that I could perceive the world with a different body, I inevitably come across the question: Could "I" be a chimp, an ameba or—to resume an example offered by Husserl—a jellyfish?[18] How does this variation, that stretches our imagination in a particularly strenuous manner, modify the phenomenological eidetic definition of "subjectivity," "thought," and "intentionality"?

It is not self-evident for all phenomenologists that non-human animals are subjects and, to employ a Heideggerian conceptuality, that they exist (ek-sist),[19] namely, get beyond themselves, question, and shape their world. As Toadvine points out, we cannot take for granted that phenomenology has "twisted free from the anthropological machine" [20] or that it can do so. I agree with Toadvine that the most promising path among the canonical phenomenologists can be found in Merleau-Ponty's work.[21] But I will also argue that Husserl's genetic phenomenology carries unborn children: It allows us to develop the concept of transcendental interanimality and substitute it for the concept of the transcendental ego (§16–19). Such a transcendental interanimality defines subjectivity—as the field without outside, that from which we must always start, the condition of possibility of every reality—as multipolar. The eidetic discourse (what is subjectivity overall?) is then still relevant but must necessarily be supplemented by a practice of empathy, transposition, and dialogue with others, including with other animals. Animal variations and the adventurous—both perceptive and imaginative—encounter with other animals thus become key to a good understanding of the world and ourselves. It is important to identify the structures of various forms of animal subjectivity, but no definition will exhaust the more existential question of what remains to be enacted, further built, and invented in the relation with other animal subjects.

Imagination and phenomenology

Without a dimension of playability and unreality, there would not be any appearing, any phenomenality. Phenomenology and existential phenomenology accordingly always attach particular importance to imagination, most often, regrettably, to human imagination. In the remainder of this chapter, I will examine why the interlacing between human and non-human imaginations must become the primary focus of phenomenology and how a phenomenology of the imaginary of animals can offer a radically new perspective on human and non-human animals.

In the first section of Chapter 2, I concentrate on the worldly side of intentionality, for it immediately breaks with an anthropocentric and logocentric perspective. Instead of starting from a well-circumscribed subject, phenomenology directs our attention to the appearing—namely, things and the world as phenomena. I investigate how phenomenology unveils a fundamental dimension of ambiguity and open interpretation in the world itself (§13), which thus proves to emerge first and foremost as an animal *Umwelt* rather than as a full-fledged rational/human reality. Drawing on Uexküll as well as recent experiments on animals' hesitative behaviors, I contend that ambiguity is a key character of animals' *Umwelten*, which challenges Heidegger's claim that animals do not exist and have no worthy-of-the-name *world* (§14). I explain in more detail what the imaginareal is, where a new understanding of the being of animals is possible and where the human imaginary of animals and the imaginary field of non-human animals enter into dialogue (§15).

The second section of this chapter focuses on the connection that phenomenology allows us to establish between human and non-human animal *subjects*. After introducing the nature of the transcendental subject in Husserl's phenomenology (§16), I study the controversy between Heidegger and Husserl regarding empathy [*Einfühlung*] and phantasy as means of relating to other subjects—in particular to animal subjects (§17). The idea of human-non-human animal intersubjectivity is fleshed out by investigating the role played by phantasy in ethology; I draw on Buytendijk and Plessner's analyses, as well as on Carlos Pereira's work with horses (§18). Further, I contend that a phenomenological approach cannot avoid acknowledging and conceptualizing an intersubjectivity of human and non-human animals, which involves a deep redefinition of the classical concept of subjectivity. Human and non-human animals are intrinsically intertwined, a relation that can be understood through the Merleau-Pontyan concept of chiasm (§19).

PART I . BEING IS PHENOMENAL: AMBIGUITY AND THE IMAGINARY DIMENSION OF REALITY FOR HUMAN AND NON-HUMAN ANIMALS

§13 An ambiguous world of appearing processes soliciting creative subjects

Reality is intrinsically phenomenal

In his ontology of dwelling, Ingold contends that it is impossible for any discourse and any idea to arise from outside the dwelling. Even in order to define a mathematical or an objective world—which allegedly preexists every subject and will continue to exist if every subject disappears—I engage, as a dweller-subject, in an interpretation and a structuration of my concrete world.

This claim also lies at the heart of phenomenology. But Husserl goes a step further: This focus on subjective engagement is not the expression of what we, human subjects, can or cannot do and think; it is, rather, an ontological stance and the very meaning of the famous motto of phenomenology "Back to things themselves [*zurück zu den Sachen zelbst*]."[22] Namely, everything, whatever it may be, is *for* a subject. In other words, phenomenality—the appearing—is an integral part of the being of things themselves.[23] Every being is phenomenal, appearances are an integral part of its nature, which allows us to overcome the idea that our conscious life is like "a pea imprisoned in a shell," in Elizabeth Costello's words, namely a private and essentially human experience.

Does this ontology of the phenomenal being take root in the apparently contingent fact that things appear to me? Certainly, but this fact cannot only mean that I have the power to relate to things and somehow to *make them* visible and thinkable. I am not relating to a mere figment of my fantasy: I experience a world and beings that resist me, surprise me, and inexhaustibly reveal new aspects to me. Beings have this ontological property of manifesting themselves, in all sorts of fashions.

One may object, in a Kantian manner, that it is important to make a distinction between the way beings appear to us and things in *themselves* [*Dingen an sich*], which entirely escape us. But Husserl responds that such an objection is self-refuting.[24] Too late, indeed: The objector has already integrated the "thing-in-itself" in the horizon of what a subject defines, conceives, and seeks to think, even though vaguely and confusingly.[25] This representation is a first tenuous link that allows further investigation and shows that beings lend themselves to such investigations.

"Everything is phenomenal" thus does not mean that everything is crystal-clear, but rather that thinking begins at the level of phenomena, in the world and not in an alleged private human mind. Things lend themselves to perception and interpretation. Even our most abstract concepts must emerge from the phenomenal world. For instance, the different imperfect lines that can be found in our surroundings refer to each other and together point toward a horizon of more and more perfect lines, straighter and straighter, thinner and thinner. The ideal geometrical line actually stands on the horizon of this world; it does not belong to the sensible world but helps us schematize and understand this world better. Similarly, modifications in our body and empathy with other living beings incite us to envisage radically different ways of perceiving the world or even of conceiving it through general ideas. As a consequence, Plato's *Ideas* are present *in the world* in the ambiguous form of *traces and hints*. As Husserl explains, "The things of the intuitively given surrounding world *fluctuate*, in general and in all their properties, in the sphere of the merely typical: Their identity with themselves, their selfsameness and their temporally enduring sameness, *are merely approximate*, as is their *likeness* with other things."[26] The objective world must be defined in terms of endless process, solicitation, dynamism, and non-coincidence with oneself. Things themselves are ambiguous and "oscillate" or "hover," they are "*in Schwebe*,"[27] which will be a key to the concept of the imaginareal. The lifeworld is multipolar, multidimensional, ambiguous, and dynamic. Such a phenomenological approach thus eschews a typically anthropocentric way of placing ideas, essences, and concepts in the reserved domain of highly intellectual and symbolic mental processes.

As a consequence, phenomenology does not reduce experiences to a private consciousness, a singular *quale*, a unique taste, what Nagel called "what it is like to *be* me." Our subjective experience involves complex and describable structures of intentionality, relations between phenomena, a world that deploys itself before us and in which emotions, interpretations, and even concepts are sensible.

A world open to interpretation

Including appearing in things themselves amounts to introducing a virus into the classical system. "Phenomenon," "fantasy" (φαντασία), and "phantom" share the Greek root, "Φα" that engendered the rich lexical field pertaining to light (φάος), its power of manifestation, and trickeries. Appearing may take the form of a glaringly obvious presence, but it also involves possibly deceptive appearances. It is intrinsically manifold. Things and persons may appear as present or past or future or absent (someone I miss, or a landscape through a photograph), or as that which is currently perceived by another (the backside of this house, the postman from the perspective of this dog), and/or as illusory (the broken stick in water), or ambiguous (this piece of wood polished by the sea ... or is it a clayey rock?).[28]

Such a manifoldness is accompanied by openness to interpretation, as a fundamental property of the world itself. The Husserlian concept of *lifeworld* is quite telling in this regard. Modern thought defines only objective and measurable entities as real; in other words, as Jonas bluntly puts it, a still-predominant ontology considers death as the only true reality.[29] This is positivism: A being is either fully what it is, or it is not.[30] It is *in fact* or it is not, and we should stick to the facts. In *The Crisis of European Sciences and Transcendental Phenomenology*, Husserl shows that, on top of making the world meaningless and oppressive, positivism is a shoddy sleight of hand that does not stand up to scrutiny. It is indeed betrayed by its ambivalence. On the one hand, the positivist claims that what truly is, is the objective world, such as it is defined by natural sciences. On the other hand, in order to speak of the world and define its fundamental nature as a whole, the positivist must regard herself, underhandedly, as an all-encompassing super-mind, and not as an objective body in this particular limited location. Only by forgetting and burying its origins can objectivism be maintained. The lifeworld must therefore precede and sustain the objective world; It precedes this uniformizing ontology and thus integrates in its very being multiple dimensions and lines of nascent meaning that may overlap but never coincide: The different facets of experience, perception through different sense organs, fantasies, emotions, objectivity, the possible perspectives of other humans, animals, aliens, or gods. Thus, for instance, irregular round shapes in the lifeworld *are* ambiguous, they *are and are not* circles: They foreshadow the ideal but in a tantalizing way.

My ability to interpret the world does not stem from an alleged extraordinary, human faculty called "imagination." In the first place, this openness to interpretation is *experienced* as a request that comes *from the world* and solicits the subjects. One may think again of the example—to which I referred in Chapter 1—of the room illuminated by daylight and incandescent light.[31] The environment usually guides my gaze and suggests, more or less strongly, which light should be perceived as the ground light. The response of my body—for instance, apprehending the light from the bulb as the ground light—is thus achieved unconsciously, through bodily intentionality. Now, if the environment is particularly ambiguous and/or if I decide to perform the epoché, I realize that I can, *to a certain extent*, play with different ways of structuring this environment and resist the most spontaneous one. I may play with accommodation processes, exploration movements, and different combinations in the use of different sense organs, and observe the way the world is resultantly modified. This play is not primarily a theoretical or even conscious activity; it is more fundamentally initiated by a practical dialogue between the sensible world and my body. I cannot shape my perceptual field arbitrarily, I depend on the world's flexible affordances, and, correlatively, the world *itself* offers leeway for my agency.[32]

With the concept of the lifeworld, Husserl suggests a profound connection between life and a meaningful world. It is remarkable that he chooses to anchor meaning and even reality in life, while he could have sticked to the paradigm of the world of mind [*geistige Welt*]. Life is directly connected by Husserl to the meaningfulness of the world, which he contrasts with dead, ossified concepts.[33] Mind without living bodies is meaningless: This is a powerful claim that has also deeply influenced Jonas' and Merleau-Ponty's approach. Some humans may dream of clear, eternal ideas, yet the latter cannot but have (literally) roots, guts, and sensitive surfaces. Meaning *de facto* unfolds as a world, it is thus essentially inseparable from a multifaceted and embodied thought, it necessitates the imbalance and impurity (non-coincidence) of opaque, vital processes, limited perspectives, and dialectic tensions between interiority and exteriority, life and death.

Two consequences follow that prove instrumental in overcoming an anthropocentric definition of the mind.

First, the fundamental, ontological structure provided by a phenomenological approach is the strict correlation between an ambiguous world and unhinged subjects. The subject is not a metaphysical entity that would be speculatively elaborated, but, rather, a feature of the world to which we belong and a dimension that pervades every recess of being. Hence, this subjective side of the world is defined in a non-metaphysical fashion. It is not understood as a soul, or a mysterious immaterial or material reality, or even a reflective consciousness, but as an intentionality, a phenomenal world we inhabit, and a set of exploratory processes through which we build and enact our perspective. As a result, phenomenology provides us with a concept of subjectivity that can be introduced in biology without turning it necessarily into an occult science or a romantic reverie.

Second, if subjectivity pervades everything as the correlate of an ambiguous meaning, it does not, however, simply consist in a vague atmosphere. Rather, subjectivity is a set of affordances and open solicitations, onto which I am immediately grafted, that I can share with other subjects and with which I can play. Phenomenology is not only a theoretical description of the world; it offers a method that leads us to an awareness of our commitment in the world and unveils the latter as that which solicits our interpretation and responds to it. Consequently, animal subjectivity and animal thinking are to be engaged through the discovery of animal ambiguous worlds and, correlatively, through the experience of an ongoing human-animal invention of new meanings, at the very level of shared interactions with the world.

§14 Non-human animals and the ambiguity of the world: Husserl, Heidegger, and Uexküll. Hesitative behaviors and depressed bees

Understanding the correlation between the phenomenal world and thinking subjects as exclusively relative to human capacities, as Heidegger does in *The Fundamental Concepts of Metaphysics*, boils down to missing one of the most daring and revolutionary aspects of Husserl's phenomenology. Phenomenology does not pertain to psychology or anthropology:[34] Husserl discovers an ontological universal structure that can be described as the correlation between a phenomenal being and intentionality. Husserl's reasoning never starts from the postulation of human freedom or the human capacity to introduce interrogation, meaning, and interpretation into the universe. Husserl's analyses concern meaning and appearing in general. There is no reason in this approach to presuppose that a non-human animal should be incapable of relating to the world and interpreting it. The key issue is, in fact, whether it is possible to separate the ability to relate to one's environment from the ability to relate to the ambiguity of said environment. Heidegger claims to achieve such a separation. Thus, after conceding that the environment of non-human animals is "somehow *given*" *to* them ("the lizard has sought out the stone, it looks for it again if we try to remove it from the stone"; it "has its own relation to the stone"[35]), Heidegger steps back. He does not simply nuance his first assertion, but reconsiders it. The nascent openness to the world (the dog eats with us ...) immediately resorbs and turns into a closure (.... and yet not). According to Heidegger, the lizard's relation to the stone could in no way be a form of apprehension of the stone as such, since non-human animals do not ever, and—an even more problematic assertion—*cannot* apprehend the many possibilities of meaning interpretation. As a result, animals are not even poor in world, they do not have any relation to the world, except through our human perspective on them.[36] Therefore, the crucial question from a phenomenological-existentialist perspective is not only whether non-human animals perceive their environment, but also, and more importantly, whether they perceive the ambiguity of the world.

Why does Heidegger deny profundity to animal perception? When looking for the foundations of Heidegger's claims in *The Fundamental Concepts of Metaphysics*, the reader is confronted with descriptions of a few animal behaviors: Bees, a moth, a lizard. One may fall under the spell of the wealth of refined and sophisticated philosophical concepts, put forward together with an intricate new terminology. However, the applicability of Heidegger's analysis to actual animals is quite rudimentary. Indeed, his approach to animality in *The Fundamental Concepts of Metaphysics* is by and large shady, and may eventually be explained only in the light of a hidden political agenda.

In *The Fundamental Concepts of Metaphysics*, Heidegger uses the animal as a contrasting marker: "In the end our earlier analysis of captivation as the essence of animality provides as it were a suitable background against which the essence of humanity can now be set off."[37] Heidegger thus wants to define humans *in contrast to* animals, while animals are not considered for themselves, but only in the framework of an anthropocentric perspective. To be sure, Heidegger develops inspiring ideas about the ambiguous transposition from humans to animals. He also mentions that the response to stimuli "brings an essential disruption into the essence of the animal," so that the latter is *"hinausgestellt"* [put outside, ejected] into something other than itself.[38]

But the entire arrangement remains, so to speak, shifty. There is a tension in Heidegger's philosophy between, on the one hand, the definition of the *Dasein* as the elusive intertwinement between authenticity and inauthenticity, and, on the other hand, the definition of the *essence of humanity*. As suggested by Adorno in *The Jargon of Authenticity*[39], Heidegger wants to have it both ways. On the one hand, Heidegger specifies that he does not mean to establish hierarchies or ethical judgements, and that a substantification of *Eigentlichkeit* would be a mistake. On the other hand, however, he demonstrates a strong tendency to betray some hints evincing defined places/beings where authenticity actually concretely lies (the cabin in the Black Forest, the German language, a certain terminology, Greece-Germany ...) or does not lie (the contemporary city man a.k.a., "the ape of civilization,"[40] communists, cosmopolitism, the Semitic nomad/*das Weltjudentum* ...). The same ambiguity appears in Gorgio Agamben's description of Heidegger's conclusions: "Only man, indeed only the essential gaze of authentic thought, can see the open."[41] It seems to me that Heidegger's approach to animals pertains to this general two-sided pattern. By contrasting animals and humans, Heidegger seeks to draw a sharp line where, in the first place, ambiguity reigns supreme. He attempts to establish human beings in an essentialized status. In place of the flickering between authenticity and inauthenticity, Heidegger settles a clear-cut boundary between humans whose access to the world is essentially guaranteed ("Philosophy has a meaning only as a human activity. Its truth is essentially that of human Dasein"[42]) and animals who utterly miss it. The phrase "the ape of civilization," as contemptuous as it gets, thus makes both no sense at all *and* total sense in the framework of Heidegger's philosophy.

However, Heidegger rightfully draws upon Uexküll's work to make his case. Considering the limited set of actions that characterizes the behavior of the tick, such as Uexküll describes in *A Foray into the Worlds of Animals and Humans*, it seems that a determinate sequence of reflexes imprisons the tick in a rigid relation to his environment. A set of fixed and regular behaviors that can be triggered by specific stimuli apparently monopolizes the animal's relation with its milieu and prevents the openness to indefinite possibilities from blossoming. As a consequence, according to Heidegger, the tick could actually never really gain access to a "world" worthy of the name. But this reading of Uexküll is reductive and fails to do justice to the complexity of animals' relations to meaning. Let us bring to foreground a dimension of interpretation and ambiguity in animals' *Umwelten* that is suggested and eventually eclipsed in Uexküll's *Bedeutungslehre,* and that contemporary biosemiotics has made one of its central topics.[43]

Uexküll's approach proceeds from the claim that animals are subjects. I will dwell more on the reasons why non-human animals can be defined as subjects in the next section of this chapter. For now, I simply want to emphasize that the relationship to meaning *entails* openness to ambiguity, *including in the case of non-human animals.*

Following a Kantian approach, Uexküll calls animals "subjects," for they are not simply subjected to the determining action of external forces, but they are the source of structuration and norms according to which a world *for them* can appear.[44] The tick focuses on a limited set of aspects of his environment (the tip of a protruding branch or of a blade of grass, butyric acid, warm blood, a spot as free of hair as possible), while the rest of his surroundings, although mechanically and chemically effective, is not selected by this acarus as an object of perception/action. In other words, the tick does not respond to these other aspects. He remains indifferent to them. The subject is the center of its *Umwelt*. The latter is structured according to the subject's interests and needs, by means of its sense and action organs. Uexküll takes up the Kantian distinction between the thing-in-itself and the phenomenal thing "as meaning." Non-human animals performatively refer to things *as* weapons, paths, obstacles, food, poisons, dangers, etc.

Several aspects of Uexküll's analyses support Heidegger's claim that non-human animals, far from dialoguing with an open-ended world, actually project a rigid univocal structure on their milieu. Uexküll thus stresses that animals are locked up in their respective *Umwelten*-bubbles. He also develops a concept of "pure" subjectivity that tends to make the *Umwelt* almost hallucinatory. Uexküll explicitly refers to what he calls the *poverty* of *at least some* animal worlds.[45] As a result, Heidegger certainly can argue that Uexküll uses the word "world" too loosely. Hence, Heidegger concludes— contra Uexküll but in a manner that is substantiated by some features of Uexküll's analyses—that non-human animals do not actually relate to a world, not even an *Umwelt*, since, when there is no leeway for other possible perspectives, *ek-stasis* cannot arise.

However, Uexküll's analyses also include elements that reveal a dimension of ambiguity and interpretation—in other words, openness—in animal *Umwelten*. Consider the following example given by Uexküll: "A toad that eats an earthworm after a long period of hunger will also seize upon a matchstick, which bears a certain similarity in shape to the earthworm. One can conclude from this that the worm it just ate serves the toad as a search image. (...) If, on the other hand, the toad satisfied its initial hunger with a spider, it possesses another search image, for it now snaps at a bit of moss or an ant, which does not agree with it [*was ihr aber sehr schlecht bekommt*]."[46] As this example demonstrates, Uexkül's claim that animals give meaning to their environment can be fruitfully connected to our previous phenomenological analysis of the ambiguous nature of appearing processes. When meaning is involved, a gap between different avatars and interpretations arises that also makes hesitations and mistakes possible. "Prey," "food," even "this specific food" always consist in a general pattern. The recognition process that connects the latter with a particular object here and now cannot but be adventurous. Uexküll explains the surprisingly underperforming behavior of the toad as follows: "Now, we do not by any means always search for a certain object with a unique perception image, (...). We do not look around for one particular chair, but for any kind of seating, i.e., for a thing that can be connected with a certain function [*Leistung*] tone. In this case, one cannot speak of a search image [*Suchbild*] but rather, a search tone [*Suchton*]."[47] A tone is a versatile indication for endless musical variations. In the same vein, Uexküll argues that a contrapuntal relation structures animals' dealings with their *Umwelt*,[48] so that the meaning of the environment is shaped and reshaped through chance encounters and variable fugue-like interactions with the milieu and other living beings. These claims demonstrate that Uexküll outlines a phenomenology of animal freedom.[49]

When, on top of that, animals have to deal with "luring" morphologies (the eyespots on the wings of butterflies, for instance) or behaviors (the broken-wing display), the ambiguity of the phenomenal world reaches new heights. Is this eye an eye? Is this a vulnerable prey? The possibly illuminating, possibly misleading, similarity is not simply made up by the subject. It takes the form of questions that arise in the *Umwelt* and wait for a response. Different responses are allowed and different individuals will respond in different ways. Some individuals may be—or have become—more skillful than others in detecting the trickery.[50]

Hesitation has recently become a topic of investigation in animal sciences. But many observational and experimental studies still focus on whether animals achieve an expected task or not. The results take the form of a set of completion rates, while, strangely enough, the question of possible forms of indecisiveness is simply not addressed. It is certainly important to know what percentage of tested animals manage to "truly" imitate an action, or to solve a problem, but we still miss a crucial piece of information: Do some of the tested individuals, or the majority of them, or all of them under some specific circumstances, achieve a given task in a more hesitant way? I first wondered whether this question was simply made irrelevant by the observational data or if it was overlooked due to a positivist bias. Do we not naïvely observe signs of hesitation in animal behaviors in everyday life? Moreover, what makes this issue particularly significant is that hesitation, agency, and even reflection are strongly connected. An animal who, for instance, approaches an ambivalent object, moves backwards, and comes closer again, is not simply slowed down by a material obstacle. She actively introduces a leeway between the environment and the response, and plays with it, according to her own rhythm and mood, by physically sketching, testing, and thus considering, different options. She enters in relation with—and specifically focuses on—different possibilities that are for a while in competition.[51] The situation takes the form of a nascent ob-ject (*ob-* "*against,*" *jacere* "*to throw*") rather than of a stimulus, since the distance between the situation and the response increases considerably. Even more interestingly, possible options themselves become objects of consideration. A series of investigations linked to the research about metacognition in non-human animals provides a much richer approach and has started to address the shortcomings I have just outlined. Not only does this new research describe hesitation and indecisive behaviors as such, but it also links them to investigations about consciousness, self-consciousness, and freedom in animals.[52]

One such study made the headlines by demonstrating that bees can be depressed.[53] The test focused on the way bees responded to ambiguous stimuli. Honeybees were trained to associate one scent with a sugary reward and another scent with bitterness. It then became possible to test their reaction to an in-between scent. Indecisiveness was observed in bees confronted with such an ambiguous stimulus. Moreover, when the hive has been shaken violently to simulate an attack by a predator, the bees become less inclined to try their luck with the ambiguous scent. The response is not invariable; and it is precisely because different responses can be displayed within the framework of a generally hesitating behavior, that the timorous response of the "shaken" bees is not described as the result of the mechanical inhibition of a reaction by a stressful stimulus. In the leeway opened by indecisiveness, individuals can adjust their original response within a wide spectrum of possibilities from extra cautiousness (and a pessimistic mood) to adventurous behaviors, through all sorts of more or less intense states of hesitation.

While the use of a binary model (task completed or not) still largely prevails, a phenomenological approach provides a general theoretical framework that affords a systematic study of animals' relationship with an ambiguous world and this phenomenological perspective incites researchers to pay more systematic attention not only to meaning and interpretation, but also to hesitation, variations in interpretation, and mistakes.

In order to fully implement the human-non-human-animal interface and to overcome the issues encountered by those who want to take the human-animal metaphor seriously, we will take a step further and examine how imagination and the imaginary are, if understood in an ontological rather than psychological approach, the most fundamental aspect of experience through which the lifeworld and ambiguous meaning are to be engaged.

§15 The imaginareal: A non-representational paradigm for imagination

If, as fundamentally stated by phenomenology, everything *is* also its ways of appearing, the power of becoming present through images and fantasies must belong to things themselves. The potency of the virus introduced in being by phenomenology reaches a climax in the imaginary field. Even absence and ubiquity are part of the very being of realities. An inquiry into the nature of the imaginary prevents the return to a definition of beings through the circumscription of what they *are* and what they *are not*, so that the question "are human and non-human animals the same *or* different?" falls short. In order to do justice to the imaginary of animals, it is crucial to overcome the dichotomy between literal and metaphoric and to rebut a representational theory of imagination that disconnects reality and its image. The concept of the imaginareal enables a radical shift of perspective in ontology and brings access to the transitional space without which the imaginative lives of animals cannot but remain largely ignored.

In a positivist approach, the focus is on the *original versus the copy*, namely on actual realities that make the representation of the imaginary object possible. While the imaginary object is absent or unreal,[54] a material picture (a photograph or a painting) makes it "as if" present, or quasi-present. Similarly, the famous imagery debate focuses on the actual neural processes, the mental pictures, the linguistic descriptions, or even the mental activity by which one refers to absent objects.[55] But the description of the positive properties of a picture or of cerebral states fails to target a crucial aspect of the imaginary: An absent or unreal object becomes quasi-present *through* a present reality. A *present* entity stands for an *absent* one, assuredly, but how are they connected and how can presence and absence, being and non-being, be united though these imaginative processes?

As Husserl and Merleau-Ponty highlight, imagination is a form of *experience* of the object.[56] There may be no connection by nature between the very being of an object and the word that designates it ("dog," "chien," "Hund"), whereas an image (a phantasy[57] or a picture[58]) has a relationship of resemblance to the object that it manages to make quasi-present. To imagine is not to abstractly conceive of something, but rather to live an "as if" experience. I physically and emotionally engage with the imagined objects. For instance, as Husserl puts it, "our fist is tightening, we are holding a high-voiced conversation with the imagined characters."[59] Imaginative experiences actually fluctuate from clichéd images or abstract re-presentations ("let us imagine a cube …") to highly vivid quasi-experiences that almost entirely absorb us and sweep the actual world away into a hazy and remote background. When the object is not quasi-present but simply absent, or if it is regarded as present without any hint of distance and ubiquity, the lived experience does not pertain to the realm of imagination. Let us therefore note that it will be convenient for our further analyses to draw a relative distinction, within the genus "image," between **copy-images** that present themselves as a dull reference to a more valuable and, so to say, *more real* original, and, on the other hand, **living-images** that carry us away and give rise to inexhaustible quasi-experiences. In fact, they are homogenous. Copy-images are never absolutely dead; living-images also possess a dimension of opacity, distance, and re-presentation, without which they would cease to be images.

Parallel to this becoming "as if" present of the absent, the presence of real things is contaminated by the imaginative experience. Such an original presence is in fact not absolutely obdurate and overwhelming since it may pale in comparison to the hyper-presence of absent objects through a work of art or an especially vivid phantasy. Even more remarkably, the objects that are actually experienced as present also appear as that which can return and haunt me in an imaginative mode of presence. They will be the same objects, but new aspects of their power to appear will be revealed. Through images, they materialize as more malleable and more intensely enacted. They manifest as an emerging being (the imaginary is always dynamic, it stages a coming into being) that navigates from death (faded images, clichés, copy-images) to a resurrection in glory (living-images, radiating images, or "icons"). Consequently, rather than a radical contrast between presence and absence, what must be understood and accounted for is a presence that can continuously vary from a perceptive actual mode to the mode of the "as if "and the "quasi."

Husserl's phenomenological explanation of imagination is complex and manifold. Two of its main features concern us here.

First, Husserl consistently refutes the idea that the origin of imagination can be defined by positive characteristics possessed by a positive reality, be it a set of propositions, neural states, or physical or mental pictures. No positive reality can be an image in itself. "If I put a picture in a drawer, does the drawer represent something?"[60] Fantasies and pictures are two-fold and ubiquitous—or ek-static—*at heart*. They refer to what could be present, but is *instead* quasi-present. Without this self-distance, images vanish; they return to pure presence or pure absence. Husserl thus asserts that what makes a photograph or a painting a picture is the *ek-stasis* of an intentionality, more precisely a specific mode of apprehension *in the subject*. I apprehend a perceptual given (the canvas, lines and patches of colors etc.) *as* the representative of—and *as* resembling—another object that is actually absent, but I could also focus on its material actual presence. The ambiguous twofoldness of presence is the key and the correlate of the subject's agency.

This first phenomenological account is yet incomplete. To be sure, the role played by the subject's creativity in phantasies and in the apprehension of a physical image is critical,[61] but, obviously, I cannot imagine whatever I want through a given material. A non-arbitrary relation between the latter and the imagined object must exist in order to make the quasi-presence *of the imagined object* possible. It is inaccurate to hypostasize a *representative* understood as an inert tool under the control of the imagining subject. The intrinsic link of this representative with the imagined object should not be overlooked. Modes of apprehension are more fundamentally in dialogue with modes of appearing of the objects themselves. In a later development of his analysis,[62] Husserl makes significant adjustments to this subject-centered theory and paves the way for a phenomenology of imagination that can integrate the role of *the imaginary*. He gives the example of theatrical performances:[63] Hamlet becomes alive on stage and we do not imagine him by *arbitrarily using* the show as a representative of something absent, but rather by being carried away by the actor's performance. The distinction between *Bild-Bewusstsein* and phantasy is fading:[64] In both cases, the "representative" ceases to be a clear-cut present object contrasting an absent imagined subject. Rather, an as-if experience blurs the boundaries between presence and absence. We see and hear Hamlet "on" and "within" the actor's body, speech, and gestures. People and objects on stage metamorphose themselves. "Certain things show themselves to be suited to excite a double perceptual apprehension [*doppelte perzeptive Auffassung*]."[65] By using the concept of *Perzeption*, Husserl underscores that imagination is a genuine experience and must be motivated by the object. *Because* the present perceptive phenomenon on stage is obviously twofold and ambiguous, it *must* be experienced in the imaginative mode. Likewise, a phenomenological understanding of the imaginative process entails that, when a real landscape inspires a painting, it *lends itself* to the creation of this new mode of presence and thus manifests its power to incarnate itself into this painting. What, for instance, is the reality of the Montagne Sainte Victoire? As essentially phenomenal, it sketches itself out through myriad fragmentary aspects: Lights, silhouettes, changing colors, smells of the scrubland. To geologists, geographers, or ecologists it reveals even more numerous facets. The mountain, *in its very being*, is not a closed ensemble of properties. What we summarize with the noun "Montagne Saint-Victoire" *is* the hovering theme of an indefinite multiplicity of contingent variations. It can therefore be genuinely *as if present* in Cezanne's painting. The paintings emerge from the real Sainte Victoire and vice versa.[66]

In *What Do Pictures Want?*, Mitchell describes the myth of "living" images as an "anthropological universal." But from what has been discussed in these pages, it follows that we can take a step further and move beyond human beliefs and the analysis thereof. Mitchell says he does not "believe" that images actually *want* things, but, he adds, "we cannot ignore that human beings (including myself) insist on talking and behaving as if they did believe it."[67] A phenomenological approach allows us to uncover the ontological grounds of the strange kinship highlighted by Mitchell between essences—*species*[68]—and images. Aristotle indeed defines species as magical images that emanate from things and, as it were, remain connected to a mysterious life of what they represent.[69] And, in effect, the original and its image have the same ontological root: An ambiguous and manifold presence-absence. The superstitious belief in "living" images draws upon this source. The concept of *zoographia*[70] (images = animals) consequently does not come from the application of the cynegetic phantasy of hunting down and capturing animals to the field of representations, as suggested by Derrida.[71] It is the other way around: The paradoxical attempt to capture (control, freeze, and, still, foster) the life of the represented object through pictures is possible only insofar as a common stream of life originally goes through images and originals. What comes first is an organic—coherent and plastic—dynamism that characterizes the growth and proliferation of hovering themes in perception and images. Recognizing the imaginary of animals and recognizing the animal roots of human imagination are absolute correlate.

The concepts of reality, pictures, and fantasies thus prove to be relative. Both real presence and imaginative presence stem from a common root: The open being of melodic themes. I have called this fundamental level **the imaginareal** because the dichotomy between real/unreal, imagination/reality does not apply to it. As a result, the distinction between the real and the imaginary will always remain approximate. The imaginareal involves two intertwined components that can be diversely emphasized either by the experience we happen to live, or through a more playful relation to the world initiated by the subject. The first component of the imaginareal is a set of themes that are guidelines around which "realities" crystallize. On the other hand, as a second pivotal component, variations and the hovering are the inner fragility and unstable life that decenters everything and makes it proliferate. In perception, for instance, a set of regular and familiar *Abschattungen* motivates the apprehension of the actual presence of a recognized reality. The majority of perceptions tend to bring a relatively stable reality to the fore. Yet, some perceptions are hazy, confusing, and "as if in a dream or a nightmare." If the presence becomes confused and ambiguous—like a cloudy sky or the clayey rock that may also very well be piece of wood polished by the sea[72]—it can foster the development of an imaginative activity. It is always possible for the subject to decide to playfully switch between a perceptual apprehension and an imaginative one (for instance: "Let us contemplate this landscape as if it were a painting"), but this leeway is provided by the world and can be spurred by some dream-like "fantasy-provoking" experiences. The distinction between reality and the imaginary remains, but becomes relative. The apprehension of reality as such and as actually present leaves the always-at-work dimension of ambiguity in a remote and hardly noticed background. The imaginative mode of presence, by contrast, invigorates this ambiguity and allows us to actively invent and play with possibilities: Things and persons reveal themselves as ubiquitous.[73]

In the imaginareal, the multipolar nature of the real becomes patent. The world is essentially undermined by gaps, distances, and haunting processes. We would not be able to imagine if beings were not simultaneously, on the one hand, a certain consistent, although imperfect and dynamic, accretion (or a melodic line) and, on the other hand, beyond themselves and partly unpredictable. Through the perception and the imagination of the Sainte Victoire, two poles—myself and the Saint-Victoire—emerge. These poles are correlated but do not coincide. The approximate delimitation of *my* situated individuality in the world is backed up by the experience of my body, the feeling of my "I can," "I move," "I desire," combined with proprioceptive sensations, and contrasted with the experience of that which escapes me and resists me. Similarly, the mountain Sainte Victoire emerges as a relatively circumscribed entity, while remaining in close connection with my exploration moves and my creative interpretations. Accretions emerge thanks to which a contingent and provisional "literal meaning" indeed crystallizes ("I," "the Sainte Victoire") and is not a complete metaphysical illusion. But new and original developments of each being are always on their way. They can surface spontaneously or may be invented in a more active manner by imagining subjects. As the next section will show, the distance between these different poles is heightened in the field of intersubjective and interanimal relations.

A metaphor is always daring and risky, but it actually conspires with the forces operating below the literal meaning. Metaphors are creative and, without contradiction, faithful to the true nature of beings. They give up grasping the latter with an iron fist and, instead, open a dialogue with it. Such a dialogue enhances the relation of an actual reality to future possibilities. Consequently, our relations with non-human animals have no room for *mere* metaphors; and it is not less irrelevant to claim that we should completely replace a metaphorical discourse with the literal knowledge of clear-cut differences and common points.

At stake here is, so to say, an iconoclast theory of images and imagination, which will prove particularly fruitful to better understanding the imaginary of animals. The model of material pictures is radically overcome, while non-visual and elusive musical themes provide a more challenging paradigm, reminding us that imagination is multi-sensorial. The unhinged center of gravity of images lies in their future potentialities, which will unfold throughout words, body attitudes, scientific metaphors, daydreams, and even perceptive experiences. The concept of the imaginareal must prevail over the concept of imagination, which designates a personal, active faculty.[74] Further, the imaginareal is not exactly an imagination of nature, in a romantic or mystical sense—or in the sense outlined by David Abram in his earthly cosmology—namely an encompassing source of community and of "plain and obvious" intuitions:[75] As a ubiquitous fundamental dimension of beings, the imaginareal keeps everything in dialogical, risky, possibly stifling, and precarious relations. As I will argue in Chapters 4 and 5, the imaginareal allows for the emergence of individual, animal imaginations whose coming-to-attunement is neither guaranteed nor ever totally impossible.

We encounter many metaphors, images, and fantasies in the lives of non-human animals, in their morphology and in their relation to their world and other animals; however, none of these metaphors, images, and fantasies will take the form of a mere picture re-presenting an absent or unreal object. A re-presentational paradigm would compel us to simply refute the imaginative nature of these phenomena. But it proves invalid even when it comes to describing human imagination. With the complex and dynamic constellation made up of the imaginareal, the flexible and relative poles of the real and the imaginary, and individual imaginations, I have now developed a more fruitful framework, in which the relation of non-human animals to imagination can be accounted for in a much more fine-grained manner.

In the next chapters, I shall expand upon how this ontological framework sheds new light on the study of our relationship with non-human animals. I will examine the extent to which a metaphorical discourse can be compatible with a scientific approach and the consequences that are entailed—on a more practical and ethical level—by the role played by metaphors in our relationship with non-human animals. At present, however, I will review the major consequences that flow from the phenomenological understanding of the imaginareal explained so far.

- **Metaphors,** as both creative and meaningful, can now be accounted for and taken seriously. They operate in the same way as Cezanne's paintings. Sustained by the open nature of beings, they can create original variations and, yet, be faithful to a theme—or, for bad metaphors, fail to be faithful to a theme. They reveal a hovering being beyond the *is/ is not* dichotomy. The abundance of human-animal metaphors thus draws our attention to an actual correlation between human and animal imagination.
- Doing justice to the imaginary of animals is also the only way of fully engaging the powers of human imagination. The latter is confined to a disembodied and eventually anodyne, recreational inventiveness as long as its origin in *zoographies*—to wit: In the life of images, and especially as I will show in §33, in the animal life of images—is not acknowledged.
- **Non-human animals possess an open-ended nature**. We will see that Merleau-Ponty, but also Bachelard, Deleuze, and Haraway in particular, have developed and fleshed out this idea.
- Through the original variations created in **the human imagery of animality**, one can trace back the themes in which the dynamic being of non-human animals consists.
- Meaning is intrinsically open to interpretation. **Animals'** *Umwelten* **also belong to the imaginareal**, they combine emerging structures and an ambiguous dimension that solicits creative interpretations.
- If every being is ek-static, *subjects* are so in a redoubled manner: They can engage the ambiguity of beings in the world, focus on it, play with it, forge images, and refer to the virtual as such. Moreover, intentionality, such as understood in a phenomenological approach, is not limited to the conscious and predicative thought. The phenomenological definition of subjectivity allows us to acknowledge and conceptualize the **subjectivity of non-human animals in its specificity and in its individual variations**.
- With regard to **animals' imagination in particular**, the phenomenological approach teaches us that (a) "to imagine" does neither primordially nor essentially mean "to consciously refer to a fictitious object as such," and (b) "to perceive a picture" does neither primordially nor essentially mean "to consciously apprehend a present object as the representative of another," while these definitions are taken for granted in many philosophical and scientific studies so that it ought to be concluded that non-human animals do not imagine. The range of forms that imagination may take is much broader and embraces ontological ubiquity, enacted ubiquity, and symbolic ubiquity (see infra, Chapter 4).

Human-animal metaphors should be taken seriously: The issue then is to determine how animal subjectivity can be integrated in a phenomenological ontology.

PART II. TRANSCENDENTAL INTERANIMALITY: ANIMALS AND THE TRANSCENDENTAL SUBJECT

§16 An issue: Is the phenomenological subject human or animal?

Phenomenology is the philosophical school that placed subjectivity, as irreducible to objectivity, at its center. But must the subject be essentially human?

Admittedly, an anthropocentric bias often infuses Husserl's definitions of subjectivity and intentionality. The clearest evidence for this is the predominant role played by the concepts of ego, consciousness, crystal-clear thought, and spirit [*Geist*] in Husserl's phenomenology, especially his earlier phenomenology and first version of the phenomenological project. However, this bias is challenged in many phenomenological analyses, including in Husserl's work, mainly because Husserl remained faithful to his promise to go "back to things themselves" and to uncompromisingly explore phenomena.

I start from a particularly daring and unsettling appendix of Husserl's later work, *The Crisis of European Sciences and Transcendental Phenomenology*. This 23rd appendix does not simply develop a generally non-anthropocentric definition of subjectivity; it states an essential relation between phenomenology and biology. "Biology's proximity to the sources of evidence grants it such a proximity to the depths of the things themselves, that its access to transcendental philosophy should be the easiest."[76] So, is the phenomenological subject animal rather than human and what does that mean?

To begin with, an orthodox disciple of Husserl might reply, the phenomenological subject is neither human nor animal: It is a transcendental ego. The latter, insofar as it is the correlate of any possible appearing and any possible world, cannot be reduced to what Husserl calls the empirical ego, that is, to *this or that* particular subject embodied in a small piece of matter within the world. The transcendental subject is much more than me (in the personal and restricted sense of the first person, as distinct from "you," it," "him" etc.). It is more than a human being, or any particular possible being. "More" and, mainly, of a different nature, for the transcendental subject cannot be assigned to any objective place. In a first idealist version of phenomenology, Husserl identifies the transcendental ego as the perfectly self-conscious subject of an all-encompassing and crystal-clear knowledge of the world. Ideally every empirical subject must, through phenomenology, "get back to its authentic and true self, its infinite self, purified from all "terrestrialities" [*mein echtes und wahres, mein unendliches, vom Irdischen gereinigtes Ich*]," it must "fulfil itself as an 'eternal I [*ewiges Ich*].'"[77] However, this claim is incompatible with the recognition of the transcendence of the world.

Appearing requires distances, obscurity, and non-coincidence. A crystal-clear world is not any longer *a world*. Therefore, Husserl's first concept of the absolute transcendental Ego enters into tension with the fundamental project of *phenom*enology, namely the project of building a rigorous and methodical science of being *as that which appears*. Hence, a second version of the phenomenological subject is foregrounded, in what has been called genetic and generative phenomenology.[78] The transcendental "ego" cannot but be torn apart between an infinity of possible subjects whose worlds do not fully coincide. It consists in a multipolar intersubjectivity undermined by blind spots.[79] Communication and transposition are always at work, but they are dynamic and endless processes. Phenomenal subjects are thus limited and situated in the world. The focus of phenomenology shifts from the absolute ego to *embodiment*, and from the world to the concrete *lifeworld*.

Thompson asserts that "the lifeworld is relationally bound to human subjectivity."[80] But the very choice of the designation "lifeworld" as well as the appendix XXIII of the *Krisis* points to a less anthropocentric interpretation. Biology, Husserl argues in this addendum, has been commonly regarded as inferior to physics, precisely because it could never fully manage to free itself from a subjective approach. It is indeed "only through empathy [*Einfühlung*]" that "animals can have sense [*einen Sinn haben kann*]."[81] Husserl, by contrast, gives biology the highest value among sciences, even the status of the foundation of all sciences,[82] precisely since it does not lapse into objectivism, which Husserl has denounced as self-refuting. Empathy is, Husserl contends, the biggest asset of biology. This intrepid claim is vehemently opposed by Heidegger in *Being and Time* and *The Fundamental Concepts of Metaphysics*. Let us examine this point of contention. The issue is whether our human being-to-the world—from Husserl's perspective: Our transcendental intersubjectivity—essentially takes the form of a human community or of an interanimality. At stake ultimately is the recognition of animal subjects.

§17 Empathy with non-human animals: Heidegger vs Husserl

In paragraph 49 of *The Fundamental Concepts of Metaphysics* and in paragraph 26 of *Being and Time*, Heidegger criticizes the concept of *Einfühlung* in general. Husserl's approach is without any doubt one of the targets of Heidegger's criticism. Husserl, although distancing himself from Theodor Lipps' theory of projective empathy,[83] frequently uses the concept of *Einfühlung*.[84] Husserl persists with this approach in *Cartesian Meditations*, giving a seemingly quite theoretical and artificial account for the encounter of the other through analogy and a form of phantasy.[85]

In *The Fundamental Concepts of Metaphysics*, Heidegger argues that the word "*Einfühlung*" implies that an allegedly isolated subject secondarily transposes itself—"feels its way"—into another, which explains nothing unless one presupposes an actually more original community on the basis of which the projection of my feelings *into* (Ein-*fühlung*) another individual may be undertaken. According to Heidegger, the question is not whether empathy is possible, but whether an original *Mitsein* exists between human beings, or between humans and non-human animals. Heidegger continues that being with one another is an always already given feature of the relationship with other humans. As a result, a process of *Einfühlung* or transposition actually only takes place occasionally, as the sign that the original *Mitgehen* has temporarily become slightly looser. With animals, there is on the contrary, in Heidegger's view, no given *Mitsein*. A transposition into animals looms as a remote and always aborted possibility: The dog eats *with* us, and yet not.[86]

Very well: A clear-cut theory indeed, but flawed for two main reasons. On the one hand, we should never take community with other humans for granted. Consequently, phantasies must, as Husserl claims, play a role in human intersubjectivity. On the other hand, the emergence of intersubjectivity unfolds through homogenous processes, whether it binds us with other humans or with non-human animals.

In *Cartesian Mediations,* Husserl acknowledges that a fundamental community exists between empirical subjects precisely in that they are all deeply rooted in the transcendental subject. If I managed to bracket every particular aspect of my particular situation, I would reach a "perspective" that everybody could share. But since the transcendental ego is eventually regarded by Husserl as a "*telos*,"[87] an ideal on the horizon, rather than an actually perfectly achieved unity, we have to take especially seriously the *problem* of our transposition into another subject's "mind." As Husserl argues, *Einfühlung* is not based on *reasoning* by analogy, which would infer a fourth term from the relations of three other given terms: a) My body, b) my mind, c) the other's body, d) the other's mind.[88] A comparison with Dennett's concept of the "intentional stance" may also be telling here. By emphasizing the concept of empathy, Husserl refuses to ascribe the encounter with another subject to the level of theoretical assumptions and explicative *tactics*. Unlike the intentional stance, *Einfühlung* is a genuine experience and a concrete encounter with the other. But like the intentional stance, it contains a dimension of speculation. How can these two aspects combine?

Einfühlung, Husserl contends, is an *analogische Apperzeption*, a genuine intuitive experience, but based on an audacious transposition and, as a result, always undermined by uncertainty. I do not *fantasize* a *copy-image* [*Abbild*] of the other.[89] I do not *invent* a fake other. I do not project into the other's mind a feeling that I would first have experienced for myself. Rather, I live an as-if experience: It is as if I were in this place over there, with this body that is *actually* not mine. "I apperceive [the other] as having spatial modes of appearance like those I should have if I should go over there and be where he is,"[90] "her motives become my quasi-motives."[91] Our previous analyses of the concept of phantasy prove decisive here: To phantasize does not mean that we simply make up an empty and arbitrary representation of the object. The empathic phantasies are fueled by melodic themes that outline themselves through the morphology, the behaviors, and the speech of the other. Differentiated rhythms, phases (for instance resting/getting moving), and repeated motion patterns are dances and "roles" that call for our taking them up.[92]

In a phenomenological ontology, these themes are an integral part of the very being of an individual. In *Wesen und Formen der Sympathie*, Scheler argues that feelings do not preexist their "expression": "When joy or love are inhibited in their expression, they do not simply remain the same from the internal point of view, but tend to evaporate."[93] Nagel wrongly presupposed that we know our feelings only through introspection. In fact, I truly find and experience somebody's shame in her blushing, and her rage in the gnashing of her teeth.[94] When I am carried away by phantasies inspired by the theme that animates the other's existence, a key dimension of the other's being travels from its behavior to mine, animates me, and enriches my perspectives on the world. But this experience cannot be defined as a simple, unproblematic being-with, neither can it sufficiently be understood as a mechanical process of reproduction or re-enactment of the same behavior, for instance, through a phenomenon of contagion, mimicry, or the firing of mirror neurons. Indeed, a theme is essentially open. Thus, when we are inspired by the theme of an intentional and subjective behavior that, for its own part, interprets the ambiguous world and hence adds ambiguity to ambiguity, ek-stasis and dynamism reach a climax. What is indeed specific in the encounter with other subjects is that the themes that solicit me can be taken up through a world-building activity, which relentlessly modifies everything by interpreting, selecting and noticing, valuing and enacting.

Empathy thus performs an investigation and a set of hypotheses about the specific tone of the other's intentionality and the world that arises through it.[95]

For instance, I am sitting on a crowded train and a man is sitting beside me with legs wide apart. He is not gesticulating randomly; he is silent and well-groomed. His eyes are riveted on his smartphone. Apparently, he has settled down and let himself go into a relaxed posture. This disposition, however, contrasts with the slight tension between his self-restraint and a contained invasion of my space. He is seemingly not preoccupied: In his world, this situation is entirely normal. And, in fact, everybody remains quiet and apparently unconcerned around us. The question of his intentionality, its nature, and the level of consciousness that characterizes it, is open. Such a question is also related to the world in which this can be normal. Do we share the same world? Not entirely.

Heidegger's idea of an obvious *Mitgehen* with my human fellows is therefore misleading, overlooking as it does power relations and ideological structures of behavior as sources of deep misunderstandings. As Derrida argues, the smooth relation with my peers cannot demonstrate that all humans belong to an allegedly unified community and refer to the same world. To be sure, such a unique world lies on our horizon, but intersubjectivity is riddled with innumerable gaps and blind spots.

Heidegger's two main arguments against the possibility of an interspecific intersubjectivity have inherent weaknesses. First, it is impossible to contrast an allegedly obvious and always already given human community with the silence and the withdrawal of non-human animals, unless one calls "human community" the comfortable relation with one's fellows. Using our phantasy to understand others is an integral part of intersubjectivity. Second, the above description of my puzzled and prospective "empathy" with what has been polemically called "manspreading" could apply to the encounter of unfamiliar non-human animals. The silent, muddled, perplexed, and still meaningful exchange that takes place in this encounter is actually better understood in terms of unhinged worlds than in terms of symbolic communication. This description in connection to interspecific empathy is actually an effective literary technique used to overcome the misleading impression that we, humans, are all sharing the same highly civilized common world. Thus, Alexandre Vialatte, in *Bestiaire* [*Bestiary*], gathers descriptions of non-human animals and humans from the inquiring and uncanny vantage point borrowed from 19th-century zoologists or entomologists. "Today's man has to be searched for in the area where he is located. At the stop of the bus 27. Under a light rain. In slouch-hat. He returns from his sad work, at the end of a monotonous day. He does not "argue," he does not "judge," he does not want to open any "case file." He wants to go back to his gray house, in his rainy suburban area, as quickly as possible. He only asks for two things: First, not making war, and second, a pay increase.[96]

§18 Scientific empathy with non-human animals? Fantasy intertwined with an objective approach (Uexküll, Buytendijk, and Plessner. Happy fish and rearing horses)

In *Wege zum Verständnis der Tiere*, Buytendijk gives many examples of the way *Einfühlung* with non-human animals can be supported by the musical themes that animate a consistent sequence of movements. Buytendijk's approach draws influence from Uexküll's work as well as from the phenomenological tradition, especially Husserl, Scheler, and Merleau-Ponty. Buytendijk, aligning with Uexküll, seeks to show that empathy and an objective approach can fruitfully collaborate in animal sciences.[97]

According to Buytendijk, who explicitly refers to Zhuangzi's famous parable on the joy of fish, when we contemplate fish swimming swiftly, wriggling and frolicking, we see their joyfulness. We immediately feel that they exult, and we do believe that they *really* exult.[98] Even so, Buytendijk asks, how exactly do we *know* that? Although there is room for this question, indeed, even in our most confident empathy with fish, Buytendijk argues that the empathetic perception of animals is based on a strong empirical given. The behavior of these fish displays a certain unity and a recognizable pattern. They play to their hearts' content, throw themselves wholeheartedly into pirouettes and frantic spins, without any obvious goal. Moreover, as far as we can see, there is not even the shadow of a threat in their environment that may give rise to concern and would cast a totally different light on the fish's agitation.

It is also on the basis of a meticulous observation of animals' behaviors that Uexküll has contended that a scientific approach must integrate the concept of animal subjectivity. Animals are selectively oriented toward certain goals: They point to them and consistently return to them, in many different contexts, while simultaneously and not less consistently overlooking many other aspects of their environment. Also, they are not simply *actually hurt* by specific situations, they flee them and do their best not to ever encounter them again. In addition, Uexküll notes, certain realities exist only for some animals and are absolutely invisible and imperceptible by others. They are not objective realities but crucial markers in the world of specific animals. Uexküll identifies them as "purely subjective realities"[99] and gives the examples, among others, of the territory, the migration routes, and the absent fly chased by a starling.[100]

Buytendijk and Plessner borrow the phrase "motion-melody [*bewegungsmelodie*]" from Uexküll[101] to describe the structure of animal behavior. A melody is a material reality that yet has the objective characteristic of outlining a dynamic theme and a horizon of future developments; in other words, a certain intentionality. Melodies and behaviors are thus located below the level where the dichotomy between psychical and physical phenomena emerges. "The living body and its forms of motion constitute a unity that cannot be said to be physical. Nor is it psychological. Although such a unity belongs to none of these two levels of being, it is nonetheless real."[102] It is always possible to break up a melody into a sequence of clearly measurable and well-circumscribed sounds, or a behavior into a set of mechanical movements; but, when doing so, we simply choose to disregard traits that are actually given in our experience. This is also why Buytendijk and Plessner consider their approach to be essentially phenomenological: We cannot choose to ignore a part of our experience for the sake of a clear, positivist ontology.

When we experience a motion-melody, for instance, with a "searching" or a "fleeing" meaning, we are inspired and carried away by the "tempo" and the melodic line that unfolds through the different trajectories, movements, and attitudes. This is an integral part of the *reality* of a motion-melody. An intentionality is therefore experienced as well. "Animals are a subjective activity, seen from the outside," writes Buytendijk, quoting von Weizsäcker.[103] A specific body, through its objective, visible behavior *demands* that we attribute an interiority to it. It is here perfectly clear that a phenomenological concept of subjectivity is at work and allows us to avoid any metaphysical speculation about mysterious entities such as souls, minds, or vital principles. As intentionality and agency, a subject exists only through the apprehension and the active interpretation of the world. In other words, it exists through behaviors and bodily exploratory movements, immediately at the level of the interaction and the intertwinement with the world (the correlation world/subjects). It is not a pure interiority, but an exteriority that diacritically refers to a dynamic and ek-static being.

Empathy is consequently compatible with a scientific approach. It is important, for instance, Buytendijk argues, to always connect the behaviors of animals to their *Umwelt* and their own apprehension of objects as prey, tools, or threats for instance. An anthropocentric observer may interpret the pirouettes of fish as jubilation because she sees as neutral—or simply fails to notice—something that actually appears as a threat to fish. Gisela Kaplan describes another striking example. She was asked to take care of an injured boobook owl. The owl was quiet; it did not moan, scream, or even tremble. The person who found it thus described the situation by saying that the bird "had a bit of an injury but otherwise it was happy." Kaplan interpreted the owl's behavior differently, first by placing it in its specific context: In the wild, "there are feelings and experiences that birds do not want to advertise to avoid attracting attention by predators."[104] Much slighter signs are yet still perceptible: "The eyelid may quiver a little or the eyes remain half-closed and, occasionally, the body posture may be a little hunched." In short, the layman certainly perceives and enjoys a melody, but a musicologist will be able to describe its structure and understand its meaning in a much more thorough manner.[105]

This form of empathy certainly goes way beyond a mere, immediate "letting oneself go" into an intuitive feeling. However, a phenomenological approach shows that, since interiority is not separate from the behaviors of a living body, the objective observation continuously morphs into an enlightened experiencing-with or feeling-in. At stake is a concept of a scientifically legitimate empathy. But are the objective and subjective sides of empathy strictly isomorphic? In other words, do I clearly know what the non-human animals feel?

Buytendijk and Plessner vigorously underline the objectivity of the apprehension of animal intentions, as if acknowledging that interpretation involves *also* imagination could ruin the scientificity of their method. They claim, for instance, that the form of a behavior is perceived [*wahrgenommen*],[106] and that, when one observes an animal, after a while, all of a sudden, the meaning of its behavior becomes obvious [*deutlich*].[107] But, interestingly enough, Buytendijk and Plessner also call the motion melody an image [*Bild*][108] and concede that misinterpretations are always possible. Moreover, to secure this dimension of objectivity, Buytendijk reintroduces the dicho-tomie between what can be known for sure and what must remain an entire mystery. In his view, we can be certain that we are perceiving joyful movements, although the unique way these fish actually experience such a joy will remain a mystery.[109] The positivist axe has struck again. As I have argued, intentionality and meaning essentially goes together with ambiguity and multipolarity: Empathy is an *experience* that includes a process of investigation and dialogue. I do *encounter* a motion-melody that initiates a process of empathy. Its characteristics can be rigorously observed and described. I *experience* the presence of an intentionality. But the latter possesses by definition a venturesome and open trajectory. I do not know *for sure* where exactly it will lead me, for no intentionality is tied to an absolutely predetermined and clear-cut goal.[110] And this empathy can certainly unveil the specific style and tonality that characterize joy or fear, such as they are actually lived by a particular individual.

As a consequence, an accurate description of animal behaviors should be achieved according to the model of *Einfühlung* defined as an experience that involves phantasy. Is this bird who is dragging its wing, limping, and squealing, wounded or is he trying to lure a predator? Does this horse who rears up want to play, feel threatened, or maybe want to impress somebody? The latter example is also interesting when considered in the context of equestrian arts. Both the trainer and the horse may—or may not—cooperate to build a common articulated language, a series of instituted signs, against the backdrop and with the support of a more syncretic affective, gestural, and behavioral communication. This requires extraordinary patience in practicing the exercises. The horse must self-discipline so that he can participate in the process. He must pay attention to the trainer, as the discrimination between different signs is subtle and must be learned by relentlessly struggling against the ambiguity that is constantly reintroduced by the background of syncretic communication. Indeed, slight inflexions in the voice or gestures of the trainers unavoidably modify the articulated signs and, obviously, each trainer performs these signs with her own unique style. The horse and the trainers work and play together in order to come to a negotiated agreement about common clear signs. This is a form of joint-invention. The regular performance of a specific reaction to a specific sign may be obtained to a very limited extent by simply using rewards and punishment. But some trainers, like Carlos Pereira,[111] argue that this method antagonizes horses and has commonly led to an escalation in the use of tools of coercion. Pereira relies on positive reinforcement and deliberately follows the model of the Primate Research Institute of Kyoto University, where the chimpanzees are free to enter and exit the lab as they see fit. Pereira thus counts more fundamentally on the interest of horses for interactions that combine play and a form of work. Such cooperation involves self-discipline on both sides and close attention paid to another living being. The rearing is a particularly impressive example. It responds to a certain specific sign performed, for instance, with two lunge whips (here only used as a signifying tool): The trainer holds a lunge whip in each hand and the sign for the rearing consists in bringing the arms and the whips behind one's back. The horse then rears in front of the trainer, both at his request and with a hint of potential display of power. And horses also play with this ambiguity. One can observe that, when the trainer is tired, horses are less inclined to concentrate. And when a less experienced or less self-confident trainer replaces a more assertive one, some horses may introduce a greater ardor, exaggeration, display, and, by coming closer to the trainer, some provocation in the rearing. What exactly do they express? Certainly not raw violence, since the play-work session may very well continue after this movement. In this dedicated space of communication, room also exists for the *staging* of one's power.

Communication with non-human animals unfolds in the same tentative way as communication with non-fellow humans. To be sure, I may try to ask another human for clarification, through one of the many already existing symbolic languages, if we happen to share one of them. But the originality of each perspective will be saved only if we also pay attention to the non-verbal dimension of our communication, and carefully undertake to define original concepts and even to coin new terms to account for this originality. What is more, interspecific dialogues intended for checking the accuracy of the comprehension of one interlocutor by the others can be performed. The provocative horse who rears up in front of the shy trainer checks what he was asked to do and with what firmness the trainer wants—and is able—to continue leading the game. Even with non-human animals, a process of institution of a set of conventional signs can be undertaken, as in the example of horses above, or in the famous experiments conducted by Roger Fouts or Sue Savage-Rumbaugh with apes. The idea that the initiative is human in such cases is of secondary significance, since non-human animals must be able and willing to take part in the process and may even compose new signs as the interaction progresses.[112] Lucy, a chimpanzee who learned Ameslan, thus famously identified a radish as a *cry fruit*, a watermelon as a *drink fruit*, and referred to a macaque with whom she had aggressive interactions by signing "dirty monkey."

§19 Transcendental interanimality and the chiasm between non-human and human animals

Admittedly, the addendum XXIII of *The Crisis of European Sciences and Transcendental Phenomenology* might very well be read as an idealistic text that submits any particular reality and, in this case, especially living beings, to the model of human subjectivity. Husserl seeks to understand that which he describes as "those variant forms that ultimately lead back to an ego and to myself, the inquirer here, as an originary mode?"[113] He also often presents the human psyche as the *normal* one, the archetype, in contrast to the thought and the world of children, "inferior" animals,[114] and cultures other than Western.[115]

Yet, Husserl's admiration for the European grand project of universal science and universal values is counter-balanced by his criticism of the predominance of the objectivist paradigm in the Western thought. The certainty of referring to the unique objective world—the real world that everyone should recognize as it is—is illusory and results from a grandiose and consistent construction, namely a questionable way of using language to ossify meaning. When I encounter another subject outside of my smooth relations with familiar peers—this man on the train, somebody who does not speak any of the languages that I know, or my neighbors' cat—we do simply try to communicate. In such encounters, the horizon of a common world is still present, but it was not brought by Europe. Indeed, tentative communication in such cases cannot take place under the banner of the imperialist injunction to europeanize ourselves and to join an already defined alleged unique world.

The smooth alterity that binds me to my fellow humans is thus a far too narrow basis for the definition of subjectivity. The more audacious the variation, the more significant and fruitful for phenomenology. I could perceive the world through the body of a close friend, but also through a bat's body or a caterpillar's body: In a phenomenal approach, these possibilities essentially belong to my subjectivity, namely, inseparably, to the transcendental subject, as the other side integral to our being, accessible only through phantasy and yet constitutive of the very flesh of the *real* world.

Consequently, it seems to me that the appendix XXIII should be read in a non-anthropocentric manner, regardless of Husserl's actual intention, which was most likely extremely ambivalent. In a genetic approach, *Einfühlung* is less a *reduction* of the other to the sphere dominated by the transcendental ego than a "*transgression*" of the very being of my monadic ego by others. Husserl's choice of the verb "to transgress" [*überschreiten*], in *Cartesian Meditations*,[116] when describing the encounter with the other, reveals a certain violence exerted on the ego, a questioning of its strict self-identity, without going as far as establishing a pure positive externality.

If decentration and phantastic transpositions constitute the ultimate nature of being, taking the subjectivity of non-human animals seriously proves crucial in the investigation of the true nature of things and persons. Phenomenology does not eventually bring everything back to a Platonic all-encompassing Subject. Rather, its ultimate principle is a world in-the-making through transpositions. *Einfühlung* is the original multiplicity of the transcendental "ego," or better, of a transcendental intersubjectivity. Empathy is the subject's diffraction in poles reflecting one another. Put another way, empathy is a chiastic relation between strands that are symmetrical and yet different.

I borrow this concept of *chiasm* from Merleau-Ponty, who particularly developed the idea of a multi-polar intersubjectivity. With the concept of chiasm,[117] Merleau-Ponty endeavors to think the deep relation between two dimensions, without reducing one to the other. More radically, the chiastic relation consists *both* in an extreme closeness and in an extreme distance, and the connecting node is also a shifting point. The structure of the figure of speech called chiasmus (ABBA) indeed always accentuates proximity *and* contrast. AB and BA are connected but through a reversal, so that the figure of speech—unlike parallelism, for instance—maintains and enhances a tension instead of reducing differences to identity or similarity. "Fair is foul and foul is fair": Fair is actually foul, but the former is not resorbed in the latter; the second strand of the chiasm ("and foul is fair") revives fairness, although it does so by connecting it tightly again with its opposite: An endless dynamic process is launched. In *The Visible and the Invisible*, Merleau-Ponty speaks of "chiasms" when referring to a series of relations such as the relation between the visible and the invisible, my body as a subject (flesh/*Leib*) and my body as an object (*Körper*), I and the other, or the objective world and the subjective world. Merleau-Ponty refuses ontologies that simply reduce one of the terms to the other. The world for me does not exactly coincide with the real world. Things and others are transcendent, so that the horizon of the real objective world emerges. Similarly, the seer both encompasses the world and is contained in it, locked-up in a tiny bit of space. My touching hand and my hand as a surface that can be touched and an object that may be squashed or severed do not exactly coincide: I do not experience these two aspects exactly at the same time. However, I can always switch from the experience of my hand as an object to the feeling of my hand as that which is animated by my "I feel," "I touch," "I can," and "I want." Each aspect is *imminent* in the experience of the other. Likewise, in a phenomenological investigation, there is always a point where thoughts turn out to be things and things turn out to be thoughts. I am not the other, but, through empathy, namely also fantasy, I can decenter myself into the other's perspective. The node of the chiasm is thus a dynamic switching operator. Each strand of a chiasm is also the reverse side of the other. Similarly, the human and non-human worlds developed by contrasting each other around poles that have woven their *Umwelt* around them and have built an interior-exterior structure; yet, other animals are imminent in me and I am imminent in non-human animals, *through* the interface of the imaginary.

The question "are animals subjects in the same sense as we, humans, are?" is badly posed. Humans are not subjects in a clear-cut sense, with perfectly rigid characteristics. And subjectivity necessarily possesses an intersubjective structure. Non-human animals are neither subjects in the same sense as we humans are, nor in a different sense. Rather, we belong to a unique and yet manifold and dynamic transcendental interanimality.

Non-human animals bring a new dimension to the transcendental intersubjectivity. By regarding them as alter egos, I gain access to original and extraordinarily diverse variations in possible body structures, perceptive systems, communication systems, and social structures. I discover, for instance, forms of expression that are not anchored in a conventional articulated language.[118] The "normal real world" then truly appears as a construct and is replaced in the foreground by a genuine *life* world. Husserl's anthropocentrism is challenged and disrupted by an interanimal *Einfühlung*. In place of the vertical structure "normal subjectivity/

variants," phenomenology should put forward a horizontal, multipolar structure. In this way the specificity of each pole is not squashed and we can relate to a *world* worthy of the name. Phenomenology needs biology and biology needs phenomenology. To be sure, the imaginareal unhinges every being, but, as the next chapters will explore, some modalities of perception, imagination, and symbolization in non-human animals do call for and sustain transpositions to an unparalleled degree. In order to flesh out this new phenomenological ontology, it is thus momentous to *explore* the imaginary of animals and to go through the wealth of concrete and challenging experiences that it offers.

Notes

1. Zhuangzi. *Wandering on the Way: Early Taoist Tales and Parables of Chuang Tzu.* Trans. Mair, Victor H.. Honolulu: University of Hawai'i Press, 1994, p.165. Buyentdijk quotes this text in *Wege zum Verständnis der Tiere.* Zürich, Leipzig: Max Niehans Verlag, 1938, p.17–8.
2. The phenomenology of animality is already a rich field of research. See several illuminating contributed volumes: Painter, C.M. and Lotz, C. (Eds). *Phenomenology and the Non-Human Animal: At the Limits of Experience.* Dordrecht: Springer, 2007. Burgat, Florence, and Christian Ciocan (Eds.). *Phénoménologie de la vie animale.* Bucarest: Zeta books, 2016. Diaconu, Madalina and Christian Ciocan. *Phenomenology of Animality.* In Studia Phaenomenologica, Vol. 17. Bucarest: Zeta books, 2017.
3. Nagel makes a first step toward this idea when he notes that perfect objectivity is actually "a direction in which the understanding can travel" (Nagel, 1974, p.443). He also specifies that when I consider the perspective of a subject who is "sufficiently similar" (442, quite a vague concept: Objectivity is vanishing into thin air ...) to me, I have access to "perfectly objective" (442) phenomenological facts. Here, Nagel contradicts his original claim (subjectivity is private) in order to stick to the facts. Indeed, when writing about what *we* can know, he must expect us to understand what the thoughts of this "we" are. This "we" is a phenomenon, and, as Husserl argues, a fundamental structure of subjectivity, which therefore is an intersubjectivity.
4. See, for instance, Harman, Graham. *Guerrilla Metaphysics: Phenomenology and the Carpentry of Things.* Chicago: Open Court, 2005.
5. See, for instance, Husserl, Edmund. *Die Krisis der europäischen Wissenschaften und die transzendentale Phänomenologie. Eine Einleitung in die phänomenologische Philosophie.* Trans. David Carr. The Hague: M. Nijhoff, 1954. Evanston: Northwestern, 1970, §41.
6. Wemelsfelder, F. "The scientific validity of subjective concepts in models of animal welfare". Applied Animal Behaviour Science 53,1997.
7. Rees, Amanda. "Reflections on the field: Primatology, popular science, and the politics of personhood." In *Social Studies of Science,* 37(6), 2007. See also Nimmo, Richie. "Animal Cultures, subjectivity, and knowledge: Symmetrical reflections beyond the great divide." In *Society & Animals,* 20(2), 2012, p.181.
8. See also on that topic Ruonakoski, Erika. "Phenomenology and the Study of Animal Behavior." In *Phenomenology and the Non-Human Animal: At the Limits of Experience.* Corinne Painter and Christian Lotz (Eds.). Dordrecht: Springer, 2007, p.75–84. And Lestel, Dominique, et al. "The phenomenology of animal life." In *Environmental Humanities,* Vol. 5, 2014.

9. Petitot, Jean, Varela, F.J., Pachoud, B., Roy, J.-M. (Eds.). *Naturalizing Phenomenology: Issues in Contemporary Phenomenology and Cognitive Science.* Standford, CA: Stanford University Press, 1999, p.2.
10. Thompson, Evan. *Mind in Life. Biology, Phenomenology, and the Sciences of Mind.* Cambridge: Harvard University Press, 2007.
11. However, the body is, according to Husserl, a "switching point [*Umschlagspunkt*] from causal to conditional process", and not a point of coincidence and identification between these two perspectives. Husserl, Edmund. *Ideen zu einer reinen Phänomenologie und phänomenologischen Philosophie, Zweites Buch: Phänomenologische Untersuchungen zur Konstitution (Husserliana IV),* The Hague: M. Nijhoff, 1952, p. 160. Trans. Rojcewicz, Richard and Schuwer André, Dordrecht: Kluwer, 1993, p.168.
12. §35 explores this subjectification process in a detailed manner.
13. Similarly, the description of such a cooperation as a relation "through reciprocal *constraints*" (Petitot et al., *Naturalizing Phenomenology*, p.67) betrays the radical freedom that is at stake with the phenomenological epoché.
14. Husserl, *Ideen II*, "Naturkausalität und Motivation," p.229 sqq.
15. See infra, Chapter 3.
16. See, for instance, *Ideen zu einer reinen Phänomenologie und phänomenologischen Philosophie, Erstes Buch: Allgemeine Einführung in die reine Phänomenologie.* The Hague: M. Nijhoff, 1950, §34 and §69–70.
17. For instance, *Ideen I*, §42–4.
18. See also Toadvine, Ted, "How Not to Be a Jellyfish: Human Exceptionalism and the Ontology of Reflection." In *Phenomenology and the Non-Human Animal: At the Limits of Experience.* Lotz, C. and Painter, C. (Eds.), p.39–55.
19. From the Ancient Greek ἔκστασις: "a standing out of one's self."
20. *Ibid.*, p.41.
21. As such, Toadvine's analyses in *Merleau-Ponty's Philosophy of Nature* (Evanston: Northwestern, 2009) led the way for my research.
22. Husserl, Edmund. *Logical Investigations.* Trans. J.N. Findlay. New York: Routledge, 2001, Vol. 1, p.168.
23. Levinas, Emmanuel. *La théorie de l'intuition dans la phénoménologie de Husserl.* Paris: Vrin, 1970, p.62.
24. *Ideen I,* §9.
25. And, indeed, Kant actually never ceased developing new ideas about things in themselves.
26. *Ibid.*, §9a, p.22, trans. Carr p.25, my emphasis.
27. *Krisis*, p.159 and p.394. See also Dufourcq, 2010, p.209 sqq.
28. Merleau-Ponty, *The Visible and the Invisible*, Trans. A. Lingis, Evanston: Northwestern University Press, 1968, p.40.
29. Jonas, *The Phenomenon of Life, Toward a Philosophical Biology.* New York: A Delta Book, 1966, p.10.
30. Positivism claims that the real is defined by its pure positivity—from the Latin *ponere*, to lay down, to place, to settle arbitrarily.
31. Merleau-Ponty, *The Structure of Behavior,* Trans. Alden Fisher, Boston: Beacon Press, 1960, p.81–2. See supra, §10.
32. Our phenomenological agency is thus illuminatingly defined in terms of enaction by Depraz, Varela, and Vermersch in *On Becoming Aware. A Pragmatics of Experiencing.* Amsterdam: J. Benjamins, 2003.
33. Husserl, Krisis, Addendum XXVIII, p.512.
34. Husserl makes this point explicit for instance in the 38th paragraph of the *Prolegomena to Pure Logic* (Husserl 1993).
35. Heidegger, Martin. *Die Grundbegriffe der Metaphysik. Welt—Endlichkeit—Einsamkeit,* Frankfurt: Klostermann, 1983, p.291–2. Trans. William McNeill and Nicholas Walker, Bloomington: Indiana University Press, 1995, p.197–8.

36. *Ibid.*, p.361. Trans., p.248. "Beings are not manifest to the behavior of the animal in its captivation. they are not disclosed to it and for that very reason are not closed off from it either. Captivation stands outside this possibility." See also trans. p.270 "The animal could only be deprived of world if it at least knew something of world. But it is precisely what we have denied in the case of animals."

37. *Die Grundbegriffe der Metaphysik,* p.408, Trans. p.282.

38. *Ibid.*, p.396, Trans. p.273.

39. Adorno, Theodor W., *Jargon der Eigentlichkeit. Zur deutschen Ideologie ist ein ideologiekritisches Werk,* Suhrkamp Verlag, 1964, p.39–45. Trans. by Knut Tarnowski and and Frederic Will Evanston: Northwestern University Press, 1973, p.50–58.

40. Heidegger, *Die Grundbegriffe der Metaphysik,* p.7, Trans. p.5.

41. Agamben, Giorgio. *The Open: Man and Animal.* Stanford: Stanford University Press, 2004, p.58.

42. Heidegger, *Die Grundbegriffe der Metaphysik,* p.28, Trans. p.18–9.

43. See, for instance, Hoffmeyer, Jasper. "Biosemiotics: Towards a new synthesis in biology." In *European Journal for Semiotic Studies.* 9(2), 355–76, 1997. See also infra, Chapter 4.

44. "To be a subject means the continuous control of a framework by an autonomous rule" (Uexküll, Jakob von. *Theoretische Biologie.* Berlin: J. Springer Verlag, 1920. Trans. D. L. Mackinnon Theoretical Biology. New York: Harcourt, Brace, 1926, p.223), the question of awareness and self-awareness is regarded as secondary by Uexküll in his definition of subjectivity.

45. Uexküll, Jakob von. *Streiftzüge durch die Umwelten von Tieren und Menschen Ein Bilderbuch unsichtbarer Welten.* Berlin: Springer, 1934, p.8 and p.60–1. *A Foray into the Worlds of Animals and Humans,* p.51 and 96.

46. Uexküll, *Streiftzüge durch die Umwelten von Tieren und Menschen,* p.81. Trans., p.117

47. *Ibid.*

48. See also Buchanan, Brett. *Onto-Ethologies: The Animal Environments of Uexkull, Heidegger, Merleau-Ponty, and Deleuze.* New York: Suny Press, 2008.

49. See also Florence Burgat's illuminating analyses in *Liberté et inquiétude de la vie animale.* Paris: Kimé, 2006, p.123

50. See infra, Chapter 4, §29.

51. Such a hesitative attitude becomes even more obvious when chimpanzees who are confronted with two alternatives in a choice situation "concurrently point to both alternatives or successively change their choice" (Suda, Chikako and Josep Call. "What does an intermediate success rate mean? An analysis of a Piagetian liquid conservation task in the great apes." In *Cognition.* Vol. 99, 2006).

52. Carruthers, P., Ritchie J.B. "The emergence of metacognition: Affect and uncertainty in animals." In Beran M et al., (Eds). *Foundations of Metacognition.* Oxford University Press, 2012. Kornell Nate. "Where is the "meta" in animal metacognition?" In *Journal of Comparative Psychology* 128, 2014. Beran, M.J. and Smith, D. "The Uncertainty Response in Animal-Metacognition Researchers." In *Journal of Comparative Psychology,* May; 128(2), 2014. Kaufman Matthew.T. et al. "Vacillation, indecision and hesitation in moment-by-moment decoding of monkey motor cortex." In *eLifeSciences,* May 5, 2015.

53. Bateson, Melissa et al. "Agitated Honeybees Exhibit Pessimistic Cognitive Biases." In *Current Biology,* June 21(12), 2011.

54. Pictures and fantasy always involve a dimension of unreality. To imagine is to refer to unreal objects (e.g., the chimera) or unreal aspects of real objects (e.g., a colleague of mine wearing a rabbit costume). Material pictures also contribute to the imaginary field by involving a reference to the unreal, although it is important to account for the distinction between the act of imagining

in the absence of any actual material pictorial basis and the act of contemplating a picture. One may contemplate a photograph and consider it to be the simple unveiling of the way an actual object in itself *looks*. In this case, the photograph is regarded as the trace left by the actual object on the paper through a sequence of physical, chemical, and digital processes. Nevertheless, this realist approach is always at least subtly unhinged by the pictorial structure itself. In an aesthetic contemplation of a realistic portrait of Napoleon, I do not simply tell myself: This is what Napoleon looked like. Pictures are not only about represented objects, but also about the secret formulas of appearing processes. In pictures, as already highlighted by Plato or Kant, the actual presence of the object as that which I can use and interact with at all possible levels is removed. What remains is the appearance. But this *reductio ad apparentia* raises questions that are usually overlooked in the everyday relation to objects: How can something manifest itself without being really present so to say "in the flesh"? How can it appear through new, diverse, and quite modest mediums such as patches of color, lines, textures, and sketches? Hence the always uncanny character of objects in pictures: They are and are not real. They are somehow phantoms, reduced to a mere set of colors and lines on a canvas, but also strangely hyper-present (the humble is thus glorified in Flemish still lives), for appearing powers (such as light effects, perspective effects, compositions, and affective powers of sensory qualities) have been worked on, enhanced, and intensified. Also, the role played by the activity of the subject in the process of appearing is foregrounded. The picture, whether a photograph or a painting, always conveys a subjective dimension: This is the way *the author* saw the object, or even intentionally wanted to present it, through unexpected forms and colors. This incites the viewer to be inventive in her turn and to engage the esthetical contemplation as a play with appearances. Hence, the enactive dimension of phenomenality is tenfold increased.

55. See, for instance, Thomas, Nigel J.T. "Are Theories of Imagery Theories of Imagination? An Active Perception Approach to Conscious Mental Content." In *Cognitive Science*. Vol. 23, 1999.

56. Imagination is an "intuition," not an empty intentional meaning [*Meinung*]: See, for instance, Husserl, *Logische Untersuchungen, Zweiter Teil, Untersuchungen zur Phänomenologie und Theorie der Erkenntnis*, 6th Investigation, §14. See also Merleau-Ponty, *L'Œil et l'Esprit,* Paris: Gallimard. 1961, p.23–4, Trans., p.126.

57. Husserl uses the word *Phantasie*. In vernacular German, *Phantasie* may mean, depending on the context, "imagination," "fantasy," or "fancy." But Husserl gives it a much more specific meaning and this is the reason why, in line with the English translation of Husserl's *Phantasy, Image consciousness, and Memory* (Dordrecht: Springer, 2005), I will use the spelling "phantasy" when I want to designate an experience in which the object is given as quasi-present, in the mode of the *as if.*

58. Husserl contrasts phantasies [*Phantasien*] with *Bilder*, pictures, or more exactly with "image-consciousness [*Bild-Bewusstsein*]" namely a certain experience in which a present object (a photograph for instance) is given to me as the representative of an absent one.

59. Husserl, *Phantasie, Bildbewußtsein, Erinnerung. Zur Phänomenologie der anschaulichen Vergegenwärtigungen, Texte aus dem Nachlaß (1898–1925)*, Den Haag: M. Nijhoff, 1980, p.42, Trans. p.45.

60. *Ibid.*, p.21, Trans. p.23. See also Evan Thompson's study of the role played by the concept of mental picture in mind sciences: Thompson, *Mind in Life*, p.267–311.

61. Sartre also puts particular emphasis on imagination as an *act* of apprehension, see *The Imaginary*, p.35–6.
62. Husserl, *Phantasy, Image consciousness and Memory*, texts No. 16 and 18.
63. *Ibid.*, text No. 18.
64. See Dufourcq 2010, p.59–77.
65. Husserl, *Phantasy, Image consciousness and Memory*, p.517, Trans. p.619 (modified translation). In this passage, it is also important to pay attention to the word "perzeptive." Husserl makes a distinction between *Perzeption*, and *Wahrnehmung*. Both are perceptions, and in both cases, Husserl specifies, the object is present in flesh and blood [*Leibhaftig*], but in *Perzeption* the belief in the existence of the object is suspended. (*Ibid.*, p.466, trans. 556).
66. See Merleau-Ponty, *Eye and Mind*, In *The Merleau-Ponty Aesthetics Reader*, Galen A. Johnson, Michael B. Smith (Eds.). Evanston: Northwestern University Press, 1993, p.130.
67. Mitchell, *What Do Pictures Want?*, Chicago: University of Chicago Press, 2005, p.11.
68. *Ibid.*, p.13 and 86–9.
69. Mitchell, *Iconology: Image, Text, Ideology.* University of Chicago Press, 1987, p.10.
70. See supra, p.3.
71. See Jacques Derrida, *Of Grammatology*, Trans. Gayatri Spivak, Baltimore: Johns Hopkins University Press, 1976, p.292.
72. Merleau-Ponty, Maurice, *The Visible and the Invisible*, p.40.
73. Husserl speaks of the hovering in phantasy ("*das Vorschweben in der Phantasie*") to describe three remarkable features of fantasies: 1) the absence of anchoring of the appearing object in objective space and time (I cannot trace a way that would start from my actual place in the world and would lead continuously to the centaur, for instance). 2) "*Das Schweben*" also describes the fleeting, evasive, and fragile nature of the object that appears in fantasies. 3) "*Schweben*" means: To hang in the air. The image may be fleeting, it is nevertheless an apparition and possesses a relative stability, its own consistency: it competes with reality without being able to be situated within it. As a result, the hovering is a "beyond," a beyond-the-actual-world, that is neither completely nothing, nor radically foreign to our world. It cannot be *fully integrated* in our world: It introduces a dimension of ubiquity, of de-centering, and of phase-difference within it. About imagination and hovering see also John Sallis, "Hovering: Imagination and the Spacing of Truth," in *Spacings: Of Reason and Imagination in Texts of Kant, Fichte, Hegel.* Chicago: University of Chicago Press, 1987, 23–66.
74. In summary, I propose the following definitions:

The imaginareal: The transcendental field that precedes dichotomies between subject and object, real and imaginary. It consists in a flow of ubiquitous sensible appearances that both echo and disrupt each other. It does not have any outside and holds together the following three dimensions:

- **The real**: A surface of crystallized characteristics that recur regularly in perception and through ossified figures.
- **The imaginary**: Possible variations in the being of things and subjects that manifest themselves in the form of haunting fantasies, recurring images, symbols, stubborn myths, and narratives. In the imaginary field, the floating dimension of beings is brought to the fore and is apprehended as such.

> The real and the imaginary are organically grafted upon one another. The boundary between them is porous. Each may shrink to almost nothing, while the other swells to overwhelming proportions.

- **Imagination** is the individual ability to be solicited by and actively and creatively play with melodic themes that stem from the imaginareal. Imagination actively enacts the passages and shifts of balance between the real and the imaginary.

75. Abram, David. *Becoming Animal. An Earthly Cosmology*. New York: Vintage books, 2011, p.305. I deeply sympathize with Abram's claim that our human imagination is sustained by the sensible world. The performative power of Abram's immersive, non-divisive approach is crucial. But it is equally important to acknowledge that the imaginareal is scattered, unfinished, and troubled, so that I will consistently avoid any prophetic and realist references to *the* Earth's mind, or any ultimate intuitions.

76. Husserl, *Krisis*, p.483, trans. Niall Keane, "Addendum XXIII of *The Crisis of European Sciences and Transcendental Phenomenology*: Edmund Husserl". In *Journal of the British Society for Phenomenology*, Vol. 44, No. 1, January (6–9), 2013, p.7.

77. Husserl, *Erste Philosophie (1923–1924), Zweiter Teil: Theorie der phänomenologischen Reduktion*. The Hague: M. Nijhoff, 1959, p.16.

78. See Steinbock (*Home and Beyond: Generative Phenomenology After Husserl*. Evanston: Northwestern University Press, 1995) and Thompson *Mind in Life*, p.33 sqq.

79. See also the concept of the *"de jure* invisible" in Merleau-Ponty's *The Visible and the Invisible*, p.254–5.

80. Thompson, *Mind in Life*, p.34.

81. Husserl, *Krisis*, p.482, "Addendum XXIII of *The Crisis,"* p.6.

82. "In fact, biology, as genuinely universal biology, embraces the entire concrete world, and thus implicitly physics too, and in the examination of correlations it becomes a completely universal philosophy" "Addendum XXIII of *The Crisis"*, p.8.

83. The psychologist and aesthetician Theodor Lipps was the first to develop a theory of *Einfühlung*, in the early 20th century (see for instance Theodor Lipps, "Einfühlung, innere Nachahmung und Organempfindungen," *Archiv für die gesamte Psychologie*, Vol. 1, 1903). Lipps claimed that empathy was not a form of inference or based on reasoning by analogy, but that it consisted in spontaneously projecting our feelings and mental states *into* others. His work deeply influenced Scheler, Husserl, and Edith Stein, although these three authors take a critical distance from Lipps' theory.

84. Zahavi, Dan. *Self and Other: Exploring Subjectivity, Empathy, and Shame*. Oxford: Oxford University Press, 2014, p.114.

85. Husserl, Edmund. *Cartesianische Meditationen und Pariser Vorträge* (1929-1931). The Hague: M. Nijhoff, 1950. Trans. Cairns, D. Dordrecht: Springer, 1960, §53.

86. See supra, Chapter 1.

87. Husserl, *Krisis*, Appendix IV.

88. Husserl, *Cartesianische Meditationen* §50.

89. *Ibid.*, p.153.

90. Husserl, *Cartesianische Meditationen*, p.146, Trans. 117.

91. Husserl *Ideen II*, p.275. Although this reference to the "as if" and the "quasi" demonstrates the connection between *Einfühlung* and phantasy, Husserl does not use the word "phantasy" to define *Einfühlung*. I have explained in a more detailed way elsewhere (Dufourcq 2011, p.157–182) why it is nevertheless accurate to describe *Einfühlung* in terms of phantasy.

92. This idea is sketched by Husserl in *Ideen II*, p. 238–40.
93. Scheler, *The Nature of Sympathy*, Trans. Peter Heath, London: Transaction, 2008, p.251.
94. *Ibid.*, p.260. See also Merleau-Ponty, Maurice. *Sense and Nonsense*. Trans. Hubert L. Dreyfus and Patricia Allen Dreyfus, Evanston: Northwestern UP, 1964, p.52–3.
95. One may object that, as Scheler has argued in *The Nature of Sympathy* (p.39), there is a difference between, on the one hand, imagining how I would react if I were in the other's situation and, on the other hand, apprehending the way the other experiences this situation. As a consequence, Scheler defines the apprehension [*Auffassung*] of the other's feelings and experiences in terms of perception [*Fremd-Wahrnehmung*] (*Ibid.*, 238). But imagination cannot be reduced to the imagination of what *I* would do. It is a field of possible decentrations. I am stirred, haunted, and carried away by a theme that I have not created. Moreover, even if, when apprehending the other's experiences and feelings, I do encounter a transcendent given, this experience is not a "perception." In effect, the other's experiences essentially consist in dynamic explorations and interpretations that develop towards an indeterminate future. The other's intentionality and agency are beyond my control. I can apprehend a theme, I can be swayed by the style in which the other plays this theme, but this experience will always have the twofold and ubiquitous structure that characterizes phantasies. And, in fact, Scheler's theory bears traces of the tension between the perceptive and the phantastic dimensions of the encounter with the other. Scheler thus does not only designate this apprehension of the other with the word "*Wahrnehmung*," but also, on a regular basis, with the word "*Nachfühlen*" [vicarious feeling]. Scheler identifies such vicarious feelings—or reproduced feelings, or feelings "after the other" — as *Vergegenwärtigungen* [presentifications or reproductions]. In other words, they are vivid experiences of an object that is, nonetheless, absent (*Ibid.*, p.9, *Wesen und Formen der Sympathie*, p.5). In Husserl's works, phantasies are defined as forms of *Vergegenwärtigung*, since there is an obvious difference between perception (where the object is simply present [*Gegenwart*], in person), and presentification [*Vergegenwärtigung*]. Scheler's phrase "*wahrnehmendes Nacherleben*" (*Wesen und Formen der Sympathie*, p.302) is therefore quite dicey and puzzling. To be sure, the words "Imagination" or "fantasy" may give the impression that a subject invents an interpretation from scratch or arbitrarily attributes feelings to another person in a completely fictitious representation. "Phantasy" is, I think, more accurate, because it designates a genuine experience but does not conceal the irreducible distance and ubiquity that undermine it.
96. Vialatte, Alexandre. *Bestiaire*, Paris: Arléa, 2002, p.35 (my translation).
97. This claim is also clearly advocated in a book co-written with Helmut Plessner (Buytendijk and Plessner. *Die Deutung des mimischen Ausdrucks: ein Beitrag zur Lehre vom Bewusstsein des anderen Ichs,* Bonn: Cohen, 1925).
98. Buytendijk, Frederik. *Wege zum Verständnis der Tiere*, p.18–9.
99. Uexküll, *A Foray into the Worlds of Animals and Humans*, p.125.
100. *Ibid.*, 102, 120–2.
101. Buytendijk and Plessner, *Die Deutung des mimischen Ausdrucks*, p.78.
102. *Ibid.*, p.83, my translation.
103. Buytendijk, *Traité de Psychologie Animale*, Paris: Vrin, 1952, p.20.
104. Kaplan, Gisela. *Bird Minds. Cognition and Behaviour of Australian Native Birds*. Melbourne: CSIRO Publishing, 2016, p.126.
105. Buytendijk and Plessner, *Die Deutung des mimischen Ausdrucks*, p.85.
106. *Ibid.*, p.82.

107. *Ibid.*, p.81.
108. *Ibid.*
109. Buytendijk, *Wege zum Verständnis der Tiere*, p.22.
110. The encounter of a subject cannot but be the experience of ambiguity. Heidegger is thus more consistent than Buytendijk and Plessner in this regard: if the orientation of an animal behavior is perfectly clear, it does not make sense to speak of *Mitsein*.
111. Carlos Pereira is the director of the Institut du Cheval et de l'Équitation Portugaise. He currently conducts research on the communication between horses and humans. I was lucky enough to spend one day with his team at Champlâtreux. I could observe the respectful and fruitful work-play that is carried out with horses, and learn more about his original approach. See also Pereira, Carlos, *Parler aux Chevaux Autrement - approche sémiotique de l'équi-tation*, Paris, Amphora, 2009.
112. See Fouts, Roger S. "Communication with Chimpanzees." In *Hominisation and Behavior*, G. Kurth and I. Eibl-Eibesfeldt (eds.), (137-58), Stuttgart: Gustav Fischer, 1975. And Lieberman, Philip. *The biology and Evolution of language*, Cambridge: Harvard University Press. 1984, p.241–52.
113. Husserl, "Addendum XXIII of *The Crisis,*" p.6.
114. Husserl, Edmund, *Zur Phänomenologie der Intersubjektivität. Texte aus dem Nachlaß, Zweiter Teil: 1921–1928*. The Hague: Nijoff, 1973. p.119.
115. See Husserl's famous Vienna Conference. "Die Krisis des europäischen Menschentums und die Philosophie", In K*risis*, p.314–48.
116. Husserl, *Cartesianische Meditationen*, §44, p.125.
117. See for instance Merleau-Ponty, *The Visible and the Invisible*, p.130–55.
118. See infra, Chapter 4.

3 Animal bodies and the virtual
Animals as real phantoms

Two and two there floated into my inmost soul, endless processions of
the whale, and,
     mid most of them all, one grand hooded phantom

Herman Melville[1]

§20 Continuism, discontinuism, and beyond

This chapter concerns itself with the ontological ubiquity of human and
non-human animals. As I will contend, living beings in general and animals in particular are to be fundamentally defined by an elusive "to be and
not to be" or "phantom-like" being, which entails their intrinsic relation to
meaning, essences, and the virtual. To make my case, I draw upon Merleau-
Ponty's concept of Gestalt in *The Structure of Behavior*. I argue that this
concept can become a key to framing the relation between the imaginary
and animal life in its most fundamental form, as a relation that pervades the
morphology of the living body, metabolism, animal attitudes, and behaviors. I also explore the topicality of this concept of phantom in contemporary biology, and show that there were scientifically relevant reasons why
Darwin's *Origin of Species* was haunted by phantoms and versatile images.
The concept of animal phantoms thus also proves highly relevant in the
framework of contemporary genomics and postgenomics.

I begin with the analysis of Merleau-Ponty's *The Structure of Behavior;* this
work is indeed of crucial importance for a research on the imaginary of
animals in that it outlines a theory of the deep kinship between the human
mind and non-human animals' behaviors.

The Structure of Behavior is a tricky book for Merleau-Ponty scholars as
well as for animal studies scholars. Merleau-Ponty's analyses in this work
draw heavily on contemporary findings in psychology, neurology, and ethology, but his approach is also critical and dismisses objectivist and positivist reductionism, even realism, while seeking to forge operative conceptual
tools for the scientific study of behavior. Even more puzzling: *The Structure*

of Behavior at once challenges and reasserts the boundary between humans and animals. In this one and the same book, Merleau-Ponty thus declares, on the one hand, that behaviors and organisms are melodies that sing themselves,[2] also in an irreducibly improvisational way, and, on the other hand, that "properly so called" knowledge and "properly so called" intelligence occur only in the human realm,[3] so that "man can never be an animal."[4]

The Structure of Behavior thus gives us insight into the difficulty of accounting for both the kinship and the differences between humans and animals. I want to show that the only way out of this issue is to acknowledge that human and non-human animals share the same imaginative being. In fact, *The Structure of Behavior* provides key tools for forming this theory but does not achieve their full potential.

The Structure of Behavior develops an original theory that claims to both bridge the gap *and* maintain a significant distinction between consciousness and life. The central focus is on the concept of Gestalt [structure, form], that Merleau-Ponty borrows from the school of Gestalt psychology, born in Austria and Germany in the early twentieth century. A Gestalt is a *new meaning emerging from a concrete material structure.* As Gestalten, living beings cannot be reduced to their material components and blind mechanisms. Indeed, a Gestalt is a form of intentionality that is immanent in a material structure.

But, after stating that mind is, as it were, already at work in life, Merleau-Ponty hastens to add that, in the animal realm, the source of meaning is "the monotonous *a prioris* of need and instinct."[5] The human realm, Merleau-Ponty contends, displays a wholly different relation to meaning: In our world, the symbolic function reigns supreme. Whereas, in a later work, the *Nature* lectures (1957–1960),[6] Merleau-Ponty put forward the daring theory of a multipolar subjectivity and *lateral*—i.e., non-hierarchical—intersubjectivity between non-human and human animals, *The Structure of Behavior* (1942) opts for a hierarchical approach and draws a distinction among three irreducible orders—the physical order, the vital order, and the human order.

As such *The Structure of Behavior* has given rise to conflicting interpretations. Toadvine and Thompson rightly emphasize the contrast between Merleau-Ponty's earlier and later approaches to life. But, whereas Toadvine suggests that Merleau-Ponty's ideas "retain the echoes of the sharp divide between human and animal"[7] *only* in *The Structure of Behavior*, Etienne Bimbenet contends that the statement "man can never be an animal" is absolutely representative of Merleau-Ponty's thought overall and deserves to be taken seriously as the inescapable triumph of the anthropological difference: "We would like to demonstrate that this statement can be regarded as the motto [*la formule*] of a consistent philosophical anthropology, to which Merleau-Ponty remained faithful till the end."[8] In the same vein, Kazuo Masuda regards *The Structure of Behavior* as the perfect illustration of the idea that the link between animals and meaning only makes sense from the perspective of humans, who remain the only true masters of the symbolic realm.[9] I want to counter this stance and show that animals, as ubiquitous phantom-like beings, are fully at home in the imaginary realm, a domain that no human symbolic superstructure *masters*.[10]

The key question turns in fact on the exact nature of structures. The close examination of the concept of Gestalt in *The Structure of Behavior* will allow us to contend that "man can *never* be animal" cannot even be regarded as an accurate summary of Merleau-Ponty's position in *The Structure of Behavior* itself, not to speak of Merleau-Ponty's explicit rejection of an anthropological perspective in his later work.[11] In fact, Gestalten can in no way be defined as positive, solid, and clearly circumscribed *beings*. They are rather defined according to the model of phantoms and works of art. Merleau-Ponty repeatedly emphasizes that what is at stake in *The Structure of Behavior* is to overcome realism:[12] I want to take him at his word and develop the concepts of phantom-bodies, phantom-humans, and phantom-animals, also on the basis of his analysis of the phantom limb in *Phenomenology of Perception*. My approach entails a much more fragile and ironical version of the three irreducible orders than the one conveyed by the emblematic quotation retained by Bimbenet. I will contend that this concept of Gestalt is to be understood as a radical challenge to a realist approach to life. The concept of Gestalt is critical in that it lays the groundwork for a theory that accounts for animals as *imaginative through and through* and as harboring the imaginary ubiquity in the very structure of their body.

The first section of this chapter (**§21–22**) analyzes the concept of Gestalt and the bases of Merleau-Ponty's "three orders" theory in *The Structure of Behavior*. The tensions that undermine this hierarchical take are studied in the second section (**§23–26**). It thus appears that living beings are phantoms and that the three orders are in fact elusive. This notion of phantom-body also resonates with the rich imagery that forms the secret core of Darwin's theory of evolution (**§25**). The crucial relevance of this concept for a scientific approach to animals will also be at stake in this chapter.

PART 1. STRUCTURES ALL THE WAY DOWN, BUT THREE ORDERS IN MERLEAU-PONTY'S THE STRUCTURE OF BEHAVIOR

§21 Structures [*Gestalten*] and behaviors

Gestalten are "total processes whose properties are not the sum of those which the isolated parts would possess."[13] In this holistic approach two processes can be identical—that is, have the same Gestalt—"while their parts compared to each other, differ in absolute size," but also, possibly, in nature.[14] In other words, structures are "transposable wholes." For instance, a group of chickens was trained to "to choose, between two equal piles of grain, the one which is signaled by a light gray (G1) and to leave aside the one which is signaled by a medium gray (G2)."[15] It turned out that, when given a new choice, between a G1-colored pile of grain and a lighter-than-G1 pile of grain (G0), these chickens, in the clear majority of cases, did not choose the reflexogenic color (G1), but the lighter-colored grain G0.[16] "The lighter" is a transposable pattern. It can shape an infinity of experiences where the concrete aspects and actual parts of the original experience are absent. Gestalten are thus essentially plastic and are always processes rather than substances.

The concept of Gestalt can be applied to physical, ecological/biological, and cultural systems. Köhler gives many examples in *Die physischen Gestalten in Ruhe und im stationären Zustand*. The charge of static electricity on a spherical conductor distributes itself "in a manner determined according to the geometrical conditions of the conductor,"[17] or its "topography." The addition or subtraction of electricity at a particular point of the system results in a new total distribution. The charge in every point of the system changes, the absolute value of the amount of the charge changes, but the relative distribution of the charges at the separate points retains the same structure. The system changes as a whole. Moreover, this structure, with its characteristic proportions, can be reproduced on different scales and with conductors of a similar topography but of different materials.[18] More broadly, Gestalt theorists stress that every elemental causal relation actually exists in the framework of a certain field, where a certain equilibrium could be struck so that relatively stable laws can come into effect. "The law of falling bodies expresses the constitution of a field of relatively stable forces in the neighborhood of the earth and will remain valid only as long as the cosmological structure on which it is founded endures."[19] What is at stake is a new overall transversal paradigm. The concept of Gestalt was inspired by Goethe's holism and his *Gestaltlehre,* and borrowed, in the first place, from the psychology of perception: It is indeed in perceptive experiences that we most commonly apprehend diffuse forms such as the global physiognomy of person, the overall *look* of a thing, or the unmistakable style of a behavior. But Gestalt theorists *integrate* this concept of form into an objective scientific approach; they want to combine observation, measures, and accounts describing the overall forms that structure phenomena.[20] Gestalt theorists claim that science must overcome an elementarist and atomistic approach that reduces every reality to discrete and localizable entities.

In biology, Gestalt theory developed in opposition to behaviorism. It also, more generally, challenges every reductionist methodology that claims to fully understand organisms, milieus, and behaviors by studying the causal role played by, on the one hand, clear-cut and localizable stimuli, and, on the other hand, specific parts of the body, areas of the brain, and pre-wired reactions.

Whereas we may naïvely believe that we turn our eyes toward a luminous spot in order to see it (for it has attracted our attention),[21] behaviorists reject any form of finality, intentionality, and meaning as purely speculative. Behaviorism replaces the phenomenal light with the "real" light that functions as a cause and *triggers* effects. Light is a *vis a tergo* that needs not be apprehended or interpreted in any manner by the subject to be effective. "If my eyes oscillate in such a way that the luminous spot comes to be reflected in the center of my retina, it is in the antecedent causes or conditions of the movement that one must find the sufficient reason for this adaptation."[22] And this must be a blind process.[23] Light is accordingly defined as nothing but a physical agent that falls onto the retina. Each point of the retina is connected, through pre-established nervous circuits to motor nerves and muscles. The light that falls on the retina triggers ready-to-work reflex arcs, so that the eye moves and the luminous impression is brought over the macula.[24] This "atomistic" approach thus presupposes a whole set of pre-existing and perfectly circumscribable reflex-arcs connecting one specific stimulus with one specific reaction. Likewise, for higher forms of behaviors, Pavlov sought to strictly define "points of innervation" that can be "marked on the map of the brain"[25] and govern simple reactions to clear-cut stimuli.[26]

As Stefan Frisch highlights,[27] a significant body of research in contemporary neurology aligns with this mechanistic and elementarist paradigm. The debate between a holistic and a mechanistic approach remains absolutely topical. Frisch shows that, first and foremost for methodological reasons, neuroscientists seek to localize functions in circumscribed brain areas, a task that imaging techniques support and elucidate. By the same token, researchers endeavor to find evidence for specific lesion-deficit correlations through "controlled experiments in which the investigated phenomenon is isolated under laboratory conditions." Such a paradigm has proved fruitful, without any doubt, but inevitably reaches its limits for being based on the careful exclusion of more plastic phenomena. As Frisch explains, "It is in fact a common advice for lab greenhorns that when conducting (neuro-)psychological experiments: (a) Never to run too many trials of the same condition in one and the same subject; and (b) never to let one and the same subject take part more than once in experiments with very similar manipulations, as effects are soon reduced or even washed out."[28] Fixed localizations are thus defined only on the basis of quite artificial experiments.

The initial motivation of Gestalt theory is precisely to account for aspects of the phenomena that an elementarist approach must artificially discard. Frisch argues that the clinical neuropsychological practice is much more influenced by the Gestalt paradigm than commonly acknowledged in a majority of scientific theorizing works.[29] Neuroimaging actually shows that the same lesion and/or activation of one specific area may lead to different outcomes in different subjects, but also in the same subject in different contexts and at different times. As demonstrated by Kurt Goldstein, one of the founders of Gestalt theory, a specific cerebral lesion results in a general modification of the behavior at all levels and, more particularly, often leads to all sorts of behavioral and even physiological compensatory reorganizations that can be experienced as "normal" by the subject and may be treated as pathological only with a great deal of caution.[30] A function that has been ascribed to a specific area may be taken up by another brain area or, beyond the modular paradigm, by a complex adaptation of the organism as a whole and as a subject. One cannot care for and heal patients, whether human or non-human living beings, without taking these holistic considerations into account. In asserting that sickness represents a qualitative, global alteration,[31] Merleau-Ponty gives the example of Lashley's experiments with rats: "The elementary movements which 'compose' the behavior of a rat do not seem to be compromised after cauterization of the central and frontal cortical regions. But the animal is maladroit; all his movements are slow and rigid."[32] Hence, sickness cannot be properly addressed through an analysis that isolates elementary reflexes/neural localizations/lesions. It requires a comprehensive approach oriented toward the general "style [*allure*]" of the behavior and its new meaning as a whole.

More broadly, Merleau-Ponty argues, the elementarist/mechanist theory cannot account for the fact that, when confronted with inner alterations or modifications in their environment, organisms reorganize their behavior and possibly their morphology globally, but also in an oriented way, to develop a system of gestures that possesses a biological meaning in their *Umwelt*.[33] These living beings adjust themselves in such a way that they re-establish a relationship with their environment that will be meaningful for them. The concept of Gestalt was designed specifically to describe such a dimension of meaning.[34]

What is a *meaningful behavior*? With "an increasing stimulation of the concha of the ear in a cat,"[35] one obtains all sorts of gestures and behaviors, but the constant tendency is that "the excitation is elaborated in such a way that at each notable increase it is translated in the motor apparatuses by new movements and is distributed among them in such a way as to release a gesture endowed with biological meaning;" for instance "the fundamental forms of the movement of walking"[36]. Similarly, hemianopic patients who, through a macular degeneration, have lost half of their visual field may nevertheless hardly notice any change in their visual experience. "The organism has adapted itself to the situation created by the illness by reorganizing the functions of the eye. The eyeballs have oscillated in such a way as to present a part of the retina which is intact to the luminous excitations."[37] The process is even more versatile than that: Actually, new "*pseudofoveas*"—parts of the retina that are not the fovea but *act as* the fovea, note the reference to the virtual, via the Greek ψεύδω, to lie—develop and vanish alternately according to each particular situation.[38] With such phenomena, the behavior's global meaning is key. The actual organs/body parts/brain areas involved can vary indefinitely, but a general form consistently guides these transformations.

Furthermore, it is possible to distinguish between, on the one hand, chaotic, isolated, blind, and repetitive behaviors and, on the other hand, oriented behaviors involved in a process of apprenticeship. Learning behaviors do not proceed through random trials and errors, but rather through meaningful attempts, and a successful gesture becomes fixed as a habit immediately after the first attempt, whereas absurd mechanical gestures become a habit only after a series of repetitions. Also, when animals are "cold, tired, or too excited," they display absurd mistakes and chaotic gestures.[39] By contrast, learning processes occur when a gesture is performed in the framework of a coherent sequence of gestures. A holistic explanation thus becomes necessary to account for actual differences that can be directly observed within phenomena. A learning behavior displays "a definite 'path,' each part of which is determined only by its relation to the direction of the ensemble."[40] For instance, "A rat introduced into a labyrinth will follow the general direction of the initial elements. Everything happens as if the animal adopted a 'hypothesis' which is not explained by the success since it is manifested and persists before the success can confirm it."[41] In general, a huge number of learned behaviors are irreducible to a rigid set of atomic gestures: They can be recognized as a set of variations on a general behavior-pattern—for example, walking, fleeing, moving toward a certain direction, or, as in the case of Köhler's chickens, "choosing the lighter gray".

Gestalten pose a daunting challenge for positivist ontologies. As an embodied but transposable pattern, a Gestalt possesses a certain generality and lies *between* the parts of the actual body or *ahead of* them, as that which consistently orients their actual transformations. The Gestalt is an original form of meaning: It is a wide-ranging and yet consistent theme. It adumbrates itself, in an impressionist manner, through many sequences of gestures/behaviors/transformations. The concept of Gestalt "renders possible the use of a finalistic vocabulary,"[42] but *not* in a metaphysical sense. Indeed, one should not picture the Gestalt as a positive force or a positive spiritual being that would be the *cause* of the observed phenomena or an enigmatic "force of attraction."[43] The Gestalt must be regarded as a descriptive concept necessary to account for observed phenomena.[44] A Gestalt is a general form, immanent in a material system: The organism as a whole—or a set of body parts as a whole—*copes* with a situation and *performs a general function*, such as walking, seeing, fleeing, or finding food. The Gestalt can never be exclusively located in this or that particular part of the body: As in the case of ubiquitous pseudofoveas, for instance, the *Gestalt* could emerge through other parts and configurations. The Gestalt is everywhere, so to say, between the different material and actual body parts, *as a virtual dimension*. When an animal memorizes her way, the

acquired skill does not take the form of "an imprint of the muscular con-
tractions that have actually been produced."[45] Instead, this acquired skill
relates to "a certain 'space traversed'"[46] which is the matrix of an infinity of
steps of another size, differently directed, and potentially performed in new
ways and shortcuts. "The organism (...) does not admit of division in space
and in time. Nerve functioning is not punctually localizable."[47] Likewise,
neurology defines "vertical localizations," namely functions that are connected to a broad variety of brain areas in such a fashion that a lesion on
a particular point compromises a general level of conduct, and not only a
specific function or reflex.[48]

Turning the concept of Gestalt into the keystone of a new non-
positivist ontology, Merleau-Ponty brings Gestalt theory to a whole new
level. He contends that Gestalt theorists were not adequately aware of the
magnitude of the ontological revolution implied by their work, and that
they often return to more traditional positivist ideas, sometimes without
realizing it.[49] But Merleau-Ponty himself struggles with the ontological
status of Gestalt. I will contend that overcoming a positivist definition of
Gestalten entails a radical critique of the rigid human/animal hierarchy.
Instead, Merleau-Ponty sets out to use the Gestaltist approach to establish
his theory of the three orders and realign with a classic hierarchy. Merleau-
Ponty's theory is certainly a quite subtle and sophisticated version of the
latter; *The Structure of Behavior*, however, never finds its balance. A new
conception of the human-animal relations, more aligned with the ontology
of the Gestalt, still lies on the horizon of the theory actually put forward by
Merleau-Ponty, waiting to be picked up and developed.

§22 The physical order, the vital order, the human order: New threshold cosmology and old hierarchy

Gestalt theorists sought to overcome a predominant version of natural
sciences that rendered the world meaningless and created a radical gap
between beings as objects of science and humans as spiritual and cultural
beings. Likewise, in *The Structure of Behavior*, Merleau-Ponty endeavors
"to understand the relations of consciousness and nature."[50] More precisely,
as put by Evan Thompson, Merleau-Ponty wants to use the concept of
Gestalt to "bridge the explanatory gap between consciousness and nature."[51]
Since Gestalten are everywhere, at every level of being, and since physical
phenomena could not be properly understood without being regarded as
inscribed in physical Gestalten, a fundamental ontological continuity must
exist between the physical, vital, and cultural realms.

With the concept of Gestalt, a form of meaning is discovered that can be
defined without referring to the activity of a subject. As a structure embodied in a material system, the Gestalt is an interior visible in an exterior
and can thus unfurl at all levels of being. Merleau-Ponty quotes Khöler:
"Each part of a form (...) 'dynamically knows the others.'"[52] Each part of

the Gestalt "refers" to the others and to a general theme that orients their mutual adjustments so that these parts form a specific whole endowed with its own norms. As a consequence, matter can no longer be regarded as pure blind exteriority. "Nature has its own interiority and thus resembles mind."[53] Symmetrically, human existence, from a holistic perspective, can no longer be separated from the human body and its environment. A human being is not a pure consciousness; she does not even *have* a pure consciousness.[54] Therefore, the reasons consciousness is usually denied to animals apply to human beings as well.[55] Many Gestalten (brain activity, reflexes, body schema, habitual behaviors, cultural institutions) sustain the activities of the human "mind"; the latter, consequently, "is not pure interiority, but resembles life."[56]

In referring to a form of analogy/resemblance between physical, biological, and cultural structures, Merleau-Ponty also casts the notion of Gestalt as a challenge to the dichotomies between things and consciousness, matter and mind, subject and object, for-itself and in-itself.[57] At stake is the possibility of conceiving of a shared thinking life between humans and animals. Beneath a surface of apparently clearly differentiated realms, Gestalten constitute a unique and common ontological root. A fundamental unity and continuity are thus discovered here, but they do not supersede the ontological ubiquity outlined by the notions of resemblance or analogy: How exactly are we to think the connection as well as, still, the difference between inert matter, living beings, and human beings? Using the concept of Gestalt, Merleau-Ponty wants to build a systematic threshold cosmology.

Drawing on Gestalt theory, Merleau-Ponty is able to discard every theory that explains the birth of humanity as a metaphysical event, the descent of a soul into flesh, for instance, or the mysterious adjunction of rationality to animality. Merleau-Ponty highlights that, with Gestalten, a new dimension appears within matter and from matter but is irreducible to the properties of the material elements within which it takes shape. Hence, Merleau-Ponty states that the realm of meaning—Gestalten—is not a new substance, a soul, or a platonic idea. In other words, between the physicochemical level and meaningful Gestalten, no causal relation can be established. As stated by Merleau-Ponty, a behavior—and I will add, more generally, a structure—is not "a mundane event, interpolated between antecedent and subsequent events."[58] Among the infinite number of possible physicochemical configurations, some configurations open up onto a new dimension: A certain number of elements interact in such a way that a whole emerges; for instance, the specific topography of the conductor in the example developed in the beginning of the present chapter. Claiming that a Gestalt has been produced by a certain physicochemical configuration is not sufficient. This amounts to overlooking the originality of what actually appeared: A whole that takes itself up, transcends elementary and actual material components, and even perpetuates itself. This relation to oneself, this circularity, cannot be understood within the elementarist and linear pattern of conceptualization that applies to the physicochemical causal chains. It must be comprehended following the model provided by the perceptual field or intellectual intuition, where a general style or idea precedes and presides over the analysis into parts. In other words, with the Gestalt, the field of meaning emerges. The relation to itself is foreign to the physicochemical level: A leap takes place to a new form of action, a new ontological pattern.

In a sense, this emergence can already be thought of as a form of *autopoiesis*, a term that Varela[59] uses to describe the emergence of life within inert matter. My contention is that it is ontologically illuminating, but also literally relevant, to apply this concept of *autopoiesis* to Merleau-Ponty's conception of the relation between the realm of the physicochemical and the realm of Gestalten. Indeed, a Gestalt has no causal antecedent, neither in matter, nor in any alleged ontological principle outside of matter. The Gestalt must be its own principle, in, so to say, a spontaneous generation. To be sure, Varela speaks of *autopoiesis* when the emerging structure is proactive and uses, or even builds, means to maintain itself and develop its own pattern of norms. But, as Merleau-Ponty emphasizes, a physical system can also, to a certain extent, actively perpetuate itself: "Doubtless certain physical systems modify the very conditions upon which they depend by their internal evolution, as is shown by the polarization of electrodes in the case of electrical current."[60] However, Merleau-Ponty also highlights that physical systems only exert a *minimal* activity, which "always has the effect of reducing a state of tension, of advancing the system toward rest,"[61] thus allowing him to define both a continuity and a difference between physical and vital Gestalten.

In contrast to the minimal activity displayed by physical systems, living structures have the ability to pre-empt external solicitations, explore, and put themselves at risk with a view to bringing still virtual norms and functions to fruition. Consider the risky phases of the development of an embryo, or think of the way cats jump and climb to high places, with varying degrees of success, when they explore a new environment. Living beings are more actively and strongly normative than physical Gestalten. The global orientation that defines them as Gestalten is focused on a virtual milieu, which they create by conjuring meanings that did not exist within physical structures (food, prey, danger, shelter ...).

I have already described two thresholds: To physical systems and to vital Gestalten. Which criteria can Merleau-Ponty follow to assert the existence of a third threshold to specifically human Gestalten?

Merleau-Ponty distinguishes between syncretic, "amovable [*amovible*]," and symbolic forms of behavior. Only the latter are, Merleau-Ponty claims, specifically human. "Syncretic forms" designates rigid reactions to determinate stimuli that take the form of reflexogenic traits (for instance, a specific color or a moving target that unfailingly *triggers* the same specific reactions): They hardly allow for learning processes.[62] A movable forms structure behavior in relation to a general goal whose achievement can take many different guises: For instance, chickens learn to choose the "lighter-colored grains," dogs or rats in mazes look for the "shortest" way, or chimpanzees use or produce whatever may help to gain access to food. Such behaviors are creative and the capacity to adapt them varies in time as well as among individuals. Symbolic behaviors address general goals, functions, and patterns of organization as such, while non-human animals relate to forms that always remain embedded in concrete situations. In other words, humans can apprehend the geometrical structure of a pile of stacked boxes, measure angles and distances, and draw up specifications for the production of tools, whereas chimpanzees find forms as they go along, directly in this or that concrete situation; as a result, their behavior is strongly determined by the particularities of the said situation. By contrast, humans, Merleau-Ponty contends, use existing structures to create new ones in a cumulative process. They vary perspectives and, accordingly, make meaning their main focus.

According to this theory, humans jump to the *systematic representation of the virtual*—for instance, they design abstract mathematical models to represent physical systems—while animals see, think, and build forms much as a tennis player "evaluates" "the direction of the ball, the angle which the trajectory makes with the court, the rotation with which the ball can be animated, the position of the adversaries and the dimensions of the court."[63] In Köhler's experiments in particular, to which Merleau-Ponty extensively refers, a chimpanzee who "had learned to manipulate cases in preceding experiments, does not use one which is offered it as long as another monkey is sitting on it."[64] Likewise, in many cases, apes who have already used rods to reach a goal that was not directly accessible do not recognize a potential instrument in a dried bush from which the branches can easily be cut, or even a rake that is not presented to them with its handle within reach of their hand.[65] They also build provisional and unsteady piles of boxes to reach food and use agile body moves to compensate for the unbalanced construction. Merleau-Ponty concludes that "the chimpanzee does not succeed in developing, in an indifferent space and time, a mode of behavior regulated by the objective properties of the instrument" and that "the field of animal activity is not made up of physico-geometric relations, as our world is."[66] In this approach, the human virtuality and the animal virtuality become incommensurable.

Merleau-Ponty here seems to fall under the spell of the positivist belief that "Symbolic forms" are *absolutely* labile. He thus refers to Hegel and claims that "a consciousness is a hole [*un trou*] in being,"[67] while "here [with animals] we have nothing yet but a hollow [*un creux*]."[68] Humans allegedly negate the concrete matter of actual situations, words, and tools to reach out toward abstract meaning and new possibilities. This radical negation would enable humans to aim at true essences, which are situated *beyond* our limited, contingent world and remain the same through their embodiment in different individuals. Correlatively, humans are skillful at considering, beyond the realm of the actual, the innumerable functions they could give to each present concrete being. In other words, humans turn the world into a spectacle. This claim is the ABC's of classical idealist and anthropocentric theories and certainly disappoints the reader who is expecting a true revolution to result from the concept of Gestalt. But let us be patient: Merleau-Ponty's text is full of trapdoors.

A radically new relationship to meaning emerges, Merleau-Ponty contends, with what thus deserves to be called "the human realm." If meaning is what matters, we must avoid simply focusing on the fact that a human is made of flesh, body tissues, bones, a heart, a brain, and so on—which, in a sense, makes her an animal. Humans are also, and, in a Gestaltist approach, more fundamentally, distinguished by a specific way of structuring behaviors and the world in such a manner that, perhaps, they are animals merely homonymously. A piece of matter, a gesture, or a behavior possesses modes of being and different meanings depending on the whole to which they belong. The heart pulled out of the body no longer functions as an organ; in other words, as Aristotle points out, it is only a "heart" homonymously.[69] The heart of a frog and the "broken heart" that, in Rimbaud's poem, "drools at the stern"[70] also have different meanings and, therefore, *are different beings*, because they too are incorporated into different worlds, behaviors, and intentionalities. Perceptions, sexual behaviors, eating behaviors, moves, and diseases acquire a whole new meaning and are simply inserted into a wholly human meaning.[71] El Greco's astigmatism thus becomes an artistic skill, transcending any biological account.[72]

To this relatively sophisticated threshold cosmology, Merleau-Ponty attaches a traditional hierarchical perspective. He regularly introduces value judgements into his description of the three orders, often without justification. It was indeed not necessary to write that "the chimpanzee *does not succeed in* [*ne réussit pas à*] developing, in an indifferent space and time, a mode of behavior regulated by the objective properties of the instrument."[73] Phrases of this sort function consequently as a symptom of ideological biases, rather than as a philosophical stance. One argument is yet put forward by Merleau-Ponty. *The Structure of Behavior* presents emergence in terms of *integration* and, from there, predictably, the human affection for system thinking becomes an undeniable advantage that ensures us a place at the center of the cosmos. A new emergent Gestalt places all the elements of the matter from which it emerged at the service of an original goal. A living being uses and interprets every possible piece of matter and physical system in connection with its vital norms. Humans, allegedly, interpret every vital process in the framework of a systematized relation to meaning as such. The degree of integration of the physicochemical random elements increases from physical to vital and from vital to human Gestalten. Correlatively, "matter, life and mind constitute a hierarchy in which individuality is progressively achieved."[74] Further, "The relation of each order to the higher order is that of the partial to the total."[75] Living Gestalten continuously and proactively gather and integrate new elements into their unity. Human beings do not simply ascribe new meaning to living processes. A greater integration is achieved by the mind and its ability to create universal ideas, thematized concepts, speculative systems, and projects with an eternity-term perspective.[76] Merleau-Ponty concedes that "integration is never absolute and it always *fails*—at a *higher* level in the writer, at a *lower* level in the aphasic,"[77] but this phrasing nevertheless implies that the highest integration possible is the absolutely valuable goal. Here Merleau-Ponty tips the scales in favor of the finality and the unity of Gestalten and departs from that which defined the fruitfulness of this concept, namely the compound of an elusive integration with versatile concrete configurations.

Merleau-Ponty strangely seeks to combine a theory of the metamorphic Gestalt with an old, ossified hierarchy: Three orders, not one more, not one less, in which individuals are stuck ("man can never be an animal"[78]). This theory cannot hold. The fragile equilibrium gives way precisely when Merleau-Ponty takes the three orders too seriously and reintroduces rigid boundaries. He then concludes without further ado that animals *"cannot"* have access to symbolic forms[79] and that our world *is made up of* physico-geometric relations.[80] In these passages, Merleau-Ponty seems to be reporting facts, when, in reality, his analysis becomes lazy and echoes tired clichés. Precisely because he cannot have it both ways, Merleau-Ponty oscillates. He thus, for instance, adds a footnote to specify that "it must be added that our world is not constantly made up of them [physico-geometric relations]."[81]

However, the last part of *the Structure of Behavior* tips the scales in favor of a focus on the fluidity of concrete Gestalten and what he will later call "a sardonic form of humanism [*humanisme narquois*]."[82] If realism is indeed to be overcome, Gestalten and organisms in particular, as well as every alleged "order" must be regarded as phantoms.

PART II. THE HAUNTOLOGY OF ANIMALS

§23 "Is there not a truth of naturalism?": Humans overtaken by the meaning emerging in living beings

Is there not a truth of naturalism? This surprising question is the title of the last part of Merleau-Ponty's *Structure of Behavior*; it introduces a significant shift in a theory that so far has oscillated between two incompatible versions. Although Merleau-Ponty does not yet abandon his model of the three orders, he clearly opts for a definition of "structure" that makes a rigid categorial and hierarchical pattern absolutely inapplicable and inaccurate.

The fourth part of the book, "The Relations of the Soul and the Body and the Problem of Perceptual Consciousness," takes the form of a tricky plot, perhaps in a manner of staging and channeling what I have identified as an authentic oscillation operating throughout the book. First, through his recurring reference to Hegel,[83] and the claim that the Gestalt is a form, a meaning, or even an object of knowledge,[84] Merleau-Ponty somewhat over-plays an idealist version of his theory. Eventually, however, he provocatively brings naturalism and its fragmented and blind processes back into play. It is therefore especially dangerous to hasten to a conclusion while reading this book.

What is the putative link between Merleau-Ponty's concept of Gestalt and idealism? The notion of Gestalt reintroduces the reference to mean-ing as decisive in physics and biology, and because this notion of meaning derives from the field of conscious thought, understanding, interpretation, and knowledge, it becomes tempting to contend that, through Gestalten, nature is subordinated to mind. An idealistic interpretation would thus maintain that Gestalten do not first exist in physical or living bodies *before* being, possibly, discovered by a mind: They are intrinsically *for* a knowing mind and the realm of nature is ontologically appendant to the realm of (human) knowledge. Naturalism tried to expunge every subjective perspec-tive and idealist metaphysical claim from science; but, as Merleau-Ponty highlights, since meaning is an integral part of physical and biological processes, it follows that "the universe of naturalism has not been able to become self-enclosed [*n'a pu se refermer sur lui-même*]."[85] Perception cannot be reduced to the result of mechanical processes in nature, for these pro-cesses can never be defined or analyzed, or simply observed, or given in any way, without their intrinsic reference to meaning. As Merleau-Ponty asserts, "It is not the real world which constitutes the perceived world."[86] The three orders may consequently be ultimately rooted in the human order, where meaning "properly speaking" resides.[87]

On several occasions, in *The Structure of Behavior*, Merleau-Ponty roughly conflates meaning and human conscious knowledge. An organism "is a whole which is significant for a consciousness which knows it, not a thing which rests in-itself."[88] "In the final analysis form cannot be defined in terms of reality but in terms of knowledge, not as a thing of the phys-ical world but as a *perceived* whole (...) A unity of this type can be found only in an object of *knowledge*."[89] The oscillation between perception and knowledge (Gestalten are "object of perception"[90]/"objects of knowledge") already indicates the fragility of this idealist take, but it is indeed tempting to give "meaning" a positive reality and to assign responsibility for its per-fect understanding to an actual rational subject. This idealist perspective of the concept of *Gestalt* matches Merleau-Ponty's claim that knowledge "properly so called" and consciousness "as a hole in the world"[91] belong exclusively to human beings. Correlatively, indeed, *Gestalt*, here under-stood as "an object of knowledge"—in other words as a *de jure* clear and circumscribed meaning—can also be defined as a "function" that has "a positive and proper reality."[92] This statement aligns with the theory of the three orders: Positive real orders correspond to positively defined functions. The function is indeed more important than its material support, as evidenced by the fact that it can even create its own body organs and improvise using whatever means at hand, as in the case of hemianopsia that I have discussed above.[93] According to this view, behaviors are governed by an intentionality towards the fixed and perfectly knowable essence of "to see," "to walk," "to flee," "to eat," and "to imitate," for instance.

But Merleau-Ponty was plotting a surprise reversal. In the last chapter of *The Structure of Behavior*, he eventually attributes the idealist claims described above to a "philosophy in the critical tradition"[94] that deeply differs from the conception of the transcendental attitude that he wants to endorse. In fact, we cannot but fail to expunge matter, blindness, and chance from the embodied meaning of Gestalten, just as naturalism never succeeded in becoming self-enclosed.[95] As I will explore, with these concerns, Merleau-Ponty's philosophy strongly echoes crucial Darwinian themes. Structures/Gestalten also evolve randomly. The concepts of design, intentionality, and essence fail to account for the always particular, concrete, and unpredictable avatars of functions, such as "to see" or "to walk" in particular individuals.

The originality of Merleau-Ponty's approach to Gestalt lies in its emphasis on the materiality of structures. "What is profound in the notion of 'Gestalt' (...) is not the idea of signification but that of structure, the joining of an idea and an existence which are indiscernible, the contingent arrangement by which materials begin to have meaning in front of us [les matériaux se mettent devant nous à avoir du sens], intelligibility in the nascent state."[96] The function is not reduced to matter, but it does not exist—as a positive idea would—outside of a concrete material structure. Indeed, the function does not subsist anywhere—in a Platonic "sky of ideas," in the realm of eternal essences, in a divine mind, in the human symbolic order, or in universal laws of nature. In effect, the overcoming of realism "in general"[97] that Merleau-Ponty calls for cannot be properly achieved if realism tips over into its doppelganger: Even when idealism does not take the form of a realism of ideas, it still postulates positive, *de jure* clear ideas as an absolute principle of everything. Merleau-Ponty's criticism of realism cannot be interpreted as a plea for a subject-centered idealism, or more broadly for any ontology with one center only. By using the pronominal form ("se mettent à avoir du sens"), Merleau-Ponty suggests, as he often does in *Phenomenology of Perception*, that the emergence of meaning in Gestalten is not initiated by the human observer and thinker, or even the transcendental Ego. Instead, this emergence of meaning takes place "in front of" this human thinker, in the sensible world.[98] I never jump to the definitive concept of a function; I perceive and discover its concrete evolutions. Merleau-Ponty emphasizes the plasticity of functions themselves and the transformation that they undergo *as functions, as meaning*, through their different concrete self-productions. "Our analyses have indeed led us to the ideality of the body, but it was a question of an idea which proffers itself and even constitutes itself in the randomness of existence [*le hasard de l'existence*]."[99]

In the case of hemianopsia, for instance, one's vision is qualitatively different from that of healthy subjects.[100] Likewise, a general impairment of the "symbolic function" will take slightly varied forms if the lesions in the patient's brain affect the "zone of language" or the "zone of perception."[101] Domestication provides innumerable examples of surprising avatars of functions. In many domesticated species, breeding can happen in any season and several times a year; lactation continues after the calf is weaned. Generation after generation the forces of domestication have selected a general ethos that is much less oriented toward acute observation, exploration, and alertness than in wild species. Then optical sensitivity changes.[102] Vision, reproduction, and lactation take on new connotations and are achieved through qualitatively new processes. The "truth of naturalism" regarding domestication could be phrased as follows: What is at stake is not the substitution of a human goal or a human design for an original "wild" or "natural" function. To be sure, domestication does imply intentional artificial selection, but it also, and more originally, took the form of a co-evolution. Early phases of cooperation—especially with regard to hunting strategies—between wolf-packs and nomadic hunter-gatherer communities[103] likewise permitted the development of humans. Even striking similarities between behavioral and social patterns in wolf packs and human societies have been highlighted.[104] These early communities did not yet have a *domus* in which pet dogs could be *domesticated*, cooped up, and systematically selected.[105] Dogs were not manufactured. Barking already exists in wild dogs or wolves, but the repertoire, the spectrum of tone, pitch, or frequency variations as well as the array of possible barking contexts are significantly narrower than in domesticated dogs.[106] Meticulous studies describe the extensive variety of barking sounds, their subtle inflexions, and the emotional meanings that human beings spontaneously recognize as expressed by them.[107] Feddersen-Petersen speculates that such an expansion of barking in domestic dogs may correlate to the companionship with and dependency on humans who happen to be predominantly verbalizing beings. Further, Molnár contends that barking has evolved into a refined language consisting in recognizable and stable patterns, and especially designed for dog-human communication. "To bark" certainly does not mean *exactly* the same for a wolf and for a dog. But humans have not taught dogs to bark. Despite our relatively good capacity to intuitively identify the contexts in which barks were made,[108] the meaning of various barking sounds (in human-dog communication as well as in the framework of still relevant intraspecific communication) and their exact connection with barks in the wild constitute a recent and scarcely explored field of investigation. As Merleau-Ponty puts it: Some nascent meaning is unfolding before our eyes. A function makes its way through evolving bodies, neotenic morphologies, new relational patterns, taking on new shapes that human breeders did not necessarily aimed at.

A function is a vague, ambiguous theme or requirement. Uexküll thus playfully imagined the surprise of the triton, a carnivore, once a graft of tadpole tissue that was implanted in its mouth developed into cartilage-like gums. The functional prompt was only relatively specific. Put into words, this prompt would be something like: "Build oral weaponry."[109] The predictability of Gestalten is essentially limited, which results from their transposability and the discrepancy between a general function and particular contingent achievements. If the function "to see" can incarnate itself in this healthy eye or in that damaged eye, its identity is essentially unhinged, thus enabling many other contingent—unnecessary and unpredictable—achievements. We have already studied the key role played by ambiguity at the behavioral level in animal life;[110] but, with the concept of Gestalt, we can take one step further: The introduction of this concept in biology turns out to be fundamentally concomitant with the radical assertion of ambiguity and interpretation in *every* life process, even at the morphological and metabolic level.

What is more, Merleau-Ponty places special emphasis on the manifoldness of Gestalten. Fragmentation is not only always possible, but also "imminent" and, I would add, concurrent with the unification process.[111] Different organs, body parts, cells, and sensory fields can contribute right now to my writing of some philosophical ideas. However, the integration of body parts in the same functional unit "does not annul their specificity": "Their imminence" is "[attested to] by the disintegration in case of partial lesion."[112] I would like to replace the concept of "disintegration" that ossifies and values one Gestalt exclusively with the notion of manifoldness or ubiquity. *While* I am writing these lines, within me, billions of cells, skin microbiota, hunger, fatigue, breathing rhythms, peripheral visual or auditory fields, budding mutations, blood clots or tumors, *simultaneously* buttress my project and live their own lives, opening, more or less silently, new developmental paths.

The question "is there not a truth of naturalism?" is not rhetorical. Although Merleau-Ponty maintains, in the last part of *The Structure of Behavior*, the concept of disintegration and its correlates, he also irremediably undermines *any* idealist ossification of Gestalten. Moreover, *a certain truth* of naturalism does not amount to *the* truth of naturalism. A materialist ossification—in other words, a reductionist form of naturalism that claims to account for Gestalten by referring only to mechanical blind processes—is equally ruled out. A certain order takes shape, some autopoietic global processes are at work: What are they if they *are* not?

§24 Phantom limbs, phantom bodies, and sardonic humanism

In order to understand Gestalten properly, and, therefore, to understand living bodies more particularly, it is necessary to draw on the ontology of the imaginareal. This condition can be demonstrated along three lines of analysis: (a) The necessity of thinking the "normal" body through the phenomenon of the phantom limb, rather than the other way around, (b) the shift towards a "sardonic humanism,"[113] and (c) the kinship between Merleau-Ponty's theory of life and the imaginary of Darwin such as Gillian Beer describes.

Phantom limbs—phantom bodies

If a Gestalt cannot be reduced to its material parts and, nevertheless, is not a positive being subsisting independently of them, it must be an open-ended project that unhinges these material parts and prevents them from coinciding with their own nature. The Gestalt is a drive that animates an open metamorphic process. It draws its invisible guideline "on the spot," depending on the constraints and contingent situations encountered; consequently, it unfolds in a recognizable but never fully predictable way. Hence, it is best defined as a form of haunting.

What does the concept of "phantom" designate? Asserting that phantoms are fictitious or illusory entities prevents us from delving into a proper phenomenological description of the experiences that are designated by this word. Also, insofar as phantoms subvert positive dualities (to be or not to be, alive or dead), they are ontologically critical entities: They can be simply discarded as divagations of our imagination or provide new paradigms beyond positivism.[114] The concept of phantom designates an entity that is given through perception: An apparition. Thus, a phantom limb is not simply imagined or fantasized: The subject actually experiences its enduring presence. And yet, this experience is also deeply tantalizing and troubling. The phantom limb does not behave like an actual limb. It does not possess the same apparent well-circumscribed morphology as an actual limb. It appears and vanishes, shrinks or expands, and passes through walls. Similarly, the dagger in *Macbeth* can be seen but not grasped.[115] As a consequence, the phantom never crystallizes in the form of a substantial real object in the world, but neither can it be easily consigned to the level of pure whimsical fantasies. It sticks to a series of actual not-perfectly-consistent perceptual manifestations (*phenomena*[116]) and must take the form of an uncertain, ungraspable entity that hovers above the flux of such momentary concrete phenomena. A form is looming, but its manifestations are not so consistent that we could simply overlook the sensible flow of its contingent appearances and directly jump to a solid reality that shows through them. The phantom never becomes an integral part of the world of real bodies. It remains superimposed on its surface, or as a chasm and a source of phase-shifting opened in its very flesh. The phantom hovers, appears and vanishes here and there, making everything uncanny. Phantoms, "revenants," are therefore essentially potentially ever-recurring. They have effective power, raise questions, open a future, and are insepara- ble from a dimension of concern and emotional attachment, which confers an extra-value on this concept of phantom in the framework of an investi- gation about living beings.

In *Phenomenology of Perception*, Merleau-Ponty surprisingly makes ref- erence to the phenomenon of phantom limbs in order to understand and define the real body. The argument goes as follows: This phantom limb phenomenon can be accounted for through neither a purely psychological explanation, nor a purely physiological one. Indeed, on the one hand, the phantom limb manifests the history of the patient: It often stays in the posi- tion occupied by the real arm when it was injured. The pain resurfaces when the patient confronts circumstances that remind him of the moment and the situation of the accident. Also, the form of the limb fluctuates according to the subject's feelings and particularly the more or less successful grieving process he is going through. On the other hand, the phantom limb is not a purely psychological phenomenon either: When the sensory conductors that run from the stump to the brain are severed, the phantom disappears.[117] The phantom limb is not a figment of the patient's imagination. It is the patient's body that grows the phantom limb. Merleau-Ponty thus borrows the con- cept of organic repression from Schilder and Menninger-Lerchenthal to describe this bizarre phenomenon. Repression is an intentional, although unconscious, process that both denies and grants an overwhelming, obses- sive presence to the repressed event. Instead of referring to speculative psychical entities (the unconscious, the id) or mechanistic, but no less spec- ulative, causal explanations, Merleau-Ponty seeks to do justice to what can be observed: The living body is able to play out a discrepancy between the virtual and the actual, the past and the present, the absent and yet mourned limb. In this way, the body can, on its own terms, express and address the paradox of absence as such. Merleau-Ponty does not tie this ability to a specifically human configuration. In his view, the phenomenon of the phan- tom limb demonstrates that, under the "actual body," a "habitual body" is at work that "acts as a guarantee for the actual body [*se porte garant pour le corps actuel*]."[118] The living body can achieve the essentially twofold repression, for it does not consist only, or primarily, in a set of objective and juxtaposed actual parts. The body is also a habitual body that exists as a whole. It implements and coordinates a complex set of consistent behavio- ral, postural, and morphological patterns that possess a certain plasticity and can adjust to new situations.[119] The habitual body is transtemporal; it addresses virtual functions, organs, and situations. Thus, in the experience of the phantom limb, the patient tries to walk and to lean on her phan- tom leg, or tries to grasp objects with her phantom hand: The habitual body "vouches" for the actual body, stands ahead of it, rushes toward the things-to-be-grasped. If an actual hand is missing, a gap can then be experienced. In other words, the phantom limb—precisely because it unveils the habitual body in an unusually conspicuous way—becomes the model according to which the normal body is best described and considered. The body is at least twofold and emerges through the interaction between actual parts and a non-localizable, general, and transposable habitual body.

Merleau-Ponty does not explicitly refer to Gestalten in this specific passage, however, the generality and transposability of the habitual body, its holistic nature, and its inseparability from the matter of the actual body make it a Gestalt, without possible confusion.

Even more remarkably, the paradigm of the phantom limb applies to animal bodies as well. In the reorganization of the eye, in hemianopsia, a phantom eye haunts the damaged parts of the actual eye. The same phantom-like structure can be found in the numerous cases of functional plasticity in animal organisms. Injured animal bodies can reorganize themselves in a way that should not be regarded as pathological but rather as a wholly new manner of achieving fundamental functions.[120] A similar plasticity can be observed via the study of animal body schemas: It has thus been demonstrated that the body schema of macaque monkeys who become accustomed to using a tool expands and incorporates this new supplement.[121] Even monkeys who were trained to "use tools under visual feedback provided through video-captured images projected on a monitor" could integrate this extension into their body schema and coordinate the unusual visual information with kinesthetic information.[122] For Köhler's chickens, "choosing the lighter" is, likewise, a phantom, and I have already indicated that, when confronted with an ambiguous situation, bees, for instance, display hesitative behaviors that are linked to the emergence of a discrepancy—and, I contend, a phantom. There are good reasons *in animals themselves* why *Moby-Dick* can be simultaneously a ghost story *and* an encyclopedic enterprise that endeavors to do justice to the protean and tantalizing *reality* of actual whales: Animals are *literally prodigious*.

As a matter of fact, Merleau-Ponty regularly uses terms related to the virtual to define Gestalten in non-human bodies and behaviors. From Buytendijk, he borrows the concept of "kinetic melody."[123] More generally, Merleau-Ponty, Koffka, and Buytendijk, among others, often turn to the metaphor of the melody—namely of an entity made up of a virtual theme and actual contingent variations—to describe Gestalten.[124]

"What regulates our motor reactions in a decisive manner is this general factor which is not necessarily tied to *any of the materials of behavior*. '... [L]ike man, animals know how to move themselves in an area of space which is not given to them in perception and without possessing signs which indicate the 'way'." Thus, animals and men react to space in an adapted manner *even in the absence* of adequate *actual or recent* stimuli.' (...) When the animal moves itself in this space to which it is adapted, a melody of spatial characteristics is unfolded in a continuous manner and is played in the different sensory domains."[125]

Likewise, Merleau-Ponty rejects the description of the nerve substance as "a container in which the instruments of such and such reactions were deposited," claiming instead that it is "the theater where a qualitatively variable process unfolds."[126] Above all, Merleau-Ponty regularly insists on the idea that every Gestalt is a creative process.[127] Insofar as a discrepancy between a general theme and contingent variations emerges, predetermining a limited set of possible actual variations in which the theme may be played becomes impossible. Structures, overall, are principles of "improvisation."[128] In the *Nature* lectures, eventually, the reference to a phantom-animal occurs once: "The directing principle is neither before nor behind; it's a phantom [fantôme], it is the axolotl, all the organs of which would be the trace: it's the hollowed-out design of a certain style of action, which would be that of maturation; the arising of a need would be there before that which will fill it. It is not a positive being, but an interrogative being which defines life."[129]

Of course, we do not know much—at least directly—about the way various non-human animals subjectively experience these phantoms, or about the disorientation, the puzzlement, and the emotions that may accompany the misadventures of Gestalten in non-human living bodies.[130] Moreover, the phenomenon of the phantom limb may very well be greatly increased in human organisms, who tend to institute and overplay an official distance between, on the one hand, a re-presentative and objectifying form of consciousness ("I" as a thinking mind) and, on the other hand, the feeling of the postures, states, and activities of one's own body (my body as a vaguely familiar background for all my thoughts). Hence, the human body becomes *something* that can operate silently, symbolically, far from "me."[131] This "repressed" body, as Merleau-Ponty calls it—officially distinct from me, but actually, secretly, the pre-personal existence from which "I" stem—can thus take responsibility for aspects of my existence that I try to shirk, for instance, the amputation and the mourning work. Correlatively, the description in terms of phantoms is laden with a rich human imagery and cannot be separated from Homer, Shakespeare, Dickens, and so many others. However, what I find remarkable in Merleau-Ponty's analysis is that the phantom limb is first and foremost ascribed to the gap between the actual body and the habitual body. Because the difference between the actual body and the habitual body does not fundamentally stem from the hypertrophy of the theoretical sphere in some human beings, but was already observed in the descriptions of vital Gestalten in *The Structure of Behavior*, it would be faulty to reduce the phenomenon of phantoms either to its literary variations or to the ubiquity of Gestalten in living beings: Both echo and inform each other.[132]

Sardonic humanism

Humanity, as a general *Gestalt*, is also a phantom. Merleau-Ponty emphasizes this idea on a number of occasions, which undermines the claim, advocated in *The Structure of Behavior*, that only human beings have access to
virtuality, truth, and freedom.[133] "Man is hidden, well hidden, and this time
we must make no mistake about it: this does not mean that he is there beneath
a mask, ready to appear. (...) The situation is far more serious: there are
no faces underneath the masks, historical man has never been human, and
yet no man is alone."[134] In his *Causeries*, Merleau-Ponty quotes Claudel who
compared "our Mother, the holy Catholic Church" to an elephant. He highlights that animals are consistently referred to as emblems of the human
and the superhuman, and calls for "a sardonic form of humanism and a
particular kind of humor" that "lay beyond Descartes and Malebranche's
reach."[135] It is not that the concept of humanity is absolutely meaningless,
but its meaning is uncertain, under construction, and never localizable in
this or that person or particular achievement.

The term "Human" is freighted with centuries of dreams, tentative concepts, ideologies, imagery, disappointments, crises, hopes, renaissances,
ideals, and practices. The human figure should not be reduced to nothing
or even to a pure chaos, but it certainly requires to be taken with a significant touch of humor. Unlike the radical annihilation of the human world
that characterizes, according to Margot Norris in *Beasts of the Modern
Imagination*, the works of Nietzsche, Kafka, Ernst, and Lawrence, Merleau-
Ponty's sardonic humanism does not intend to *"strip"* human existence from
every meaning;[136] nor does it ultimately identify humans with mere brutal
and irrational forces.[137] The very idea that animality is essentially beastly
and feral is in fact more anthropocentric than it first may seem. A poetic
relation to the virtual, meaning, and ideas arises in human and non-human
animals, but this relation is *in both cases* phantom-like. In other words, it
goes together with a wobbly structure and entails endless misadventures,
some stupendous, others risible.

The concept "human" naturally tends to puff up, for it is based on
auto-deleting structures. *The Structure of Behavior* provides striking illus-
trations of this phenomenon: Merleau-Ponty goes too far when he claims
that our world is made up of [*fait de*] physico-geometric relations[138] or that
"a consciousness is a hole [*un trou*] in being."[139] With a bit of disappoint-
ment, we have to observe that the human body's perceptual capacities do
not allow us to scan our environment systematically in order to foreground
its geometrical structure. Some robots and prosthetic devices can do so, but
they reveal at least as much what certain humans would like our world to
become, as what our world actually "is." "Made up of [*fait de*]" is, in any
event, too strong a characterization, and the remark appended in a footnote ("it must be added that our world is not constantly made up of [such
physico-geometric relations]"[140]) is an understatement. If I come to believe

that the world, which *may* indeed be geometrically schematized, is *made up of* geometrical relations, I negate the idealization process that is at work in geometry and is also inseparable from obscure and debatable inclinations, emotions, and ambitions. Admittedly, this idealization process tends to spontaneously recede to the background, since geometrical drawings, words, and formulas lend their solidity to concepts while becoming invisible mediations. But, when we jump to the geometric idea and overlook its material conditions of emergence, we do nothing but fantasize ourselves as pure minds.

Merleau-Ponty consistently mocks the arrogant "pensée de survol [high-altitude thought]"[141] that reduces the world to a transparent object and *its own* domain. And, in *Phenomenology of Perception*, he clearly contradicts the claim put forward in *The Structure of Behavior*: "Thus I am not, to recall Hegel's phrase, a "hole in being," but rather a hollow, or a fold that was made and that can be unmade."[142]

The paradigm of the hollow relates to a form that arises *within* a material body, *from* a process that takes place *in* this body, for instance, *from* the "invagination"[143] of an organic tissue. Matter, in living beings, has a certain way of folding and structuring itself that creates a new space and a new framework where original functions can develop. Human animals are no exceptions. They *always* apprehend meaning in a concrete context that binds their interpretation—as the Milgram experiment, for instance, shows: Ideas coming from a stern man in a gray laboratory coat carry more authority than those of anyone else.[144] Our most ambitious and "spiritual" ideas bear the seal of our emotions, our background, and concrete contexts, including the morphology of words and layers of significations that sedimented in them. Hence, we never break from the performative thought of the tennis player and of the chimpanzees who build provisional and unsteady piles of boxes, restore their balance acrobatically, and invent new solutions along the way. Symbolic thought is illusory and dangerously authoritarian when it ceases to regard its material basis as akin to—although relatively more rigid than—such unsteady piles of boxes. Chimpanzees, tennis players, and theorists can thus come together for the greater good. Structures are not rooted in an exclusive human symbolic realm, but in animal flexible behaviors. Eventually, Gestalten do not have their root in the human mastery of the symbolic realm, as claimed by Masuda, but in amovable living structures.

§25 The imaginary of Darwin: The introduction of phantoms in life science

The suggestions that living beings are phantoms makes Merleau-Ponty the heir of Darwin—more precisely of what Gillian Beer calls "Darwin's plots"[145]—to a much greater extent than he himself acknowledged or believed.

In fact, the figure of phantom animals has sustained life sciences since Darwin.[146] Contemporary science, art, and literature found a fiery heuristic power in the official marriage of modern science and the imaginary of metamorphoses, such as played out by Darwin.

As Charles Lyell bluntly phrases it in a letter to J.D. Hooker: "I fear much that if Darwin argues that species are phantoms, he will also have to admit that single centres of dispersion are phantoms also."[147] To this remark, Darwin replied that he "could not conceive why his theory would invalidate specific centres;"[148] but his approach certainly allowed for such an interpretation. Darwin's theory indeed introduces phylogeny in ontogeny and therefore makes individuals inseparable from a past history of innumerable and yet largely unknown divergent modifications.[149] By the same token, living beings carry with them the open future of an infinity of unpredictable transformations.[150] Darwin's theory is thus famously based on the dismissal of any form of essentialism. Darwin was remarkably talented at conjuring up spectacular metamorphoses and fast-motion sequences, such as the transformation of a swim-bladder into an air-breathing lung,[151] of a nerve into the perfect structure of the eagle eye,[152] or, mostly famous, of a bear into a whale.[153] It is consequently not surprising that Darwin's *Origin* caused such a commotion in his contemporaries. It proposes both exciting and deeply frightening ideas, and suggests even more startling possibilities, while recasting and reviving familiar narrative tropes:[154] My ancestry is not what I believed, past history haunts the present and goes as far as haunting my body through vestigial features. I am not whom I believed myself to be. And I may suddenly go through bizarre and distressing metamorphoses and be confronted with all sorts of ambivalent animals, including previously harmless, now threatening, companion animals, like Alice confronts with a giant puppy in Wonderland.[155]

More than 160 years after Darwin's *Origin*, animals are more phantom-like than ever: Developmental systems theory, epigenetics, and "postgenomics" have rejected the one-to-one correspondence between genes and phenotypic features. They challenge the idea of a perfectly determinate "program" engraved in genes and on the basis of which a "large enough computer" could compute the organism.[156] The role of *pseudo*genes for instance—yet another phantom-like ψεῦδος—has proven crucial. An abundant stock of genetic resources flourishes in the background of expressed genes. These pseudogenes are not available to the transcription process. They are not subject to natural selection and undergo wild mutations, but they may resurface in the genome after centuries of latency. What is more, particular experiences undergone by individuals can deeply modify the way information coded in one's DNA and in the DNA of one's offspring will be expressed. The genome is far from univocal or purely actual. It does not constitute a sufficient determining cause. As such, may it still be regarded as a set of well-circumscribed, inert tools that are selected through a blind mechanism? This assumption is challenged by epigenetics and developmental systems theory. Animals actively shape their environment on the basis of a set of global functional patterns, those that Merleau-Ponty precisely defines as Gestalten. A recent study demonstrates that the offspring of mice who were trained to fear the particular smell of a chemical called acetophenone not only produce more smell receptors that can detect acetophenone, but are also more nervous, overall, after having been exposed to acetophenone.[157] While the parents were trained to associate acetophenone with a foot shock, the offspring overreact to loud noises when they follow the exposition to acetophenone: The inherited pattern should not be described as a label placed genetically on the process of synthetizing a specific smell receptor but can rather be phrased in the following way: "When you smell this, be prepared for—broadly—bad things to happen."[158] Moreover, if the environment makes this or that morphology more relevant—adaptatively speaking—than another, it should not be forgotten that such a relevance is not a positive and purely blind datum. "Adaptation" will vary according to the individual's and a group's modifications of their environment: They can create *niches* where the factors of selection are original. Once again, the ideal of essences, program, and circumscribable species or individuals fades away.

Let us return to Darwin's imaginary. Why can it be claimed that, with Darwin, the archaic imaginary of metamorphoses is incorporated into a controversial but undoubtedly foundational scientific text? Admittedly, to be precise, evolution is not metamorphosis. Whereas, in Ovid's poem, transformations are rapid and random, everything changes and nothing dies, Darwin's evolution, for its part, is a slow and gradual process. Through this evolutionary process, to be sure, everything may eventually become anything, but evolution as Darwin defines it actually gives birth to a *de jure* traceable genealogy of species: The stately tree of evolution. Darwin carefully draws a distinction between observable, superficial analogies and confirmed homologies. He seeks to move from metaphors to "plain signification."[159] Still, Darwin also consistently stresses the irreducibly conjectural dimension of his theory. The first person and its assumptions, speculations, and wild guesses play a conspicuous role in the first edition of *The Origin*. Darwin gives pure contingency a key role. The free association George Eliot makes between music, *Tales of Arabian Nights,* and Darwininian evolution thus proves meaningful.[160] New species emerge through an erratic process. Darwin does not define one unequivocal causal law; rather, he intertwines several principles. Natural selection is the result of the interplay between a mad hyperproductivity of differences and a drastic selection, two processes that are absolutely not coordinated by natural harmony. Sexual selection introduces a new logic that takes esthetics and a form of subjective appreciation into account. Furthermore, the concept of struggle for life enters into tension with the consistent emphasis Darwin puts on interdependence and cooperation.[161] Darwin thus provided us with "conflicting narratives:"[162] Necessity vs chance, progress vs degradation and loss, joyful admiration for nature vs our horror when we consider profusion, struggle, destruction, death, and extinction in evolution. Further, Darwin is torn between, on the one hand, an "unresolved trouble,"[163] a tendency to pity individuals, to irresistibly join their perspective, to fear for them, and, on the other hand, the denial of their significance, the disregard of their personal agency or even their suffering on the "true" scale of evolution.[164]

Crucially, this imaginary dimension of Darwinism is intimately entwined with his will to account for life in all its complexity. "Imaginary," in this case again, should not be understood as contrary to "real." Darwin refuses to "invoke a source of authority outside the natural order,"[165] and wants to start from the description of an overabundant life. Darwin, like Milton or Wordsworth, was overwhelmed and astonished by the multifariousness of natural forms, a regard that decisively shaped the style of *The Origin*.[166] Darwin sought to prevent a human or allegedly divine conceptual pattern from molding the inspirational contact with nature in advance. To be sure, there was no question of wallowing in an endless description of details.[167] But in the radical absence of a *de jure* conceptual pattern—since life is inseparable from free profusion and possible monstrosity—imagination becomes pivotal and gains a new epistemological legitimacy. The scientist must observe and map congruences between (a) particular characters and—in order to understand centuries of divergent mutations—(b) unstable relations of cooperation between individuals, as well as (c) modifications in the environment possibly induced by the individuals' activity. She can only daringly venture synoptic insights based on shortcuts, far-reaching associations, and schematizations.[168] "'Relative similarities,' 'graduated differences'—these are the major topics of *The Origin*."[169] Darwin's *Origin* is thus driven by a tension between the will to establish a positive system and the description of life's profusion, surprising descents, and minuscule, innumerable, unpredictable changes.

This strange combination of positivity and imaginative hovering particularly manifests in the famous tree diagram, which possesses all the features one can expect from a phantom. Darwin worked on this diagram for years and sketched many other versions that substantively differ from each other[170] before finally opting for this one. These revisions and adjustments are all the more significant in that, eventually, Darwin wanted the final version to bear the mark of a certain dynamism. As he explains, "I must here remark that I do not suppose that the process ever goes on so regularly as is represented in the diagram, though in itself made somewhat irregular."[171] Darwin certainly intended to give a synoptic, clear, and solid representation of his theory; and he explicitly validates the correlation of the diagram to a tree.[172] But a comparison between this 1859 diagram and the much more literal and substantial version presented by Ernst Haeckel, for instance, is particularly telling (see Figures 3.1 and 3.2). Darwin's "tree-diagram" in *The Origin* has no trunk. It is at least as much coral-like as tree-like.[173] It is structured around "imaginary

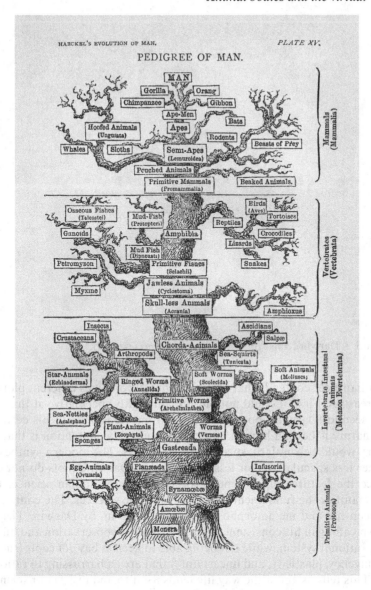

Figure 3.1 Haeckel's Tree of Life in *The Evolution* of *Man* (1879).

breaks"[174] and is entirely drawn in dotted lines. Darwin's diagram points to a positive model and, because of its vertical orientation, even to an interpretation of evolution in terms of progress, but it certainly does not coincide with them. As the dots refer to invisible, long since disappeared, ungraspable lines of descent,[175] they are the very incarnation of speculation and undecisive compromise.

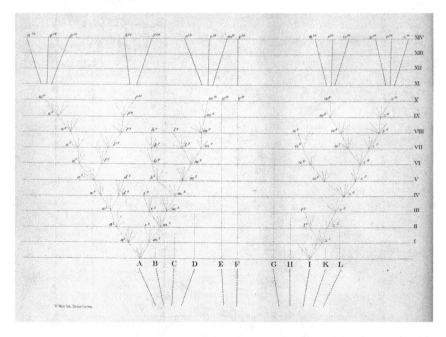

Figure 3.2 Darwin's diagram in *The Origin of Species* (1859).

Gillian Beer contends that Darwin's theory combines different plots for two reasons: First, it could not but remain highly speculative at that time and, second, Darwin's positivism entered into tension with his certitude that nature is beyond human understanding.[176] My contention is that this theory would be similarly speculative nowadays. The modern synthesis is neither less speculative, nor less questionable. Darwin's plots do not only evince the features of one epoch, or the limits of human science, they also manifest the fundamental structures of life sciences. The connection between life and metamorphoses was not made up by Darwin. There is deep meaning in his consistent effort to begin with observation and build a truly rational system, while not being able to keep at bay concepts (such as contingency, plasticity, and uncertainty) that are embarrassing to rationalism. This tells us about the way life lends itself to and resists our scientific endeavors, constraining us to invent hybrid phantom approaches.

In the *Nature* lectures, Merleau-Ponty first and foremost underscores his disagreement with Darwin:[177] Darwin's theory has no room for meaning, intentionality, and, therefore Gestalten. Fair enough: The introduction of the concept of Gestalt undoubtedly allows for a further and more radical acknowledgement of the phantom-like dimension of living beings. Indeed, this concept of Gestalt gives a full-fledged status to ubiquity, and it directly accounts for the coexistence between the level of the actual parts of the Gestalt and the virtual structure that gather them into a whole. But

Merleau-Ponty missed the imaginary of Darwin. He mainly reads Darwin through neo-Darwinism and the modern synthesis. This approach is regrettable, for it cuts short the opportunity to undermine these purely positivist interpretations of Darwinism. It is crucial to show that, precisely through their tie to Darwin as their leading light, neo-Darwinism and the modern synthesis must be still haunted by ambiguity, conflicting plots, hovering "essences," and even the issue of meaning and agency in animals. Developmental systems theory, epigenetics, and "postgenomics"—which contend that selection cannot be regarded as a purely blind and mechanical process that individuals passively undergo—thus have the possibility to strategically claim the legacy of Darwin.[178] "Evolutionism has been so imaginatively powerful precisely because all its indications do not point one way."[179]

§26 Perceiving animals as images

Animals are phantoms. In other words, they first and foremost consist in an effective, metamorphic, and recurring melodic meaning. Does that mean that they are images or that they imagine?

When a phantom-like theme governs morphological changes, like in hemianopsia, when the progression of rats in a maze manifests the general virtual pattern that guides them (for instance, "looking for the shortest way"), the body and the movements of animals let a reference to the virtual show. Gestalt essentially consists in the discrepancy and the systematic interplay between the virtual and the actual, which always comes across through the living body. This is also why non-human animals spellbound our imagination: Observing their movement or interacting with them inspires inexhaustible fascination. In this regard, non-human animals imagine. As autopoietic beings, they consist in the active and effective reference to a specific virtual theme operating within oriented ontogenetic, phylogenetic, and behavioral processes.

However, the gestures toward virtuality can be at work in animal bodies without necessarily being intentionally played out. The goal of the race of rats is the actual exiting from the maze, not the *expression* of the general theme "finding the shortest way." As we describe these Gestalten and learn to perceive animals as phantoms, we certainly bring these virtual themes to the fore. We express them, and, in the same process, we turn non-human animals into images: We aim, through them, at a virtual theme.

"Image" in this context does not mean copy-image (*Abbild*), namely the reproduction of the appearance of a model. The movements of rats do not mimic the appearance of a concrete reality. But human images are never essentially copy-images.[180] The original "Hamlet" does not exist. The original "Montagne Sainte Victoire" is not reproduced but reincarnated into a new appearance by Cézanne. The body of the actor and Cezanne's painting point to and depict a phantom-like theme. All of them are images, however, and not signs, since there is no hypostasized signified here, no abstract meaning, and no conventional signification, but only an endless series of particular sensible appearances.[181]

It is therefore no surprise that Plessner and Buytendijk describe animals and their behaviors as "images [*Bilder*]"[182]: "The dog, with his head on the ground, who strains forward, restlessly runs to and fro, stops and sniffs now here now there, and returns to his starting position, gives us the typical image of 'to look for.'"[183] An image is indeed a sensible layout that essentially presents itself as a whole and points to a meaning that can be reduced neither to this actual appearance, nor to a conceptual definition. Behaviors and morphologies also present themselves as wholes, or Gestalten: We certainly can break them into pieces (body parts, molecules, isolated frozen positions, isolated knee-jerk reactions), but this would be to miss their intrinsic meaningfulness.

An enthralling reversal of the myth of living images becomes thus possible. Mitchell outlined its possibility in *What Do Pictures Want?*: Not only are images *as if* living beings, but "living things themselves were always already images in one form or another."[184] Ontology and biology have long grappled with the kinship between life, images, and essences (*species*, namely kinds *and* visible forms, from *specere*, to see[185]) and the paradigm of absolute rigid pictures, forever shining in the sky of forms or in enlightened intellects, largely prevailed. But species were already also defined as forms that give themselves to perception, through shapes, patterns, and likeness. They were images at least as much as immaterial and intellectual essences. As a result, Mitchell claims, ontology and biology were always a form of iconology. In fact, contemporary research in life sciences brilliantly confirms this daring idea: We are now dealing with "pseudospecies"[186] in asexual living beings, but also more broadly, as I have argued above, with phantom-like kinds. The analyses developed in the present chapter allow us to flesh out a crucial aspect of this uncanny knot between ontology, biology, and iconology. Indeed, in living beings, and particularly in animals, the unfolding of essences is brought to a unique level of *autopoiesis* and active play—which entails stretching and yet recreating the moving boundaries of phantom-like forms. As a consequence, animals are and cannot but be perceived as images. A piece of the puzzle is still missing, however. It will be investigated in the next chapters: Non-human animals are also actively and *subjectively* engaged in a relation to appearance and "likenesses." They take it up and start dealing with recognition, pretend, symbols, and their ambiguities.

To a certain extent animal bodies and behaviors are what I have named living-images,[187] but—to put it in phenomenological terms—my apprehension-act of the animal body *as* an image plays a key role here, even though this active apprehension would be powerless without the support of ontological ubiquity in non-human animals. But it is another matter to investigate the processes whereby non-human animals can themselves perform behaviors or metamorphoses that bring such virtual themes, in other words, meaning, to the fore.

Several issues follow: (1) Is the meaning that surfaces in the body of non-human animals only *for* us, human animals, or do non-human animals themselves thematize it, and, if in the affirmative, how do they thematize it? (2) How are we to conceive of the relation between, on the one hand, the cultural understanding of phantoms, sustained by a rich literary, artistic, and even scientific imaginary, and, on the other hand, the imaginary of animals (subjective genitive)? When I think of an animal as a phantom, or a melody, or when the hissing becomes a haunting theme and the very incarnation of evil in horror movies, do animals imagine or is it only us, human beings? (3) Should we not abandon the conceptuality of imagination and the imaginary altogether in order to fully acknowledge the ontological ubiquity of animals, from a more literal perspective? The first two questions will be addressed in Chapter 4, the third in Chapter 5.

Notes

1. Melville, Herman, *Moby-Dick; or, The Whale.* New York: Harper & Brothers publishers, 1851, p.7.
2. Merleau-Ponty, *The Structure of Behavior,* Boston: Beacon Press, 1960, p.159.
3. *Ibid.,* p.126 and 188.
4. *Ibid.,* p.181.
5. Merleau-Ponty, *The Structure of Behavior,* p.162.
6. See Chapter 4.
7. Toadvine, *Merleau-Ponty's Philosophy of Nature,* Evanston: Northwestern, 2009, p.83.
8. Bimbenet, Etienne. "L'homme ne peut jamais être un animal", In *Bulletin d'analyse phénoménologique* VI 2, 2010, p.164–5, my translation.
9. See also Masuda, Kazuo, "La dette symbolique de la Phénoménologie de la perception," in Heidsieck, François (Ed.), *Merleau-Ponty: Le philosophe et son langage.* Paris: Vrin, 1993.
10. Chapter 4 shows that the human symbolic realm never really breaks away from this animal imaginary.
11. See, for instance, Merleau-Ponty, *Le visible et l'invisible,* Paris: Gallimard, Tel, 1964, p.322, Trans. A. Lingis, Evanston: Northwestern University Press, 1968, p.274. See also p.177, 312, 315 (Trans. p.136, 264, 267).
12. For instance, Merleau-Ponty, *The Structure of Behavior,* p.182: "It is realism in general which must be called into question."
13. Merleau-Ponty, *The Structure of Behavior,* p.47. Merleau-Ponty refers to Köhler's *Die physischen Gestalten in Ruhe und im stationären Zustand. Eine naturphilosophische Untersuchung,* Braunschweig: Friedrich Vieweg und Sohn, 1920
14. *The Structure of Behavior,* p.47.
15. *Ibid.,* p.106.
16. *Ibid.*
17. Petermann, Bruno. *The Gestalt Theory and the Problem of Configuration.* London: Routledge and Kegan Paul, 1932, Reprint 1999, p.69.

18. See Köhler, *Die physischen Gestalten in Ruhe und im stationären Zustand* p.78–9; Petermann, Bruno, The *Gestalt Theory and the Problem of Configuration*, p.68–70; And Humphrey, George, *The Nature of Learning: In Its Relation to the Living System,* London: Kegan Paul, Trench, Trubner & Co, 1933, p.35.
19. Merleau-Ponty, *The Structure of Behavior,* p.138.
20. See, for instance, Koffka, Kurt. *Principles of Gestalt Psychology.* New York: Routledge, 1935. See also Vatan, Florence. "L'obscur attrait des formes: Wolfgang Köhler et la catégorie de Gestalt." *Revue d'Histoire des Sciences Humaines* (5), 2001.
21. Merleau-Ponty, *The Structure of Behavior,* p.7.
22. *Ibid.,* p.8.
23. See also *Ibid.,* p.35.
24. *Ibid.,* p.33.
25. *Ibid.,* p.60–1.
26. *Ibid.,* p.60.
27. Frisch, Stefan, "How Cognitive Neuroscience could be More Biological—and What it Might Learn from Clinical Neuropsychology," In *Frontiers in Human Neuroscience,* Vol. 8, 2014.
28. *Ibid.,* p.4. See Bates, Elizabeth and Dick, Frederic. "Beyond Phrenology: Brain and Language in the Next Millennium." In *Brain and Language,* 71, 2000.
29. Through the concept of degeneracy, nevertheless, the plasticity of the brain is accounted for "rarely but increasingly" (Frisch, "How Cognitive Neuroscience could be More Biological," p.4).
30. *Ibid.,* p.4 and 6.
31. Merleau-Ponty, *The Structure of Behavior,* p.65.
32. *Ibid.,* p.67.
33. Merleau-Ponty points out another symptom of this inadequacy of the mechanistic approach, namely the inflationist reference to new *ad hoc* hypotheses to account for behaviors, when an alleged pre-wired reaction fails to be triggered by its alleged specific stimulus. It is always possible to presuppose that an inhibiting mechanism is at work and that another pre-wired connection has taken over (see for instance *The Structure of Behavior,* p.19). But this explanation *forges* the idea of inhibition and posits the permanent existence of an extension reflex "which can in no way be observed in the adult and normal subject" (*The Structure of Behavior,* p.19). Moreover, an indefinite number of inhibition mechanisms and alternative reflexes must then be presupposed to stick as closely as possible to plastic behaviors.
34. As shown by Ash, Mitchell G. (*Gestalt Psychology in German culture,* p.290) and Vatan, Florence ("L'obscur attrait des formes" p.98), Gestalt theory can be tied, although in an ambiguous manner, to the critique of positivism that developed at the beginning of the 20th Century and pointed to a "crisis of science." The traditional scientific methods were then reproached for turning every reality into blind mechanisms and plunging us into a meaningless and disenchanted world. Gestalt theorists nevertheless always "couched their position in language specific to the scientific community" (Ash, p.7). They also introduced in science a concept that reaches back to esthetics and history (Vatan p.99–100) and regularly emphasized that their approach allows science to eventually incorporate a dimension of meaning into its objects.
35. Merleau-Ponty, *The Structure of Behavior,* p.25 (SC p.25).
36. *Ibid.*
37. *Ibid.,* p.41.
38. Frisch, "How Cognitive Neuroscience could be More Biological," p.10.

39. Merleau-Ponty, *The Structure of Behavior*, p.98. Merleau-Ponty refers to Köhler, *Intelligenzprüfungen an Menschenaffen*, 2nd ed., Berlin: Springer, 1921, pp.140 sqq. (Cf. English edition, The Mentality of Apes, trans, by E. Winters, London, Routledge & Kegan Paul, 1925, pp.194 sqq).

40. Merleau-Ponty, *The Structure of Behavior*, p.100.

41. *Ibid.* Merleau-Ponty refers to Tolman's experiments (E.C. Tolman, "Sign-Gestalt or Conditioned Reflex?" Psychological Review, XL 1933) and to Guillaume, Paul, *La Formation des habitudes*, Paris, Alcan, 1936, p.69.

42. Merleau-Ponty, *The Structure of Behavior*, p.51.

43. *Ibid.*, p.160.

44. I will return to the significant and tricky issue of the ontological status of the Gestalt in the second section of this chapter.

45. Merleau-Ponty, *The Structure of Behavior*, p.30.

46. *Ibid.*

47. *Ibid.*, p.155.

48. *Ibid.*, p.73.

49. Kurt Koffka's *Principles of Gestalt Psychology* is particularly representative in this regard. Koffka makes a case for a science that would integrate meaning and a holistic perspective, but he *ultimately* places his trust in mechanistic explanations (for instance p.48: "I admit that in our ultimate explanations, we can have but one universe of discourse and that it must be the one about which physics has taught us so much"). The risk is indeed to lapse into vitalism: If meaning is not reducible to a set of material parts and causal chains, how can science avoid being torn apart between two incompatible forms of knowledge, namely mechanism and vitalism? Merleau-Ponty suggests that a much more refined ontology is required to achieve the full revolutionary potential of Gestalt theory. Merleau-Ponty's approach to science in general and Gestalt theory in particular is dialogal. By referring to Gestalt theory, Merleau-Ponty avoids criticizing science from an external purely philosophical point of view. He demonstrates that, within contemporary science, precisely through the concept of Gestalt, behaviorism can be overcome and, for some scientists, is in fact overcome ("c'est un fait que la théorie classique du réflexe est dépassée par la physiologie contemporaine" *La structure du comportement*, Paris: PUF. 1942. Reprint Quadrige 1990, p.8, *The Structure of Behavior*, p.9–10). Thus Merleau-Ponty lets Gestalt theory take the floor but also responds to it. He seeks to make the radical shift of perspective implied by Gestalt theory explicit and to manifest its ontological significance. Merleau-Ponty underlines that we do not only have to choose between atomism and gestaltism, but also between realism and gestaltism (*The Structure of Behavior*, p.152 and 182), the latter point being usually overlooked in the existing Gestalt theory (p.132–4 and 219).

50. *Ibid.*, p.3.

51. Thompson, *Mind in Life. Biology, Phenomenology, and the Sciences of Mind.* Cambridge: Harvard University Press, 2007, p.78.

52. Merleau-Ponty, *The Structure of Behavior*, p.143. I have modified the translation. "Chaque partie d'une forme 'connaît dynamiquement' les autres" (*La structure du comportement*, p.174).

53. Thompson, *Mind in Life*, p.78.

54. Merleau-Ponty, *The Structure of Behavior*, p.126.

55. *Ibid.*

56. Thompson *Mind in Life*, p.78. The concept of resemblance still needs to be questioned, but it rightly echoes the concept of "analogy" used without further comments by Merleau-Ponty in *The Structure of Behavior*, p.47 and 229.

57. Merleau-Ponty, *The Structure of Behavior*, p.126–7.
58. *Ibid.*, p.133.
59. Varela, Francisco J. "Organism: A Meshwork of Selfless Selves," In *Organism and the Origin of Self*, Tauber, A. (Ed.). Dordrecht: Kluwer, 1991, esp. p.80–88
60. Merleau-Ponty, *The Structure of Behavior*, p.145.
61. *Ibid.*
62. *Ibid.*, p.104–5.
63. *Ibid.*, p.115–6.
64. *Ibid.*, p.114. Merleau-Ponty refers to Köhler, *Intelligenzprüfungen an Menschenaffen*, 1921.
65. *Ibid.*, p.114.
66. *Ibid.*
67. *Ibid.*, p.126 (*La structure du comportement*, p.136–7). I have modified the English translations since "a penetration in being" does not render the idea of radical negativity that is at stake in the French text ("est un trou dans l'être"). In this passage, Merleau-Ponty's use of the concept of consciousness proves shaky and ambiguous. Indeed, Merleau-Ponty has claimed that consciousness is somehow everywhere. This stance aligns with Hegel's theory, but Merleau-Ponty oscillates between two meanings of "consciousness" and awkwardly tries to get off lightly by introducing a hierarchy between true consciousness and what must then appear, negatively, as a mysterious gray zone, the "consciousness, although not-true consciousness" of nature.
68. *Ibid.*
69. Aristotle, *Parts of animals*, 640b.
70. Rimbaud, *Lettre à Izambard du 13 mai 1871*, in *Œuvres Complètes*, Paris: Gallimard, Pléiade, p.2009 p.340. See my article "Is a World without Animals Possible?," Trans. by Ramon Fonkoué. In *Environmental Philosophy*, Vol. 11, Iss. 1, 2014, p.79.
71. For instance, Merleau-Ponty, *The Structure of Behavior*, p.181: "Reorganized in its turn in new wholes, vital behavior as such disappears. This is what is signified, for example, by the periodicity and monotony of sexual life in animals, by its constancy and its variations in man." See also p.202–3
72. *Ibid.*, p.203.
73. *Ibid.*, p.114.
74. "se réalise toujours advantage" Merleau-Ponty, *La structure du comportement*, p.143 (trans. p.133).
75. Merleau-Ponty, *The Structure of Behavior*, p.180. See also p.184.
76. *Ibid.*, p.203–4.
77. *Ibid.*, p.210, my emphasis.
78. *Ibid.*, p.181.
79. *Ibid.*, p.116 "In other words, the animal cannot at each moment adopt a point of view with regard to objects which is chosen at its discretion." See also p.118. The problem lies in the use of the verb "cannot." The latter is indeed imprisoning, strictly speaking unjustifiable, and typical of the great divide as a fundamental, brutal *decision*.
80. *Ibid.*, p.114.
81. *Ibid.*, p.239.
82. Merleau-Ponty, *Causeries 1948*. Paris: Seuil, "Traces écrites," 2002, p.42. Trans. O. Davis, New York: Routledge, 2004, p.77.
83. For instance, *The Structure of Behavior*, p.162.
84. *Ibid.*, p.143.
85. *Ibid.*, p.145 (*La structure du comportement*, p.157).
86. *Ibid.* p.88.
87. *Ibid.*, p.202.

88. *La structure du comportement*, p.159.
89. *Ibid.*, p.143.
90. Ibid., p.143, p.144.
91. *Ibid.*, p.127.
92. *Ibid.*, p.88.
93. See supra p.96.
94. *Ibid.*, p.206, Merleau-Ponty mentions Léon Brunschvicg rather than Kant himself, whose *Critique of Judgment* overcomes a purely intellectual perspective.
95. *Ibid.*, p.145. See also the Preface of *Phenomenology of Perception*, Trans. D. Landes. New York: Routledge, 2012 p.lxxvii: The transcendental reduction must fail.
96. Merleau-Ponty, *The Structure of Behavior*, p. 206–7 (*La structure du comportement*, p.223).
97. *Ibid.*, p.182
98. See *Phenomenology of Perception*, the chapter "Sensing," for instance, p.222. Perception cannot be properly understood as the projection of a conceptual meaning on raw meaningless data. Indeed, perception can never occur without a cooperation between our body and a sensible environment that solicits us and guides the way we explore it.
99. Merleau-Ponty, *The Structure of Behavior*, p.210 (*La structure du comportement*, p.227).
100. *Ibid.*, p.41.
101. *Phenomenology of Perception*, p.146.
102. See for instance Roth, Lina and Lind, Olle. "The Impact of Domestication on the Chicken Optical Apparatus." In *PLOS One*, 2013.
103. Schleidt, Wolfgang M. and Shalter, Michael D. "Co-Evolution of Human and Canids: An Alternative View of Dog Domestication: Homo Homini Lupus?" In *Evolution and cognition* 2003, Vol. 9 (1). See also Olsen, Stanley, *Origins of the Domestic Dog: The Fossil Record*, Tucson: University of Arizona Press, 1985, and Botigué, Laura et al. "Ancient European Dog Genomes Reveal Continuity since the Early Neolithic," In *Nature Communications* 8, 2017.
104. Schleidt and Shalter, "Co-Evolution of Human and Canids," p.59–61.
105. *Ibid.*, p.65–6.
106. Feddersen-Petersen, Dorit Urd. "Vocalization of European Wolves (Canis lupus lupus L.) and Various Dog Breeds (Canis lupus f. fam.)," in *Archives Animal Breeding, Archiv Tierzucht*, 43, 2000.
107. See Pongrácz, Peter, Molnár, Csaba, Miklósi, Ádam, and Csányi, Vilmos. "Human Listeners Are Able to Classify Dog (Canis familiaris) Barks Recorded in Different Situations." In *Journal of Comparative Psychology*, Vol. 19, No. 2, 2005.
108. Ibid., See also Molnár, Csaba et al. "Classification of Dog Barks: A Machine Learning Approach." In *Animal Cognition*, 11, 2008.
109. Uexküll, *A Theory of Meaning* (in *A Foray into the Worlds of Animals and Humans. With A Theory of Meaning*, 2010, p.154).
110. See supra, Chapter 2, §15.
111. This idea is developed in *Phenomenology of Perception*, but not yet in *The Structure of Behavior*. The theory of the three orders is still tightly connected to the conception of fragmentation in terms of disintegration, failure, and illness.
112. Merleau-Ponty, *The Structure of Behavior*, p.207.
113. Merleau-Ponty, *The World of Perception*, p.77.
114. See Derrida, Jacques. *Spectres de Marx*. Paris: Galilée, 1993. I borrow the term "hauntology" from Derrida, but I derive this idea of an ontology centered around the model of phantoms from a phenomenology of the imaginary and of transcendental interanimality.

115. Shakespeare, *Macbeth*. Cambridge, New York: Cambridge University Press, 1997, II. 1.

     "Is this a dagger which I see before me,
     The handle toward my hand? Come, let me clutch thee.
     I have thee not, and yet I see thee still.
     Art thou not, fatal vision, sensible
     To feeling as to sight?"

116. Husserl also defined a perceptual phantom as that which never ossifies beyond the flux of *Abschattungen*. See my article "Vies et morts de l'imagination: La puissance des actes fantômes", in *Bulletin D'Analyse Phénoménologique*, 13 (2), 2017.
117. Merleau-Ponty, *Phenomenology of Perception*, p.79.
118. *Ibid.*, p.84 (*Phénoménologie de la perception*, p.98).
119. There is an essential link between this habitual body and the body schema. The latter can be defined as a set of postural and behavioral functions, capacities, and habits that make coordinated movements as well as the maintenance of body postures possible. The body schema is a form of self-knowledge possessed by the body, a practical and non-representational knowledge.
120. See, for instance, the study of complex reorganization processes in cortical visual areas in cases of early-blindness (Renier, L., De Volder, A.G., and Rauschecker, J.P. "Cortical Plasticity and Preserved Function in Early Blindness," In *Neuroscience & Biobehavioral Reviews*, Vol. 14, April 2014).
121. Maravita, Angelo and Iriki, Atsushi, "Tools for the Body (schema)." In *Trends in Cognitive Sciences,* Vol. 8, No. 2 (79–86) 2004. See also Holmes, Nicholas P. and Spence, Charles. "The Body Schema and the Multisensory Representation(s) of Peripersonal Space," In *Cognitive Processing*, Jun; 5(2), 2004; and a study about the plasticity of the body schema in rats: Khvatov, I.A. et al. "Body Scheme in Rats Rattus norvegicus." In *Experimental Psychology*, Vol. 9, No. 1, 2016.
122. Maravita, Angelo and Iriki, Atsushi, "Tools for the Body (schema)," p.81.
123. Merleau-Ponty, *The Structure of Behavior,* p.107.
124. See also *The Structure of Behavior*, p.155 ("Nerve functioning is not punctually localizable; a kinetic melody is completely present at its beginning and the movements in which it is progressively realized can be foreseen only in terms of the whole"), p.159 ("Every organism," said Uexküll, "is a melody which sings itself"), and p.173.
125. *Ibid.*, p.30 (my emphasis). Merleau-Ponty quotes Buytendijk: "Versuche über die Steuerung der Bewegungen," *Archives néerlandaises de physiologie de l'homme et des animaux*, XVII, 1, 1932, p.94. See also *The Structure of Behavior* p.110.
126. Merleau-Ponty, *The Structure of Behavior,* p.68–9.
127. *Ibid.*, p.14, 25, 32, 52, 105.
128. "It becomes necessary to renounce conceiving of nerve activity in its most essential character, not only as restricted to certain determinate pathways, but even as a choice between several pre-established pathways. (...) The metaphor of a switching station is not applicable since one cannot find out where it should be situated, since it would be a station which receives its instructions from convoys which it is charged to direct and which improvises the pathways and the switches according to their indications." Merleau-Ponty, *The Structure of Behavior,* p.32. See also p.38, 40.

129. Merleau-Ponty, Maurice. *La Nature. Notes de cours au Collège de France (1957–1960)*. Paris: Seuil, 1995, p.207. Trans. Robert Vallier. Evanston: Northwestern University Press, 2003, p.156.

130. Phenomena of phantom pain in non-human animals have been observed, although through an objectivist approach. These studies connect such phantom-limb-like behaviors and the plasticity of neural system. See for instance Marshall Devor and Patrick D. wall, "Plasticity in the spinal cord sensory map following peripheral nerve injury in rats", in The Journal of Neuroscience, Vol. 1, No. 7, 1981. See also Cook, Alison J. et al. "Dynamic Receptive Field Plasticity in Rat Spinal Cord Dorsal Horn Following C-Primary Afferent Input," In *Nature*, Vol. 325, 8, January 1987; as well as Katz, Joel et al. "Injury Prior to Neurectomy Alters the Pattern of Autotomy in Rats: Behavioral Evidence of Central Neural Plasticity," in *Anesthesiology*, 75(5) (876–83), 1991. And Li, Jianguo et al. "Alteration of Neuronal Activity After Digit Amputation in Rat Anterior Cingulate Cortex," In *International Journal of Physiology, Pathophysiology, and Pharmacology*, 5(1), 2013.

131. This transformation that affects the body constitutes, Merleau-Ponty claims, an original "repression" (or "repression as a universal phenomenon," *Phenomenology of Perception*, p.86) that makes human existence possible; this passage is fascinating albeit still obscure. Who exactly represses what, if "I" and "the body" are the result of such an original repression? The text is not entirely consistent on that point, and calls for a tricky but tenable circular interpretation. Another difficulty is to determine whether every human being achieves this original repression: there are always, as Merleau-Ponty highlights, moments when the bodily perception of life and the subject become almost one, when impulses and "my" decisions merge. In this regard, a dialogue with Michel Henry's phenomenology of life would be illuminating, as it fleshes out this reference to invisible life as a new form of phenomenality. Moreover, it should be recalled that the body-mind dualism does not belong to *human* thought overall, but pertains to a particular cultural interpretation of existence, among many others.

132. I will examine more thoroughly in the next chapter how to conceptualize this nature-culture, human-non-human relation.

133. Merleau-Ponty, *The Structure of Behavior*, p.122.

134. Merleau-Ponty, *Signes*, Paris: Gallimard, 1960, p.38–9, Trans. Richard McCleary "Preface to Signs," In *The Merleau-Ponty Reader*, Toadvine and Lawlor (Eds.), Evanston: Northwestern University Press, 2007, p.347. See also *Notes de cours sur L'origine de la géométrie de Husserl*, Paris, PUF, 1998, p.34.

135. Merleau-Ponty, *Causeries 1948*, p.42, *The World of Perception*, p.77.

136. Norris, Margot. *Beasts of the Modern Imagination*, London: The Johns Hopkins University Press, 1985, p.157. See also p.83 "The gift of Nietzsche's thought (...) is a gift stripped of all reciprocal function, of all economic motive, of all mediative power and social utility."

137. See, for instance, Norris, *Beasts of the Modern Imagination,* p.4: "the animal surrenders to biological fate and evolutionary destiny." Norris' analyses provide fascinating insight into the way Darwin, Nietzsche, Kafka, Ernst, Lawrence, and Hemingway connect animality to the destruction of human symbolic systems.

138. Merleau-Ponty, *The Structure of Behavior,* p.114 (*SC*, p.124).

139. Merleau-Ponty, *The Structure of Behavior,* p.126 (*SC*, p.136–7).

140. *Ibid*. p.239.

141. Merleau-Ponty, *Le visible et l'invisible*, p.97, Trans. p.69.

142. Merleau-Ponty, *Phenomenology of Perception*, p.223.

143. Merleau-Ponty, *The Visible and the Invisible*, p.152.
144. Milgram, Stanley. "Behavioral Study of Obedience." *Journal of Abnormal and Social Psychology.* 67 (4), 1963.
145. Beer, Gillian. *Darwin's Plots: Evolutionary Narrative in Darwin, George Eliot and Nineteenth-Century Fiction.* Cambridge: Cambridge University Press, 1983, reprint 2004.
146. Obviously "Darwin" is also the name of an epoch and a rich cultural context. Gillian Beer thus demonstrates that Darwin was influenced by poets as much as by scientists. In fact, he "drew on familiar narrative tropes" (*Darwin's Plots*, p.xxiv), poetic descriptions, and already trendy scientific metaphors to write his *Origin of the Species,* while simultaneously providing, in his turn, an overabundant source of inspiration to Victorian imagery.
147. Lyell, Charles, letter to J.D. Hooker, 25 July 1856, quoted in *The Correspondence of Charles Darwin,* Burkhardt, Frederick and Secord, James A. (Eds.), Vol. 6, 1856–1857, Cambridge University Press, 1990, p.195. Lyell feared that Darwin's theory may destroy every solid marker, including, on the one hand, species and, on the other hand, an identifiable, localizable, original group from which all the subgroups of a species spread out.
148. Letter to J.D. Hooker, 30 July 1856. *Ibid.,* p.193.
149. Darwin. *On the Origin of Species,* p.92.
150. "Which groups will ultimately prevail, no man can predict" (*On the Origin of Species,* p.97).
151. *Ibid.,* p.153.
152. *Ibid.,* p.140–1.
153. *Ibid.,* p.138.
154. Beer, *Darwin's Plots,* p.xxiv.
155. Carroll, Lewis. *Alice's Adventures in Wonderland.* London: Macmillan and Co, 1865, p.54. Regarding the influence of Darwin on Lewis Carroll's work, see William Empson, "Alice in Wonderland – The Child as Swain," in Phillips, Robert (Ed.), *Aspects of Alice: Lewis Carroll's Dreamchild As Seen Through the Critics' Looking Glasses,* New York: The Vanguard Press, 1971; Lovell-Smith, Rose, "The Animals of Wonderland: Tenniel as Carroll's Reader." In *Criticism,* 45.4 (2003); and Talairach-Vilmas, Laurence, *Fairy Tales, Natural History and Victorian Culture,* Basingstoke: Palgrave Macmillan, 2014.
156. Lewontin, Richard. *The Triple Helix,* p.10. See also Charbel Niño El-Hani, "Between the Cross and the Sword: The Crisis of the Gene Concept," in *Genetics and Molecular Biology,* Vol. 30, No.2, 2007; and Hoffmeyer, Jesper, "The Semiome: From Genetic to Semiotic Scaffolding," In *Semiotica* 198, 2014. For the analysis of the connection between genetic scaffolding and Merleau-Ponty's ontology of the imaginary, see Westling, Louise, "Deep History, Interspecies Coevolution, and the Eco-Imaginary," In Calarco, M. and Ohrem, Dominik (Eds.), *Exploring Animal Encounters: Philosophical, Cultural, and Historical Perspectives,* London: Palgrave Macmillan, 2018.
157. Dias, Brian G. and Ressler, Kerry, "Parental Olfactory Experience Influences Behavior and Neural Structure in Subsequent Generations," In *Nature Neuroscience,* 17(1) December 2013.
158. Hekman, Jessica Perry, "The Epigenetics of Fear," 2013, In the Blog "Dog Zombie," http://dogzombie.blogspot.com/2013/12/the-epigenetics-of-fear.html
159. Darwin, *The Origin of Species,* p.357; Beer, *Darwin's Plots,* p.49.
160. Beer, *Darwin's Plots,* p.7.
161. Beer, *Darwin's Plots,* p.93. See Darwin, *The Origin of Species,* p.51: "I should premise that I use the term *Struggle for Existence* in a large and metaphorical sense, including dependence of one being on another". See also the famous description of the "entangled bank" (*The Origin of Species* p.360) and p.60

"The dependency of one organic being on another, as of a parasite on its prey, lies *generally* between beings remote in the scale of nature. This is *often* the case with those which may strictly be said to struggle with each other for existence, as in the case of locusts and grass-feeding quadrupeds. But the struggle *almost invariably* will be most severe between the individuals of the same species, for they frequent the same districts, require the same food, and are exposed to the same dangers. In the case of varieties of the same species, the struggle will *generally* be almost equally severe," my emphasis.

162. Beer, *Darwin's Plots*, p.xix.
163. Beer, *Darwin's Plots*, p.35.
164. Ibid., p.12, 35, and 108. See, for instance Darwin, *The Origin of Species*, p.62. See also Gillian Beer's comment upon this passage (*Darwin's Plots*, p.35).
165. Beer, *Darwin's Plots*, p.12.
166. See for instance *Darwin's Plots*, p.32: "Milton gave Darwin profound imaginative pleasure – which to Darwin was the means to understanding."
167. *Ibid.*, p.73. Beer refers to Darwin, Francis and A.C. Seward (eds.). *More Letters of Charles Darwin: a record of his work in a Series of hitherto unpublished letters*, 2 vols. London: John Murray, 1903. 1, 176. To W.W. Bates, 22 November 1860.
168. See, for instance Darwin, *The Origin of Species*, p.62: "It is good thus to try in our imagination to give any form some advantage over another. Probably in no single instance should we know what to do, so as to succeed. It will convince us of our ignorance on the mutual relations of all organic beings; a conviction as necessary, as it seems to be difficult to acquire. All that we can do, is to keep steadily in mind that each organic being is striving to increase at a geometrical ratio; that each at some period of its life, during some season of the year, during each generation or at intervals, has to struggle for life, and to suffer great destruction."
169. Beer, *Darwin's Plots*, p.90.
170. Julia Voss (in *Darwin's Pictures: Views of Evolutionary Theory, 1834–1874*. New Haven and London: Yale University Press, 2010) and Horst Bredekamp (*Darwin's Korallen: Die frühen Evolutionsdiagramme und die Tradition der Naturgeschichte*, Berlin: Klaus Wagenbach Verlag, 2005) have thoroughly investigated the genealogy of Darwin's diagram.
171. Darwin, *The Origin of Species*, p.92.
172. *Ibid.*, p.99.
173. This idea was mentioned by Darwin himself: "The tree of life should perhaps be called coral of life" (Notebook B, 20 and 26, in Barrett et al. (Eds), *Charles Darwin's Notebooks, 1836-1844: Geology, Transmutation of Species, Metaphysical Enquiries*, London: British Museum (Natural History) and Cambridge University Press, 1987)
174. Darwin, *The Origin of Species*, p.92.
175. Julia Voss, *Darwin's Pictures*, p.90: "Geographers used dotted lines on maps to indicate the location of coral reefs that formed the foundation of an atoll under the water's surface while remaining invisible to the human eye".
176. Beer, *Darwin's Plots*, p.1, 6, 54–8, 92. Beer nevertheless also considers a less relative approach, p.6.
177. Merleau-Ponty, *Nature*, p.309, 317–8 (Trans. p.243–44, 250–1).
178. See for instance Stephen Jay Gould and Richard C. Lewontin, "The Spandrels of San Marco and the Panglossian Paradigm: A Critique of the Adaptationist Programme", in *Proceedings of the Royal Society of London, Series B*, vol. 205, No. 1161, 1979. See also Jesper Hoffmeyer "The semiome: From genetic to semiotic scaffolding", in *Semiotica*, 198, 2014, p.11: struggle for life irrepressibly points out to agency.

179. Beer, *Darwin's Plots*, p.6. See also p.3: "Darwin was telling a new story, and (...) [this] story itself proved not to be single or simple. It was, rather, capable of being extended or reclaimed into a number of conflicting systems."
180. see supra, chapter 2, §15.
181. Unlike perceptive appearance, although akin to it, images (pictures or phantasies) present themselves as contingent, hovering, whimsical manifestations, with which imagination can play endlessly (this could be...). Images bring phantoms to the fore, while perception hides them behind a surface of habitual "reality."
182. Buytendijk and Plessner. *Die Deutung des mimischen Ausdrucks*, Bonn: Cohen, 1925, p.77–8, for instance.
183. *Ibid.*, p.79, my translation.
184. Mitchell. *What Do Pictures Want?*, Chicago: University of Chicago Press, 2005, p.13.
185. See supra, p.68.
186. *What Do Pictures Want?*, p.92.
187. See supra, §15.

4 They talk the way we dream
Animal communication and the symbolic realm

§27 Animal expression and the dream comparison

What was at stake with the concept of phantom is the ontological ubiquity of animals. In this chapter, I investigate phenomena that fall within the categories of what I have called, in my introduction, "enacted" and "symbolic ubiquity." I will examine how the virtual is thematized—addressed as such—in behaviors. Even more radically, it will appear that symbols emerge in the animal realm. Namely, material and objective mediums arise that provide the thematic reference to the virtual with a more constant and solid presence in the world, in the same way as our pictures or symbolic systems do.

Human language mainly takes the form of conventional sign systems that give the virtual, including abstract concepts, an identifiable anchor in reality, as well as steady and clear limits. Which form does expression take in the non-human animal realm? Of course, this question is too general: Expression actually takes many forms in non-human animals' existence, including chemical, visual, or auditory signals, but also, in some cases, the form of innate or acquired sign systems, possibly syntax-like communication systems,[1] and even of symbolic-representational signs, as demonstrated, for instance, by Sue Savage-Rumbaugh's experiments.[2] In fact, communication begins immediately with the formation of made-to-be-perceived surfaces and cues: A part of animal morphology is systematically oriented towards appearing, although the targeted receivers are often not clearly defined.

By all means, I do not intend to exhaustively analyze the different forms of signs in animal life. I will focus on the general role imagination plays in non-human animals' expression. By imagination, more specifically, I mean a form of thought that proceeds poetically through associations, analog (vs digital) codes, and plastic symbols. We will see that such animal imagination can also be compared with the form of messages that our dreams deliver, as suggested by Portmann, Bateson, and Maran. This will also elucidate Derrida's concept of animals' poetry and, eventually, give ground to a theory that ascribes the origin of human imagination to archetypes in animal communication.

Human communication is dominated by and centered on the super-structure of connected symbolic systems—natural or constructed languages and institutional translation tools such as bilingual dictionaries or grammar books. By contrast, such symbolic systems at best only appear marginally in non-human animals' existence and never take the form of a cumulative superstructure. Investigating animal communication amounts to venturing out of the comfortable gated community built on the system of our symbolic languages. Animal communication is broad, manifold, and elusive also because it involves an open interspecific dimension. Even within the limits of intraspecific exchanges, it must thematize and express a meaning that is nowhere tied up with conceptual delimitations. Our clear-cut concepts hold through the grace of written definitions and agreements: We have commonly stopped believing that God placed eternal ideas in our mind. Humans may certainly do what humans do and try to establish a nomenclature that clearly connects animal signals to a specific meaning or function. For instance, the eyespots on the wings of butterflies are pictures of owls' eyes and could be defined as a signal used to deceive and frighten predators. But how may such a rigid signification come to being if no formal agreement and no written contract ever took place among animal emitters and receivers? It has thus been noticed that the predators of butterflies are often other insects and creatures too small to be deceived by this pattern. Moreover, several studies show that even birds and lizards are not systematically deflected by eyespot patterns.[3] And eyespots may also operate as "satyric mimicry,"[4] when bearing a distorted resemblance to actual eyes and combined with other patterns that obviously conflict the owls' eyes illusion. As monstrous and ambiguous "signs," eyespots may then be regarded as startling but not exactly deceiving patterns. Hence, time, hesitation, possible learning, and negotiation processes become part of the picture. In the realm of phantoms, an eye is and is not an eye.

I will consequently turn my attention to the studies of forms of expression in animals that highlight the critical role played by interpretation in non-human animal communication instead of exclusively devoting themselves to mathematical modelling or to the definition of functional typologies (a classical approach in evolution theory: What is it fit for?). In other words, I will focus on authors and scientists who do not simply study signs as objects to be analyzed and classified, but cast them as intentional relations between subjects. These researchers accordingly examine how animal signs "aim at" a meaning that emerges and evolves through concrete, subjective expression acts.[5] Our companions and guides in this chapters will thus be a phenomenologist (Merleau-Ponty), an unorthodox zoologist (Portmann), biosemioticians (Hoffmeyer, Maran), and, in a more ambivalent way, the anthropologist and founder of an *ecology of mind*, Gregory Bateson.

As Jesper Hoffmeyer claims,[6] eyespots cannot be reduced to a genetic code (digital code). They indeed consist in appearances in the world and are given as wholes together with their context. They are to be perceived as such by subjects who possess interpretive skills. Non-human animals do not simply determine each other through cues and signals. They observe each other, try to lure each other, and endeavor to ferret out the deceit, with varying success. Our reveries on eyespots are thus not a layer of subjective impressions superficially laid upon what would *per se* be nothing but a digital code: I will argue that they are an integral dimension of the analog nature of animal camuflage. Animal communication must accordingly be regarded as an art rather than a computer program. This compels us to qualify Uexküll's claim: Animal communication is a baroque performance, rather than the score written by a great composer.[7]

In this approach, the meaning for emitters, as well as for direct or indirect receivers must be investigated. This also involves, as an integral part, the human apprehension of animal signs. In a beautiful book about the esthetic sense of animals.[8] Etienne Souriau pointed to the fascinating kinship between human art and the prolific animal play with forms, rhythms, and colors. Animals dance, draw, paint, strut, and want to impress, even sometimes "dress" to impress, like peacocks or like Julie, the original designer of the "grass-in-ear" trend in a group of Zambian chimps.[9] Animals put themselves on display, as if they had a theatrical instinct.[10] The realm of animal forms also supplied heraldry with a whole set of conspicuous and easily recognizable attributes: Claws, horns, antlers, paws, tentacles, wings. But are such attributes ready-to-use symbols or are they turned into symbols by humans?

As I have already emphasized, when a subject-centered approach of animals is at stake, we must embrace the role played by *our* subjectivity in our understanding of other living beings. We cannot gloss over what it is like *for us*, human beings, to use a sign, use our physical appearance to communicate with others, or interpret signs. In the case of animal language, however, the difficulty is especially acute. To what extent can we think *as if* symbolic language were not the obdurate medium through which everything appears to us? Even insights, emotions, and alleged raw sensations are somehow marginalized by discursive thought. This is a classic issue for mystical philosophies: They cannot but be formulated through words and must therefore find a way of running, so to say, against the tide, of playing against the inevitable discursive medium that constitutes their very flesh. Animal philosophy, biology, biosemiotics, and cognitive ethology encounter a similar problem.

In order to investigate this issue, the common thread of this chapter will be the analysis of a strange metaphor that several authors put forward: Non-human animals think, communicate, and, overall, refer to meaning in the same way as human beings dream.

This comparison is ventured by Portmann, who seeks to consider the appearance of animal organisms as a full-fledged object of biology and as a form of expression. The colorful, sophisticated, and highly diversified animal "costumes" are meaningful for human imagination and in the framework of animals' perceptive *Umwelt*. From both perspectives, animals' visible appearance is not easily reducible to a clear and unique signification. Portmann thus observes that "at times, the sight of these organic forms makes us feel as if we are faced with the uncanny materialization of our dream life, the products of our fantasy."[11] I will examine why this comparison is not simply incidental or ornamental in the framework of Portmann's endeavor to introduce a form of phenomenology in biology.

The dream metaphor is also a recurring theme in Merleau-Ponty's *Nature* lectures: "What the animal shows is not utility; rather, its appearance manifests something that resembles our oneiric life."[12] Merleau-Ponty shows special interest in the way arbitrary signs emerge through imitations, ritualization, and role games in the animal kingdom: A diffuse intentionality proceeds through accidents and ubiquitous structures, developing a nascent capacity for representation.

Interestingly enough, similar metaphors reappear in 1969 under Bateson's pen, apparently independently of any direct influence from Portmann's or Merleau-Ponty's work. Bateson consistently carries out the dream comparison in his metalogue "What is an instinct?": "We've found a whole lot of things in common between dreams and animal behavior. They both deal in opposites, and they both have no tenses, and they both have no 'not,' and they both work by metaphor, and neither of them pegs the metaphors down."[13] This assumption was first figured out by the daughter:[14] "It looks as if we are going to be anthropomorphic in one way or another, whatever we do. And it is obviously wrong to build our anthropomorphism on that side of man's nature in which he is most unlike the animals. So let's try the other side. You say dreams are the royal road to the other side."[15] And the ensuing discussion fleshes out her hypothesis. Nonetheless, opposing dream to reason and drawing a list of what dreams cannot do may be a way of yielding too complacently and unnecessarily to anthropocentrism. Bateson's theory is without any doubt provocative and intended to be taken up playfully. This theory is inspiring and fruitful, but it will take, as I will contend, some significant adjustments to allow the dream comparison to become the principle of a genuine decentering process and of a deeper encountering with animal signs.

Bateson's hypothesis is momentous also in that it proved programmatic for a branch of contemporary biosemiotics that challenges a computational paradigm and intends to acknowledge iconicity, ambiguous codes, and open interspecific communicational networks as genuine and crucial forms of expression in nature. Bateson's "What is an instinct" was first published in Sebeok and Ramsay's *Approaches to Animal communication* (1969), a foundational book for biosemiotics and was once again incorporated in Timo Maran, Dario Martinelli and Aleksei Turovski's *Readings in Zoosemiotics* in 2011. Timo Maran's painstaking study of mimicry[16] will help us understand why the dream/imagination metaphor can actually be fruitful and illuminating in biosemiotics.

This chapter investigates the exact meaning of the dream comparison and its limits. I will contend that this comparison has two main heuristic advantages.

First, the dream comparison gives a relevant characterization of animal expressiveness as a general form of inchoative symbolism. As this chapter will demonstrate, speaking of the production of images or symbols in animal communication remains insufficient as long as one sticks with the model of the *Abbild* (copy-image defined by an identifiable model and its reproduction) or of the arbitrary symbol referring to a specific object. Forms of expression are diverse and versatile in the animal realm and, moreover, only a *specific* type of images/iconicity can provide a relevant paradigm to understand the overall structure of animal signs. The dream comparison aims to overcome this difficulty by presenting animal communication as essentially and ultimately defined by the following traits: Agency but no mastery; no *re*-presentation and, nevertheless, a representation; no clear-cut signified, and yet the emergence of an endless quest for more meaning, a hunger for meaning.

Dream is not *any* form of image, imagination, or fantasy. It does not *posit* the world *as an object that is aimed at via* sensations, body movements, and representations. But it does not simply present us with the world as a given. The world, placed at a distance by sleep, becomes much more plastic than in wakefulness. Personal themes, obsessions, and patterns take over perceptive receptivity, but certainly not as platonic ideas. They do so rather through sensible experiences and metamorphoses. The dream does not *know* where it is heading and what it aims at. It gropes its way towards virtual and underdetermined new options of existence: I wake up wondering why aunt Jeanne was fighting against a whale, or I find myself suddenly resolved to fly to Eugene and meet my friend Molly. Bateson and Merleau-Ponty both have in mind a conception of dream inspired by a long tradition and which makes it the disclosing principle of a side of meaning that logical thought and symbolic language structurally miss or tend to conceal. However, this duality itself is problematic and must be challenged, so much that, as it will turn out, *we humans also talk like we dream.*

This last point is linked to the second heuristic asset of the dream comparison, namely that it provokes us to investigate how we can understand animals' language phenomenologically, in the first-person mode. From an objective perspective, I create mathematical models that describe, from the outside, communication in animals cast as a remote "*them*." By contrast, the dream comparison claims that *we*, non-human and human animals, "dream" signs. This comparison describes a way of experiencing meaning that is *shared* by human and non-human animals. Through our dreams, we can imagine how it is like for a non-human animal subject to talk and think. However, the dream comparison operates as a *provocation* to investigate animals in the first-person mode, because this comparison brings trouble rather than it easily and unconditionally offers a satisfactory model. If the dream comparison amounts to the assertion that non-human animals *simply* dream but do not reason, that they sense meaning but do not grasp it, that they *cannot* say "no"[17] and are benumbed subjects, then its subversive virtue is lost. It depersonalizes animals at the very moment it claims to grant them with a subjectivity akin to ours. As we will see, Bateson here comes close to Rainer Maria Rilke—who famously writes in the *Eighth Duino Elegy* that animals "gaze into the open (...) the Nowhere without the Not"[18]—Heidegger, Lacan, and, to a certain extent, Agamben. We will also find this version of the dream comparison in Jean Christophe Bailly's *Le dépaysement*. But is dream really the opposite of *logos*? Merleau-Ponty's conception of the dream comparison, his non-hierarchical and non-dualist approach in the *Nature* lectures will allow us to understand why human and non-human animals always dream-talk together.

I will first study Portmann's theory of animal appearance and show that animals essentially consist in an intentionality that turns their own appearance into an expressive surface. Animals are therefore images, so to say, in and for themselves and not only for us (§28). I will then examine the reasons why signs in the animal kingdom always go together with a dimension of underdetermination, plasticity, and open creation (§29). Biosemiotic analysis of mimicry and meaning, but also interesting developments about the role of learning in communication, animal humor, lies, play, and their invention of symbols, at the evolutionary or individual levels, will buttress our reflections. The third part of this chapter (§30) will be devoted to the analysis of Bateson's approach to the dream comparison. I will eventually (§31) confront Bateson's conception of the said dream comparison with Merleau-Ponty's philosophy of the lateral, i.e., non-hierarchical, relation between non-human and human animals, images, and words. The phenomenology of the imaginareal that I have developed in Chapter 2 will help us take the dream comparison seriously.

§28 *Self-depiction* in Portmann's zoology: Animals turning themselves into images

Organs to be seen

As Portmann highlights in *Die Tiergestalt*, modern biology has been driven by the belief that the riddle of life shall be solved by the penetration further and further into invisible components and microscopic structures.[19] Portmann does not dispute the importance of this orientation. But he regrets that the macroscopic level and the phenomenal dimension of animals have been more and more consistently neglected by zoology.

If we analyze the structure of a feather, break it down into its physico-chemical components, and try to find which genes code for it, an important aspect of the feather will be left aside: The apparent pattern. While one side of the feather is black or light-grey, the visible side displays an iridescent, colorful, and symmetrical pattern. "It is as if a painter's brush had passed lightly over the ends of these insignificant-looking, mouse-colored feathers so as to give them a beautiful sheen, an 'outward show.'"[20]

The difficulty is that the latter description sounds brazenly anthropomorphic. When we define feathers and more generally animals as visible wholes, don't we subject them to the standard of our gaze and scale? However, the strength of Portmann's approach resides in the rigorous argumentation through which he demonstrates that macrophenomenal appearances stem from animal bodies themselves. Our perception of animals can and must, under certain conditions, be incorporated into a scientific perspective in order to do justice to a perceptible form displayed *by animals*. Portmann introduces the concept of "*Gestalt*" to designate this perceptible appearance as a whole, in contrast to "*Form*" that designates the structure of the organism, the objective arrangement of its outer and inner, macroscopic and microscopic, parts. In German the word "*Gestalt*" connotes an overall appearance. Like the English word "figure," it can designate the recognizable physical appearance of a person for instance or a character in a novel. In line with the Gestalt theorists, Portmann aims at a global plan that rules the morphogenesis of animals and consistently produce the same particular patterns, among all the patterns made possible by physicochemical interactions. Portmann takes one step further by emphasizing that such recurring patterns always involve an orientation towards the formation of a perceptible, recognizable, and expressive outward appearance.

Portmann lets phenomenality enter objective zoology. In this regard the concept of "authentic phenomena [*eigentliche Erscheinung*]"[21] (that he also calls "organs to be seen [*Organe zum Geschaut werden*]"[22]) is pivotal in his work. This concept enables Portmann to stress that a rigorous discrimination is possible between what simply *appears to us as* a figure, as an artistic and meaningful appearance, and what *actually is* such a figure. By following objective criteria, Portmann argues, we can identify a global, consistent, and sophisticated "strategy" of appearing that is at work in the animal realm. Portmann thus delineates, in animals' appearance, manifestations that cannot be only in the eye of human observers. Such *authentic phenomena,* having their source in the ontogeny of animals, must be recognized as an integral part of the being of animals themselves.[23]

How can we detect the presence of such "authentic appearances"? The first criterion is that one and the same form constitutes a consistent appearing whole, while being the result of a cooperation between diverse chemical processes, organic processes, and, possibly, behaviors. Portmann gives the example of the famous Oudemans' principle. At the natural resting position, many butterflies' hindwings are almost entirely concealed by the forewings; only the tip is still visible. Hindwings are covered with vividly colored patterns, whereas forewings display a cryptic pattern. But the little tip of the hindwings that is still visible at the resting position also displays the same cryptic pattern that can be seen on the forewings, in such a way that it exactly complements the motif appearing on the forewings and composes with it a seamless design. These parts make a whole, which, Portmann insists, is all the more surprising that forewings and hindwings stem from two separated ontogenetic processes.[24] Diverse vital operations here contribute to one and the same oriented process, thus giving birth to a visible figure.

The second essential characteristic of *eigentliche Erscheinungen* is that they are always situated on the outer surface of living bodies. They cannot be found on non-visible parts of the body, for instance, on the reverse side of feathers. This appearance also displays clearly-structured, eye-catching patterns: Contour effects, strong contrasts, non-natural colors (yellow, red) or, at the other extreme, incredibly mimetic/cryptic patterns. Portmann points out that, in animals whose external membrane is opaque, there is a striking contrast between the chaotic appearance of the hidden organs and the much more "readable" visible appearance. Even an expert will have difficulty recognizing the species of an animal when contemplating its entrails, whereas a child easily recognizes a giraffe or a lion on the basis of their outward appearance.[25] Strikingly, in transparent animals, organs are arranged in a symmetrical and clearly structured fashion.[26]

Among an immense number of appearances made possible by the physicochemical processes in living beings, only a few specific forms are stubbornly reproduced by each organism. Portmann therefore contends that a Galilean science cannot adequately study these phenomena.[27] Indeed, in an analytical approach, the appearance does not possess any meaning in itself and is nothing but the secondary product and, at the most, the indication of underlying physicochemical and metabolic processes. The cock's comb is thus understood as the "manometer in the machinery of hormones."[28] Acknowledging authentic appearances as such entails the definition of a certain autonomous oriented activity; moreover, these appearances, insofar as they are destined to be perceived, only fulfill their function through their reception by a perceiving subject. Authentic phenomena do not possess an immediate mechanical efficiency: They are to be apprehended. Hence, the fake eyes displayed on the wings of butterflies may certainly possess an immediate frightening power, but, as the next part of this chapter will show, they can also be foiled by a predator and lose their effect.

How expressive is this destined-to-be-seen outer surface? It is in fact not always easy to tell. The meaning of the stripes of the zebra is still considered a riddle nowadays. A great number of hypotheses have been investigated:[29] Are these stripes in fact a visual cue that confuse predators, deflect flies, and/or enable intraspecific individual recognition? Or are they above all a temperature regulation system?[30] To complicate matters, nothing excludes that the stripes may be multifunctional and serve simultaneously expression, communication with conspecifics, and mechanical purposes.

Animal-visible Gestalten display many recognizable patterns and play with striking colors, contrasts, and effects of symmetry. But another difficulty is to determine whether other animals apprehend these Gestalten as such. Portmann suggests that there is a connection, although not absolutely systematic, between the development of sense organs and observation in various species, on the one hand, and, on the other hand, the increasing complexity and expressiveness of animal appearances within such species. These appearances are apprehended as means of recognizing individuals, but also as means of gaining information about their mood.

Animal recognition of animal appearances and portraits galleries: Becoming a face

Portmann points out that a correlation seems to exist between the development of optical sense in vertebrates and mammals and the proliferation of conspicuous and possibly impeding traits, such as bulky antlers, rams, or scrotal sacs.[31] Furthermore, the visual appearance of animals plays an important role in mate recognition, as well as in predator, conspecifics, and individual recognition. Quite often, for both vertebrates and invertebrates, chemical cues are not the only basis for recognition. Many non-human animals proved extremely talented in recognizing human or non-human faces.[32] Sheep or horses for instance can also recognize human faces from two-dimensional images.[33]

Liostenogaster flavolineata female wasps can discriminate between aliens and nest-mates using facial patterns or chemical cues in isolation. They let many outsiders in the colony when they rely on chemical cues only and make fewer mistakes when they memorize the unique individual facial patterns of their nest-mates.[34] It is thus possible to compose a portrait gallery of wasps,[35] whereby the remarkable diversity of individual features becomes patent. And these experiments show that wasps *patiently* observe such apparent differences, through a remarkably time-consuming and fallible process that strangely prioritizes the mediation of perception and interpretation over more efficient instinctive recognition mechanisms.

To be sure, visual recognition can be achieved according to an innate search image. Thus, a cuckoo can recognize conspecifics without having ever met them,[36] and male sticklebacks attack rival conspecifics, but also all sorts of crude dummies as long as their underside is brilliant red.[37]

However, in many cases, the recognition process can incorporate a significant dimension of learning and combines attention for visual as well as auditory, olfactory, and chemical cues. For instance, jumping spiders

Lyssomanes viridis males "examine features on both the face and the legs in both the presence and absence of pheromones, and female pheromones tip the balance in favor of a female identification when a male is unsure how to categorize an incongruous set of visual features."[38]

It should be noted that Portmann's emphasis on visibility must not be interpreted as exclusive, or else it would become naïvely anthropocentric. In fact, Portmann extends his theory to other perceptible appearances such as bird songs. "Through hundreds and thousands of structures and movements, scents and sounds, creatures speak to creatures, to members of the species, to enemies, sometimes event to 'friends' from other species. All that speaks and is seen, heard, smelt or otherwise 'comprehended', can create significant relationships between one life center and another."[39] Actually, nothing stops us from imagining chemical or olfactory Gestalten. To be sure, the reference to visual appearances is especially evocative for us of holistic apprehension, observation, and possible misrecognition. Vision is indeed the sense of distance and overview. It enables a subject to apprehend a remote whole that may be misleading, further interpreted, and explored. Nevertheless, expression through scents, sounds, or pheromones also implies a mediation and harbors the possibility of mistakes, misrecognition, hesitation, and the production of deceptive replicas in defensive or aggressive mimicry, for instance.[40]

Moods: From organs-to-be seen to signs

The arising of expressive interfaces also offers scaffolding for the further emergence of expressions of mood, emotions, and intentions. The delicate transparency of the skin, the lability of its colorations, the sophisticated network of nerves, ligaments, and muscles of the face provide a versatile medium for the expression of changing moods. Stress, fear, curiosity, aggressiveness, submission, pain, playfulness, or schadenfreude show through them as much as through the body attitudes and the style of behavior.

A study on Tilapia *mossambica* fish demonstrates that the color patterns on the surface of their skin do not only change during the life of the fish but also with the individuals' mood. Specific color patterns correspond to aggression, arousal, and fear, for instance.[41] Mice and many other mammals grimace when in pain, a hint that has been meticulously studied and led to the establishment of "grimace scales" which human experimenters can draw on in further developing their own sensitivity and interpretative skills.[42] There are also many examples of animal appearances that manifest transitory aggressive states and accompany threatening behaviors. Such is the case, for instance, of the bristling of the hair on the back of cats and dogs,[43] or the ruffling or smothering of the raven's head feather, or also the spreading of gill covers whereby two dark spots are unveiled that look like an extra pair of eyes in jewel fish. Correlative understanding behaviors in non-human animals have been observed and studied: Many animals indeed adapt their behavior when observing others' facial expressions or even pictures of others' facial expressions.[44]

Some more firmly circumscribed and intentionally displayed cues—thus more worthy of the name "signs"—can stem from this rich repertoire. As famously pointed out by Darwin in *The Expression of Emotions in Man and Animal*, showing fangs, for instance, can be used, in place of an actual bite, as a warning. Likewise, the "play face"—in which "the mouth is held partially open with the unretracted lips only slightly exposing the unclenched teeth"—serves to make clear that aggressive behaviors that will follow are not for real.[45] The highly disarming "puppy look" is intentionally displayed by dogs when humans look at them,[46] and it may be used in a more or less histrionic fashion by playful dogs, for instance. This sign has been taken up by humans as well, and lives a life of its own through cartoons, movies, and everyday scenes of social relations. Promoted as a meme, in a more recent mutation, it has gained the status of an institutional symbol. So when and where exactly does the symbolic realm arise? I will return to this question in the next part of this chapter.

Self-depiction

As we will see, it is difficult to clearly circumscribe the function of this emerging expressive interface described by Portmann: It may serve to impress, seduce, startle, lure, or frighten, for instance. What is certain, however, is that this outer surface functions as the medium for a manifold surge of expressiveness.

Portmann primarily insists on defining this expressive surface as a fundamental *Selbstdarstellung* [self-representation] consubstantial to animal life. *Darstellen* in German means to display, to represent, to depict. A *Darstellung* may—but does not necessarily—take the form of the production of a picture or a medium existing independently of the presented subject. More fundamentally, it always involves an expressive process. Dar-stellen means to lay something *there*, to bring something to the fore, to make it appear. The term *"Darstellung"* can also be used to designate the performance of a role, an impersonation, or the relation between a painting and its subject (*"dieses Bild stellt Athena dar"*). According to Portmann, acknowledging the existence of to-be-seen organs amounts to asserting that animals display an essential tendency towards self-depiction. Through the appearance of each singular animal, Portmann claims, an at least twofold identity is represented. On the one hand, the identity of the species shows through in a systematic manner: Its manifestation is reproduced over and over again through the ontogeny of each individual. By means of *Selbstdarstellung,* the species would thus present itself in the same way as a human clan identifies itself with a blazon.[47] On the other hand, a more or less individualized, but possibly highly individualized, *Selbstdarstellung* occurs throughout the animal kingdom.

In fact, more generally, *Selbstdarstellung* is *per se* the arising of the self. Portmann goes as far as to speak of the expression of the "interiority" of animals. Such an interiority never coincides with this or that organ or group of organs,[48] but is rather an active principle of auto-delimitation and organization. A self arises through autopoietic, oriented, and selective processes, as already pointed out by Uexküll. Similarly, Portmann aims at a *subjective interiority* of non-human animals. This "metaphor" makes all the more sense in that individuation goes together with a selective "hiding-letting appear" process: An organic interiority is combined with the ability to electively *keep to oneself* or *display on the outside*.

What is at stake with *Selbstdarstellung* is thus the formation in animals of a material expressive envelope. Not only do animals tend to define themselves through the emergence of a membrane that, as beautifully put by Hoffmeyer, "keeps the world away in a physical sense but present in a psychological sense,"[49] but, what is more, they turn it into a way of selecting what they will keep hidden, at least by default, and what they let appear, what they let the others perceive. Chemical or olfactory features, for instance, can likewise contribute to form a multisensory expressive interface whereby the self and the other are dynamically delimited. An individual and the interface for its communication with other individuals emerge simultaneously and correlatively. There is consequently a continuity between this first fundamental expression of the self and traits or behaviors through which animals more actively camouflage themselves and lure others.

Moreover, interiority as a creative principle is even more manifest when the oriented autopoiesis squanders vital energy to produce a riot of blazing and sophisticated patterns. Animal psychology and its consubstantial theatrical nature arise here, even "below" the behavioral level.

With this concept of self-depiction, Portmann daringly states that turning oneself into an image is "a basic property of life."[50] In this way, he also bypasses the tricky question of the exact and narrowly utilitarian function that should be ascribed to appearing. *In any case*, animals consist in self-depiction. They must appear. An animal presents herself to the face of the world, and an infinite number of receivers will deal with this nascent meaning: Interaction begins, theater begins.

Animals are natural born fantasies, in a sense that combines, on the one hand, an irrepressible surge of expressiveness that crops up wherever animal life develops, and, on the other hand, the inchoative agency that gives itself the realm of meaning to play with. But I am hurrying through this: We have just had a glimpse of this dimension of interpretation, and we have to study it in more depth in the next paragraph. Animals turn themselves into images … or may it be "into symbols," or "signs"? It is precisely the intertwinement between these possibilities, the meta-stability of each of them, that brings meaning to the dream comparison.

§29 Plastic meaning, mimicry, symbols, signs, and play

The absence of fixed function

According to Portmann, any functional study of animal appearances will fall short. Reducing expression in animals to one function is as improper as trying to determine why animals—or humans—play: This is an onto-logical misplacement. Indeed, applying, a rigid pattern of characterization to that which arises from self-distance, irony, and ubiquity simply does not make sense. Proliferating creativity cannot be confined within a clear-cut definition.

Admittedly, one can define a univocal relationship between some particular to-be-seen organs and certain very specific addressees. And, indeed, Portmann actually studies animal appearances in the framework of the Uexküllian theory of the functional circle [*Funktionskreis*]:[51] The appearance is designed *for* the eye of *this* animal (or species), whether a predator, a prey, or a mate. The meaning of such an appearance can then be connected to a determinate apprehension. Eyespots and owl faces drawn on the wings of butterflies, for instance, when displayed suddenly, very efficiently set off a momentary state of stupor in predators.[52]

Nonetheless, there are motives to question the model of the functional circle. Portmann consistently rejects a Darwinist approach: Interpretations that aim at the specific function of animal expression "belong to a wide-spread conception of the living world which contends that conservation functions, through the mutation game of genetic inheritance, determine the survival and the evolution of species."[53] Portmann contrasts "fitness for survival" with "becoming a spectacle." In effect, "We cannot adequately understand the structure of such objects as feathers and fur in all their finest details unless we assume that this outward appearance has been designed for something more than those functions which we know are necessary to preserve life (...) We must assume that they have also been designed in a very special way to meet the eye of the beholder."[54] But Portmann strikingly implies that "designed to meet the way of the beholder" is not (or is more than) a matter of preservation. Isn't communication absolutely useful for survival? Portmann highlights several aspects of animal appearance that challenge a reductive approach.

First of all, the appearance of an animal often possesses a partial and debatable usefulness. Bulky adornments (the peacock's tail, the deer's antlers ...) may be useful to be chosen at the end of a courtship perfor-mance, but they are considerable handicaps in daily survival.[55] Natural weapons lose their efficacy *as such* to become spectacular and, thus, more effective *as signs*. In other words, sexual selection enters into tension with fitness to fight. Likewise, why do *Coatonachthodes ovambolandicus* beetles, integrated as guests into termites' nests, develop on their back a large pro-tuberance that imitates the form of the termite?[56] They inhabit an inner space of termite nest where normal termite workers can never see them. What is more, the appearance of termites is not a very interesting protection: Termites are very attractive food for several predators. "We must go beyond the functional conception, which judges only according to purpose and performance, and arrive at a concept of the animal which, while never ignoring the functional point of view, yet for that reason also realizes how much wider and greater is the full significance [*Bedeutung*] of the animal form."[57]

It is always possible to claim that the animals who eventually survived, through the selection process were, overall, the fittest and that they reached the best balance of costs and benefits. But this would be a way of reintroducing essentialism in Darwin's approach: "The fittest" at best means "that which actually survived at the time, in those circumstances." As Marjorie Grene argues, "usefulness" is essentially relative and cannot constitute a rigorous concept.[58] There is no Great Calculator of *the* best cost-benefit balance. Rather, a prolific production of incredibly diverse appearances is at work independently of a benefit-cost calculation. A general function or set of functions can in no way account for the exuberance and the complexity of colors and forms among living beings:[59] They fail to explain its extravagant spending of energy. A warning or protective function would be much more efficiently achieved if all the dangerous species were signaled by white polka dots on a red background for example.[60]

Further, Portmann stresses that authentic phenomena can also be found, for instance, in animals who do not possess eyes or live in the dark and cannot be seen by any mate, prey, or predator.[61] Such appearances are, in his words, "unaddressed phenomena [*unadressierte Erscheinungen*]," "sent 'into the blue' ['*ins Blaue*' *gesendet*]." Actual addressees are nowhere to be found; possible receivers are innumerable and the meaning of the "signal" becomes virtually multiple.[62] In this case, Portmann argues, the selection of colorful and symmetrical to-be-seen patterns cannot be ascribed to the eyes of *actual* receivers. Hence, Portmann rebuts even a non-essentialist version of Darwinism according to which these Gestalten would be produced through a totally random set of mutations. There is an original and consistent appetite for self-manifestation in living beings. Self-depiction is a self-sustained, inflationist, and risky creative process. But it is a phantom rather than a substance: Although not random, this appetite cannot be assigned to a specific function.

Portmann thus compares self-depiction to human fantasy: "How often does it seem to us as if roving fancy had been at work; sportiveness, the capricious free play of creative force, comes to mind rather than a technical necessity."[63] Self-presentation is autotelic, like play. When presented with unaddressed appearances, the observer "can begin sensing that this gratuitous self-structuration [*Selbstgestaltung*], this self-depiction [*Selbstdarstellung*] of plasmatic being may very well, ultimately, be the first and supreme sense [*Zinn*] of the living appearance."[64] According to Portmann's most daring hypothesis, organisms appear for the sake of appearing.

Nonetheless, when it comes to the meaning of self-depiction, Portmann's work becomes quite oneiric: There is a tension and a shift between this pure gratuity (*Selbstdarstellung* regarded as a form of fine art, depiction for the love of plastic creation) and communication (namely, self-depiction as the presentation of the species through its emblems and as the expression of the interiority and the mood of the subject). Which function prevails? Portmann's texts are not entirely clear in this respect,[65] but, in the end, what is at stake are precisely and correlatively (a) a form of communication that is a kind of art and (b) art as rooted in animal communication.

Playing with appearances: Interpretation and the war of wits

An additional fundamental reason why expression in the animal realm cannot be assigned to perfectly determinate functions is that images are not only sent into the blue, they are also sometimes used to deceive. This implies that animal appearances are taken up, interpreted, and reinterpreted by a *de jure* infinite number of conspecifics and heterospecifics. Following Plato's nomenclature,[66] mimetic costumes are realities of the third degree, copies of mere appearances. In other words, they are quasi-beings; their efficiency is not simply mechanical but based on apprehension, mistake, and illusion.

This aspect of animal communication is meticulously studied by several biosemioticians, in particular in the Copenhagen-Tartu school (Claus Emmeche, Jesper Hoffmeyer, and Kalevi Kull) and the Prague school (Karel Kleisner, Anton Markoš). As Kleisner clearly explains,[67] the distinction between function and meaning is crucial, not only because presupposing that the appearance is destined for one predetermined rigid function is arbitrary and ideological,[68] but also because one cannot reckon with the dimension of apprehension involved by appearances without considering how images are received, understood, and possibly interpreted by other animals.

And first of all, emitters definitely produce images that give receivers food for interpretation. Through the complex phenomenon of mimicry in particular, the realm of animal appearances becomes increasingly bewildering and multilayered. As Timo Maran highlights, it is not accidentally that mimicry has always resisted classification: "Mimicry is not a solid class of entities with a common origin that have developed from the same initial conditions, or follow the same biological laws."[69] Rather, mimicry "emerges in very different conditions" and at various organic levels. It takes on different forms. Mimicry can indeed be purely morphological or reinforced through a specific behavior; it may also be in the form of imitation or abstract signs. Further, Maran importantly points out that such phenomena appear when animals venture into risky untrodden ways. "Think for a moment about the acts of laying eggs in the nest of another species

or trying to copulate with a plant instead of a member of one's own species. Such cases appear to expel rationality of the biological or evolutionary
processes and are rather based on occasional mistakes or on the creative
interpretation on behalf of the animal subjects."[70] "Rationality" may be
too anthropocentric a word, but the metaphor is meaningful. It certainly
can be claimed that mimicry would never occur without the emergence of a
relatively reliable set of recognizable specific appearances, a "rationality"
that, assuredly, mimicry exploits and undermines. Mimics boldly *bet on*
signs and deceit, instead of actual flight or attack. They disseminate specific traits and appearances, blur the boundaries between species, aim at
the perspective of others, and make appearances dubious.

However, doesn't Müllerian mimicry consist in the emergence of reliable signs? In this case, indeed, different unpalatable species develop a
similar warning pattern, which, according to Müller, helps the predator
learn more quickly and efficiently to avoid pursuing these animals.[71] Each
bad experience strengthens the deterrent force of the pattern. Yet, many
cases of Müllerian mimicry are ambiguous, for the degree of toxicity of
the co-mimics is variable and covers a broad spectrum. Several of the
individuals or species involved actually parasite the process.[72] Moreover,
poly-morphism also exists between unpalatable Müllerian co-mimics,
which makes the pattern less clear and less easy to learn.[73]

Another objection to the claim that mimicry thematizes and plays with
appearances may be that the purpose of mimicry could be defined in many
cases as "becoming invisible."[74] But the recognition process is commonly
based on a complex set of cues,[75] when mimicry only displays some of them.
Even in cryptic mimicry, imitation is imperfect. The eyes on the wings
of butterflies are certainly hypnotic, but they cannot entirely conceal the
characteristic form of the butterfly. As a whole, the wing is also quite conspicuous: Predators are presented with a twofold troubling appearance,
and, as we will see, they do not all react in the same way.[76] Likewise, the
Coatonachthodes ovambolandicus beetle certainly looks like a termite from
above, but, from the side, it appears as a weird hybrid, a costumed beetle.
Additionally, in the recognition process, behaviors, attitudes, and movements often play a key role. Thus, unsurprisingly, morphological mimicry may be reinforced by mimic behaviors. The fake twig must stand still
and spiders who resemble ants tend to walk in a zig-zag ant-like pattern[77]
and to wave their first pair of legs in the air, thus mimicking the antennal
movement of ants.[78] Morphological mimicry, like every copy, is perfectible
and can benefit from a behavioral support, while the latter concomitantly
introduces a reinforced dimension of ambiguity. Indeed, a behavior can
be launched with a bad timing. It is performed with a certain malleability,
in response to different circumstances. Moreover, individuals can be more
or less skillful at performing a behavior or can even learn to become more or
less skillful. For instance, cuckoos can learn to mimic the begging call of

host young, but the accuracy and the speed of the learning process varies from individual to individual.[79]

Still on the emitter side, mimicry gets even more ambiguous when it becomes "abstract" or "satiric."

Timo Maran has coined the name "abstract mimicry" to designate mimicries that display one stereotyped trait (eyespot, yellow and black stripes, fangs, opened jaws, body-lifting/bloating, a shadow from above ...), which is isolated from a recognizable specific appearance as a whole and cannot be ascribed to one unique model-species in particular. Moreover, Maran contends that, in such cases, the image produced by the emitter does not essentially refer to a model in the objective world, or even in its own *Umwelt*, but rather to a certain schema or search image—understood as search-tone[80]—that comes so to say from the fantasy of the receiver, from the *Umwelt* of the receiver. Schematization is in fact a standard psychological process, and, as the previous chapter indicated, non-human animals also commonly perceive Gestalten, namely overall structures, in their world. Uexküll described the key role played by innate or acquired search images in the behavior of animals: A general perceptible pattern is or becomes sufficient to release a response, so that the animal may choose a human as a *Kumpan*,[81] a bicolor crude dummy as a mate, or a matchstick as a prey.[82] Thus, through abstract mimicry, the mimic *gets into the other's head*. In effect, the mimic displays the image corresponding to a schema and a search tone that characterize the subjective intentionality of other animals, *their* perception of their *Umwelt*. In addition, the images—or rather, perhaps, symbols—used in abstract mimicry also seem to be particularly connected to strong emotions: Their schematic character buttresses and/or is buttressed by the inarticulate and non-observational nature of fear, for instance: Eyes glowing in the dark or a shadow from above are also for us schematic, unobserved, and fearsome. Such are the secret roots of monsters.

Abstract mimicry thus gives flesh to what was only the virtual correlate of a subjective intentionality. Accordingly, it adds a second-degree layer to the realm of animal expressive appearances. Instead of being the *Selbstdarstellung* of a species, abstract mimicry represents one aspect of the mental universe of animals who *perceive and interpret* these appearances. In this way it can more effectively capture their attention, shock them, and move them. Through this adaptive strategy, by introducing the mediation of the other's mental universe, abstract mimicry further distances the sign from what it represents and takes an extra step towards heightened symbolism: Less resemblance, more schematization, more abstraction. This strategy also amounts to relinquishing a slavish imitation of a concrete model. Abstract mimicry fosters the development of monsters and chimeras. Think of the escalation in the number of eyespots on peacocks' tails. Eyespots or snake-like motions that come out of the blue and become incorporated into an organism that has kept its general shape are also highly chimerical. Moreover, abstract mimicry makes the perception of the *Umwelt* through general schemas dangerous and unreliable.

Satyric mimicry is a grotesque assemblage of motley features. Like abstract mimicry, it also targets the understanding of images in others. Satyric mimicry brings together features that usually belong to different images and forges a chimera or, as put by Howse and Allen,[83] a *surrealistic piece*. Satyric mimicry may also exhibit a specific trait in an unfamiliar context, instead of taking up already existing expressive forms. It creates an intentionally "monstrous" pattern that blurs and disrupts the normal interpretation of appearances in receivers.[84] Satyric mimicry is thus akin to what has been called "protean behaviors," namely the intentional use of unsystematic, unpredictable patterns—for instance, sudden changes of colors, zigzagging or creating a random run as evasion strategies—in order to thwart the expectations of the most cunning predators.[85] Here again, mimicry proves to be far advanced in what could be called a war of wits.

On the side of receivers, a correlative process of adaptation and learning takes place. Certain appearances do trigger strong immediate reactions, but, in many cases, hesitant behaviors have been observed, for instance, in lizards,[86] fish,[87] and birds.[88] The responses to a specific mimic are not uniform.[89] In addition, over time, individuals do learn to discriminate between mimics and models.[90] As mockingly pointed out by Caillois, the stomachs of predators are full of mimetic animals.[91] At the evolutionary level, inbuilt biases against repeated patterns of deceit have developed in the tricked species.[92] Of course, such biases, like all search images, can be used to the detriment of these animals by new mimics. But it also seems accurate to assume that more observant and discerning animals will be promoted through evolution, while more sophisticated and more puzzling mimics will prosper.

To be sure, "mimicry is a learning device"[93] for emitters and receivers, but the war of wits is *de jure* endless. Hence, the distance between expressive appearance and what is expressed through it increases. Elements of stability—quasi-symbols—develop. Stubborn search images and abstract symbolic traits—that Kleisner and Markoš call "semes"[94]—emerge. Yet, simultaneously, new tricks can always appear. Machiavellian strategies arise and appearances are more and more intricate and treacherous.

Further, as Maran stresses, the approach in terms of clear-cut emitters and receivers is schematic and limited. Many animals are sensitive to the eyespot pattern and, through these various receivers, a mimic does interact, regularly or occasionally, with several species.[95] Similarly, the termite-costume may be a way of coaxing termites and "paying" them with signs to obtain their protection,[96] but it simultaneously makes the beetle more vulnerable to a huge number of predators.

Kalevi Kull defines "Ecological codes" as "the sets of (sign) relations (regular irreducible correspondences) characteristic of an entire ecosystem, including the interspecific relations in particular."[97] But, when using this concept, one must take the term "code" with a grain of salt and bear in mind that such "ecological codes" cannot be linked to universal clear-cut meanings: They are interspecific, and, as such, evolve through different *Umwelten*, in connection with different sense organs, search images, interests, emotions, and learning processes. As Maran points out: "It is plausible to assume that codes on the ecological level are not strict regulations, but rather ambiguous and fuzzy linkages based on analogies and correspondences."[98] For instance, what do eyes mean?

Excursus: The rise of symbolic eyes in the imaginary of non-human and human animals

Eyespots somehow turned the appearance of the eye into a conventional—ritualized—symbol. And yet, these patterns are ubiquitous at heart. They can be puzzling, dreadful, attracting (used as adornments by peacocks for instance) and, moreover, they are somehow linked to actual eyes, namely eyes such as they appear at the already meaningful level of *Selbstdarstellung*.

This "eyespot" symbol has in fact been taken up by many species, including humans, which demonstrates that an emergent interspecific meaning and some common highly telling features must connect many individuals beyond the limits of more clearly defined *Umwelten*. But such interspecific images cannot but be loose, elusive, and plastic means of communication.

Eyes "are" signs of the presence of an observing subject, a subject that is withdrawn in its *Umwelt* and engaged into an endless play with appearances, delusion, and unmasking processes. Eyes materialize distance and mediation. These eyes *over there* point to a scanning operation that will have consequences *here* for the observed animal: The observer may recognize the vulnerability of the observed and prepare an attack, but it may also opt for flight or perhaps identify a possible ally. A decision is maturing in a mysterious *there*, which may be salutary or fatal *here*. Eyes "are" thus ambiguous *evil or lucky* eyes, like in Greek or Turkish amulets, surrounded by an emotional overwhelming aura.

Were eye patterns selected by the process of formation of animal appearances because they are perfect symbols of a scrutinizing unpredictable subjectivity? Or is it the reverse? Did eye patterns become such perfect symbols because this process of formation of animal appearances actually isolated eyes, schematized them, and multiplied this pattern in the appearance of a huge number of species? In fact, this should not be regarded as a dilemma. A symbol always involves a dimension of arbitrariness and metonymy as well as a contingent accidental "choice," but it also builds on connections and analogies that were operative beforehand. There is no *reason* that justifies the emergence of the eye symbol, but it would never have flourished without an intrinsic suggestive power.

We do know such symbols viscerally and feel their mysterious potency. Human animals take them up abundantly in visual arts. Close-ups on eyes and particularly on non-human animal eyes constitute a classic motif in films.[99] As pointed out by Jonathan Burt in *Animals in Film*, film directors particularly like to isolate the human or non-human eye. Separated from the rest of the body, the eye becomes less a situated finite body—although still an objective body [*Körper*]—and more the paradoxical opaque entrance to a parallel world. The sudden appearance of an eye on the screen hinders the scopic drive of the spectator and her tendency to overlook her own finitude. It decenters her. The close-up on the eye often suggests that we are going to see the world through a different perspective—it is therefore also the symbol of the very essence of cinema—possibly the perspective of a non-human animal. But, in addition, the close-up on an eye displays a tantalizing distance, for the image of the opaque orb remains so to say in the way, as an uncanny, obscene, enigmatic, and vulnerable body. Eyes in films are always akin at once to the all-seeing eyes of Lumet's *Equus* and to the woman/ox's eye cut by a razor in Buñuel's *Un Chien Andalou*. It is precisely the contrast between the resistance of this opacity and the concomitant powerful evocation of a marvelous new world of appearances and emotions that forces us into an oneiric mode: We quasi-enter a sacred, awe-inspiring, and unfathomable otherworld.

In *The Night of the Hunter*, the symbolism of animal eyes particularly draws on the complex layer of meaning that emerges in the war of appearances between animals. Charles Laughton's movie is without any doubt a tale of predator and prey. It tells the story of aggressive mimicry and shows the difficulty to unmask the wolf in sheep's clothing. Harry Powell, the fake preacher, is always on the hunt. He seems to never sleep, appears out of nowhere, fades into thin air, and lets out a horrific beast-like scream when he fails to capture the children, which singularly contrasts with his usual honey-sweet voice. The comparison is almost explicit in the film: Like the owl that patiently observes a rabbit before sweeping down and capturing its prey—"It's a hard world for little things"—Powell relentlessly stalks John and Pearl. He lays siege to Rachel's house, filling the air with a Christian song turned into a hypnotic, melodic ambush, waiting for Rachel's vigilance to slacken. In one of the most famous scenes of the film, John and Pearl, tracked by Powell, flee down the river, at night. In an unsettling expressionist style, the scene shows their small craft in the background, while a spider web appears in the foreground and magnified animals on the bank seem to watch over them: A toad, an owl, a turtle—"They make soup out of them. But I wouldn't know how to get them open," John comments—rabbits and sheep. All the familial and social protections in the life of these children collapse. John and Pearl drift, unmoored, thrown into a nightmarish world, where it has become impossible to rely upon any existing codes, norms, and conventions. Even words are weapons and disguise. In these circumstances, non-human animal perception, fully geared towards such a war of appearances, is spurred on and must develop to the highest degree of acuity possible. At dawn, a dog howls and heralds the approaching threat: She could sense Powell's presence before John. In this oneiric world, where every shadow has become frightening and where nothing any longer possesses a fixed meaning—which is also suggested by the poetry of night and water as essentially fluid images[100]—it becomes more crucial than ever to *watch out*, to check, through all the senses, who is watching and guess what moves are being plotted other there, behind a gaze. A tensed exchange of glances takes place throughout the film between Powell and John, each one trying to read the other's mind and to conceal his own intentions. Hence, the eyes of the animals showcased in the river scene foreshadow our quasi-entering into the world of animal communication. The latter remains mysterious, uncanny, and ambiguous, for each subjectivity fundamentally both hides and uncovers itself.

Signals, images, symbols

Before studying the dream comparison in further depth, let me draw up a glossary of the terms that are the most suitable to describe the emerging signs in animal communication.

In the absence of a conventional solid system of symbolic signs, there cannot be such a thing as an objectified reality, namely a reality radically independent of our perspectives and emotions. As a consequence, the strong objectifying act of apprehension that takes a material reality as that which stands for a clear-cut, independent reality or idea does not play a central role in non-human animals' relation to images, symbols, and signs. This act of apprehension has a more significant role in the human world, or at least, more specifically, in a rationalist approach to pictures. Rationalism has indeed consistently denied—rejected as a superstition and repressed—the relations of affinity and contamination between the picture or the sign and that which it represents. A rationalist take thus tends to conceive of the apprehension of images as a form of active and enlightened judgement. However, as I have argued in Chapter 2 (§15), the core of imaginative phenomena does not lie in the clear-cut distinction between the representation and that which it re-presents. And, in humans, the recognition of pictures *as such*, namely the distinct recognition of what pertains to the image and what belongs to the original, requires some training. In fact, even the effort to make such a distinction implies an attempt to impose an analytical and impoverishing pattern on an integral phenomenon. Moreover, even the recognition of images as such can develop in some animals. Several studies show that some animal individuals can—or can learn to—relate to signs or pictures in a re-presentational way. In other words, they do not mistake the two-dimensional picture with a mediocre exemplar of the real object, and they apprehend the picture as referring to the real object.[101]

Non-human animals do not simply mechanically react to **signals**. Nor do they simply feel independently of any mediation. The contrast—although not the opposition or the *caesura*—between interiority and exteriority, what appears and what is kept hidden, is consubstantial to animal life. At stake in animal images and symbols is consequently the description of a **non-re-presentational expressiveness**. The expressed dimension is not turned into a separated object, but it is not simply one with the appearance. For that reason, I find Peirce's concept of **icon**, as pertaining to "firstness"— namely the "in itself" "perfectly simple and without parts"[102]—to be misleading. Secondness and thirdness are not any more helpful here. Ubiquity and its unhinged and dynamic nature do not fall within firstness, nor do they belong to secondness or thirdness. However, animal appearance is undoubtedly related to the form of appearing that Peirce describes as typically iconic. An *image* looms through an oneiric experience that does not posit any solid ob-ject, any solid reality, or any well-circumscribed relation between a representative and a represented. It oscillates between the shock of a raw quality and a form of nascent recognition.[103]

Such a non-representational expressiveness is exactly that which we experience in fantasies and image-representations, but also to a certain extent in symbol-representations: When a too clear-cut distinction between the copy and the model arises, the appearing of the object *through* a set of states, emotions, and sensations becomes impossible. A certain appearance suggests—but does not fully embody—the presence of a virtual being such as the king here portrayed, a chimera, or the life of Emma Bovary. I have described this experience as *ubiquity* (§§3 and 15): It is both a state and a nascent activity of play with the thus-opened possibilities. We are not fully aware that we are currently imagining, but our actual unhinged and uncanny state encourages a more or less conscious and more or less active imaginative activity. Likewise, in the realm of animal appearances, the war of wits emerges continuously from the to-be-seen organs that are first images sent into the blue.

On the basis of a non-representational paradigm[104] of imagination, images, and symbols, it is legitimate to assert that animals produce images, apprehend them, and create symbols as well.

It should be noted here that the distinction between **image** and **symbol** is relative, but not entirely meaningless. No image boils down to the mere reproduction of a concrete individual: Images essentially open the realm of the virtual and adumbrate the innumerable possible metamorphoses of the represented reality. They already possess a dimension of generality. If one defines a symbol as that which evokes innumerable associations, emotions, and even abstract themes, then the terms "image"—as I define it—and "symbol" become synonyms. However, the concept of symbol usually aims at representational devices that especially emphasize the distance and the discrepancy between the given and what it stands for. An image can be roughly regarded as the image *of* a particular person, whereas a symbol—e.g., the scale of justice, the Maltese cross, the evil-eye amulet—only vaguely relates to the perceptible concrete appearance of a specific object. "Symbol" may thus also designate an arbitrary sign. A symbol can subsist in an autonomous manner, as a token that stands for a certain meaning. It then takes the form of a crystallized, circumscribed being and a reproductible body that will be transmitted from individuals to individuals in order to designate a fixed and clear-cut meaning.

I thus contend that, via the production of "semes" or the process of ritualization, animals create **symbols.** Through ritualization, some specific behaviors that could be serious are played out with exaggeration, ostentation, and, possibly, a certain attention paid to the others' reactions.[105] These behaviors become stereotyped and are taken up repeatedly by many individuals. They then function as easily recognizable signs. Narrowing the eyes or baring one's fangs are classic examples of such ritualized behaviors. Food-begging or food-offering as part of a courtship and leaning away as a sign of submission have also gone through a similar ritualization process. Ritualization operates as sign formation, with a dimension of convention, on the scale of evolution. A certain gesture for instance proclaims "this behavior is similar to a real attack, which, in fact, it is not:" An increased distance from oneself forms together with a threatening sign. Bateson argues that, with such histrionic behaviors, as well as with play behaviors, a metacommunicative level is reached and non-human animals "discover that their signals are signals,"[106] or, rather, that their images are symbols. In such cases, indeed, it makes sense to claim that animals produce symbols and not only images, whereas, again, the distinction must be regarded as relative, especially since the term "symbol" here does not essentially designate a completely arbitrary sign that belongs to a grammatical system. By using the term "symbol," we can stress the step further towards symbolism that semes and ritualization constitute, although the said step is made in the continuity of the process of image-creation, rather than by breaking away from it. Symbols crystallize in the animal realm as well. A **symbolic ubiquity** emerges that is not yet endorsed by an analytical mind.

In general, animals produce **images** that correspond to the level of **enacted ubiquity**. Self-depiction is a form of performance and theatricality, even though this image already becomes available as a striking appearance and a set of recognizable patterns that may be isolated and taken up as **symbols.** Animals develop an intentionality that actively and consistently draws a contrast between, on the one hand, appearances and, on the other hand, what appears only through them. The represented is not a concrete model or an idea, but a transposable theme and the endless series of possible similar appearances. For instance, in *Selbstdarstellung*, the species of the lion does not appear as an essence, but as proclaiming itself to the world through a concrete emblem.

Not describing these phenomena in terms of image and imagination would be reductive. It would lead us to overlook the original theatrical nature of animal life and dismiss the overflowing of play with seeming and symbols that inspires the animal realm. These phenomena force us to focus on a non-logocentric dimension of our own imagination: this is where the dream comparison comes into play.

§30 Animal communication and dream in Bateson's "What is an instinct?"

The dream paradigm

Bateson explicitly refers to Freud:[107] Dream is a means of communication, since it thinks through me and speaks to me. I paradoxically seek to communicate something to myself, which means that I am outside of myself, grafted upon a source of meaning that is beyond my mastery. I have not entirely vanished. I feel, struggle, and try to find solutions. I even may make some decisions, but ek-sistence and the self have become loose. I feel that I am just a part of a tumultuous meaning-unfolding process, mainly swayed around by it. This estrangement is the fundamental form of dreams, so that it does not make sense to wonder if I dream because I lost myself or if I lost myself because I dream. No organizing center decides what must be prioritized. No authority determines what is important and has to be kept in the background, what deserves to be observed, what belongs to the superficial flux of appearances, and what is real. A dream is an atmosphere. As in images, fantasies, and the imaginary overall, the whole is given before the parts and the parts only arise from the whole. Things, characters, landscapes, and qualia are imbued with the overall mood. They morph into each other and are eventually swallowed up into the matrix. Dream communicates "in images and feelings."[108] It is thoroughly analogical. As a result, there are no "labels," no statements, no possible disambiguation in dreams. A dream may have a moral, but the latter "is not stated in the dream."[109] Somebody can firmly make a statement in a dream or voice a crystal-clear "yes" or "no," but the words are engulfed with a swarm of connotations, associations, and emotional undertones, so that they can never denote a clear-cut idea. No "not" can be found in dreams.[110] Words lose their ability to clear holes in being. Like an existentialist version of Midas, dream turns everything into self-otherness.

While a fantasy is a specific representation among other possibly heterogenous acts of apprehension—such as perceptions, memories, and judgments, for instance,—a dream is a universe. *Dream* is a general modality that contaminates every possible intention, object, sensation. I do not forge this dream atmosphere, I find myself caught up in it, even sometimes during my waking life, when everything happens *like in a dream* or a nightmare. The dream thus provides a paradigm for a general way of dealing with meaning.

Are non-human animals caught up in a dream? This phrasing is precisely what makes the comparison between animal mental life and human dreams possibly misleading. But let us see first why, in Bateson's view, non-human animals may indeed have such an oneiric fashion of dealing with meaning.

Animals' dream-like communication

In "What is an Instinct?," Bateson seeks to overcome a simplistic explanation of animal behavior. "Instinct" is a paradigm for lazy scientists. It is not true, Bateson highlights, that animals always blindly react to stimuli. But why would dream be a better paradigm?

Bateson endeavors to account for specific behaviors that express a floating meaning. The examples he brings to the fore in "What is an Instinct?" are akin to those placed at the center of the 1954 essay "A theory of Play and Fantasy." A puppy lies on his back and presents his belly to a bigger dog, both inviting the latter to attack him and preventing him from doing it.[111] Similarly, a lot of fights "end up in some sort of peace-making:"[112] Animals negotiate via ambiguous serious/playful attacks and submissions. In "A theory of Play and Fantasy," Bateson uses similar examples to contend that non-human animals can handle metacommunicative frames; in other words, they can realize that their signals are signals and aim at the virtual (the nip says "I *could* bite you…") while, on the other hand, they are able to navigate from seriousness to play and back, a leeway that cannot be maneuvered without active and constantly re-enacted interpretation. Play would not be play and a sign of submission would not be a sign without the subjects switching to the right mode of apprehension, to the right *frame*. This communication also requires the expression of regular reminders ("yes, this last stronger nip was still play").[113] Every party keeps checking and reassuring the others, when the situation can quickly get out of hand.

The paradigm of instinct falls short here. Bateson therefore suggests the need to abandon the "black box" hypothesis and to move to a paradigm that allows us to acknowledge a proportion of interpretation in animal behavior. Dream is then regarded as a relevant model, since this form of interpretation is never a clear *discrimination*.

According to Bateson, non-human animals cannot say "I am not biting you."[114] The "not" belongs to verbal language,[115] and non-human animals "do not have language."[116] To support Bateson's stance, one may add that non-human animals do not have an authoritarian ego who makes clear-cut decisions in order to assign a well-defined meaning to a sign. Such decisions as well as the figure of the sovereign ego have no solid reality outside of the superstructure of language. The word "not" hardens and martializes, so to say, what first arises as a mere reluctance. However, non-human animals may *somehow* say "I am not biting you," yet only via what Bateson describes as a "*reductio ad absurdum*"[117] or as "dealing in opposites."[118] In other words, animal individuals actually start a fight and "demonstrate" through their actual behaviors that *it is not* a fight. The demonstration is never complete: Misunderstanding may always occur and the interplay is a field of endless negotiations.

"Father: (...) does the big dog know that the little dog is saying the opposite of what he means? And does the little dog know that that is the way to stop the big dog?
Daughter: Yes.
F: I don't know. I sometimes think the little dog knows a little more about it than the big dog. Anyhow, the little dog does not give any signals to show that he knows. He obviously couldn't do that.
D: Then it's like the dreams."[119]

Bateson stresses that dreams and myths—"iconic communication in general"[120]—also "deal in opposites," to wit: They say "the opposite of what they mean in order to get across the proposition that they mean the opposite of what they say."[121] This description is, however, still logocentric and somewhat misleading. A dream does not simply say the opposite of what it means: How would we know? It says something—intrinsically ambiguous— in a weird, absurd, or ridiculous manner. In Freud's "Irma's injection" dream,[122] the injection given by Otto to Irma cannot reasonably be regarded as the cause of her infection; this dream thus suggests a question: Who am I secretly holding responsible for Irma's disease? Freud describes many bizarre and troubling associations in dreams (for instance two persons become one, "my friend R is my uncle"[123]) that make the meaning slippery. I do not realize that I meant "the opposite," but, rather, I *wonder* what this dream means and I oscillate between an indefinite number of possibilities. Likewise, non-human animals communicate within a hovering field of expression, deception, ritualization, and play, where no rest is allowed.

Bateson's comparison between animal communication and dream can be given its full scope by the theory of animal appearances such as developed in the previous paragraphs of this chapter. The dream actually begins at a more fundamental stage than suggested by Bateson. The relation of animals to meaning develops through the realm of expressive appearances. It forms as the result of a combination between functional aspects and a tendency towards expression that either takes advantage of existing morphological traits (the zebra's stripes may very well also function as a temperature regulator) or creates new ornaments. Images and symbols thus arise that bear the mark of a half-intentional, half-contingent process; they cannot but be polymorphous, evolving as they do through interindividual and interspecific interactions.

Even when, through training, with some unquestionable success, human and non-human animals cooperate to institute better-circumscribed conventional signs, the interactions these individuals engage in remain grafted upon the immense analogical, ambiguous field of communication so that one never fully knows in advance how signs will evolve.[124] Vicki Hearne describes for instance how her dogs may take up what was supposed to be the behavioral response to a "command" (the "sit" attitude, the retrieving of a ball) and turn it into a creative humorous new sign: "Salty (...) now performs recalls with passionate energy and precision. She is a pretty sight indeed. My friend glimpses this from the study window and comes out to watch her work. Salty the next time I call her, performs a swift and flawless recall, a straight sit and an unbidden but accurate finish – but to my friend, who is idling about smoking his pipe. This is an ancient form of animal humor."[125] Some phantom-like, relatively consistent images and symbols do form, but everything remains possible. The enterprise whereby non-human and human animals co-create signs is compared by Vicki Hearne to the relation between Prospero and Caliban in Shakespeare's *Tempest*. To embark on such an endeavor is like summoning rough magics and releasing meaning processes that no subject can maintain under its full control.[126] This brings about miracles as well as evil outcomes,[127] so much that magicians will be often tempted to give up.

Questioning Bateson's approach to the dream comparison

The dream comparison such as developed by Bateson has its limitations. Bateson's "What is an Instinct?" strangely combines logocentrism and its overcoming, maybe a bit too cunningly.

On the one hand, claiming that dreams provide the best paradigm to describe animal communication is clearly a way of emphasizing that non-human animals are not stuck in a life of immediate and blind reactions. More than that, their relation to meaning implies a dimension of virtuality and metacommunication that lies at the origin of fantasy, poetry, and art in general, as also highlighted in the metalogue "Why a Swan?"[128] Dreams place the world and the urgency of action at a distance. The dream comparison thus states that non-human animals somehow *contemplate*. Moreover, dream is a rich and fruitful form of awareness. It gives access to aspects of meaning that a rational objectivist approach simply misses. And, indeed the daughter proposes to use the dream as a paradigm precisely because she and her father realize that the objectivist project of finding out what things *really* are and where the cause *really* lies is totally inadequate to understand intrinsically ambiguous phenomena. "It's difficult to be objective about whether a rat is really exploring or really playing."[129] Language and tools are the royal road to objectivity. Therefore: "Daughter: (...) what is the royal road to the other half? Father: Freud said dreams."[130]

On the other hand, in many respects, this approach is tainted with logocentrism. The contrast between the waking life and dream, between objectivity and oneiric thought, and between non-human and human communication is exaggerated by Bateson. On several occasions, Bateson puts forward cut-and-dried assertions. Animals *cannot* be sarcastic or ironic;[131] they *cannot* "say" I will not bite you, or "I am not biting you"[132] since "tenses are *only* possible in language;"[133] and the puppy "*obviously couldn't* give any signals to show he knows" that he is saying the opposite of what he means.[134] All these passages bear the "strictly speaking" certification mark. Bateson also claims—I should say "the father claims," which, as we will see, makes things a bit more complex—that humans are "split"[135] into the objective side and "the other half,"[136] namely "the animal part in us,"[137] while non-human animals are not.[138] Bateson, in this metalogue, does not reject the concept of objectivity overall, he rather asserts that it is easier to be objective about some aspects of reality than others,[139] about sex, for instance, and not about love. But isn't this last example so incredibly naïve?[140] Dream is also presented as the realm where primary processes rule. In dreams we can only find a form of "archaic" logic that is entirely foreign and "inaccessible"[141] to the realm of consciousness and verbalized thoughts.[142] Bateson characterizes dream by the *absence* of the "as if."[143] The artist, the dreamer, and the pre-human mammal or bird share the same "limitation:" They can use *only* iconic communication. In short, in Bateson's approach, dream remains something we describe retrospectively, and, therefore, non-human animals are described as *them*.

But where exactly does the cut-off lie between these alleged two halves? These supposed two realms? Claiming that animals communicate the way we dream may suggest that the waking life is our prerogative. Yet, non-human animals also create better-circumscribed and relatively stable signs. They set markers. They do not especially love wallowing in ambiguity. As Bateson himself highlights in "A theory of Play and Fantasy," specific conspicuous signs indicating that "this is play" (the play face, the play bow etc.) do exist in the animal realm. Further, the claim that "tenses only come with human language" comes across as a surprising breakout of ingenuous logocentrism. Many non-human animals relate to the future.[144] Declaring that, strictly speaking, they cannot say "I will not bite you" tarnishes the promising complexity of the dream comparison. Non-human animals do "sort of"[145] say "I will not bite you." By returning to the "strictly speaking" perspective, Bateson, like so many before him, backs off, digs his heels in, and refuses to embrace the metaphor.

Non-human animals do not *know* if they will be heard, understood, and believed, but neither do we. "D: The animals would know that they bared their fangs in order to say, 'I won't bite you.' F: I doubt whether they would know. Certainly, neither animal knows it about the other. The dreamer doesn't know at the beginning of the dream how the dream is going to end."[146] Since none of us *knows* at the beginning of a day, a conversation, or any interaction, how *this is going to end*, we should rather deduce from Bateson's argument that we, human and non-human animals, all live in the realm of dream.

In the same vein, Bateson presents the advent of metacommunication as a radical novelty, a metaphysical event, a break away from blind and automatic reactions.[147] "The drama precipitated when organisms, having eaten of the fruit of the Tree of Knowledge, discover that their signals are signals."[148] And he refers to a specific "stage of evolution."[149] By the same token, in "What is an Instinct?," Bateson only gives examples of histrionism, deceit, and play, which suggests that what is at stake is a set of quite particular animal behaviors, usually regarded as the cutting edge of the evolutionary process. By contrast, Portmann's perspective is revolutionary in that it obliges us to consider much more fundamental forms of the imaginary of animal. Over the paradigm of the Fall, I will prefer Merleau-Ponty's model of institution.

But before we move to this last part, I must point out that a non-divisive perspective *also* develops, implicitly, in Bateson's "What is an Instinct?," as the oneiric counterpart of the aspects that I have just analyzed. This has to do with the form of the text: Bateson ensures that the reader will not be able to rest in a literal reading. A *metalogue* is the very enactment of the ubiquity described by Bateson as this sort of being that objectivity always misunderstands. The *metalogue* does not only examine a topic, but also the way we speak of the said topic. To the question "what is an instinct?," the father responds: "It is an explanatory principle"[150] and the subjective choice to avoid considering a set of too complex phenomena (in other words, the creation of a "black box").[151] Science thus begins with fiction.[152] Moreover, a metalogue, like Platonic dialogues, thinks through a fictitious conversation. The quest for an ultimate message, Bateson's theory, is satirized by the very form of the discourse and every assertion in this context becomes unhinged. Irony, reflection, and fiction dig a chasm—or what Derrida would describe as the "foliated structure of the abyssal limit"[153]—in every concept. At work here is not only play, as that which is able to draw a limit between reality and fiction, but oneirism as a principle of alienation that haunts every certitude. Although Bateson says that, even in a metalogue, we will have to be objective,[154] he actually philosophizes in the same way as we dream. Here the model of the two halves collapses, or, more exactly, becomes one voice among others. In that sense, "What is an instinct?" may truly be a study of animals from a first-person perspective.

§31 Lateral kinship: The dream comparison in Merleau-Ponty's *Nature* lectures and the animal roots of human imagination

In the *Nature* lectures, Merleau-Ponty works toward a theory of the lateral kinship between humans and other animals. "Laterality" means that each human and non-human animal has a perspective irreducible to mine but exactly of the same value as mine, so that I always decenter myself to dialogue with it. To be sure, the way animals communicate is oneiric, in that they ordinarily do not use a superstructure of symbolic systems. However, taking up Merleau-Ponty's sparse suggestions and drawing on his theories of language and institution, I will argue that this animal oneiric form of communication is in no way heterogenous to the human one. Human and non-human animal ways of communicating contain the same components, possibly in different proportions. Thus, we should not simply *describe* the animals' dream-like communication, but rather dream and imagine *with* them, or, better, realize that human and non-human animals never really stopped doing so.

The oneiric, diacritical nature of language

Merleau-Ponty has developed a systematic theory of the emergence of language through the corporeity of speech and the poetical expressiveness of sounds, rhythms, intonation, mimicking, and resemblances.[155] The digital never becomes autonomous from the analog.

Bateson contends that the acted-out *reductio ad absurdum* is "a cumbersome and awkward method of achieving the negative."[156] But waking thought deludes itself if it believes that the word "no" provides a straightforward and safe manner of achieving the negative. Not less cumbersomely, in 2016, it appeared necessary to the German Legislature to enact a law in order to make clear that "no means no." Without empowerment, without political struggle, and without the awareness that "no" means "yes," for instance, in many romantic movies and in a still stubborn collective imaginary, the achieving of the negative is highly precarious. And woe to her who thinks that the struggle has come to an end. Is this unfortunate? The preceding example shows that ambiguity is not always valuable, but the irreducible ambiguity of expression does not make a process of relative disambiguation impossible. And we have seen that non-human animals do also create more stable and easily recognizable signs.

In addition, Merleau-Ponty stresses that there would not be anything to communicate if the ideal promised by prosaic language was achieved. A set of perfectly defined and mastered ideas would become meaningless. Indeed, in a worthy of the name *world*—namely: Opaque and transcendent— Platonic "eternal ideas" become incarnated into an indefinite series of *singular* individuals, which entails that such "ideas" must be intrinsically unhinged and metamorphic. We need to draw a line to understand what a line is,[157] and we need to experience—or *as-if* experience—how our ideas are connected to the sensible field of concrete phenomena, otherwise we will not *know* anything, we will only confirm conventions made with ourselves. The difference between a text and a musical piece is not absolute. Beyond the "conventional means of expression" ("which only manifest my thought to another person because, for both of us, significations are already given for each sign"[158]) and beyond what the other and I already know, a "genuine communication"[159] takes place through the unique arrangement of words, the style of a speech, and the new connotations thus smuggled. Expression requires that "the signs themselves externally induce their sense."[160] Communication at the purely technical level becomes dull and empty: Signification is mastered, and meaning vanishes.

Words and concepts develop conjointly as diacritical entities. Merleau-Ponty takes his inspiration in Saussure,[161] and more specifically in the chapter of *Cours de Linguistique Générale* devoted to "The linguistic value." Signifiers and signifieds emerge through the interaction between sounds and shapes in the flow of speech. There are many different ways of pronouncing words, depending on individual or local idiosyncrasy, and, likewise, we come across different typewritten or handwritten versions of a word. But, in this Heraclitean flow of sounds and shapes, relations of resemblance and difference, as well as unstable associations emerge. Individual signs do not preexist this dynamical sensible system: Signs are *values*. A value is not a positive entity; it is only an open set of singular sounds grouped together by virtue of rough resemblance and because of their common difference from all of the other sounds. Each value (for instance, the signifier "sheep" as gathering an infinity of spoken or written versions of this word) exists only relatively to the rest of the system (the difference with "ship" is for instance relevant in the English system, although the dividing line between some versions of "sheep" and some versions of "ship" can be fuzzy and the various contexts make the whole system precarious). Correlatively, each signified occupies the field left by the rest of the system. The signifier and the signified are both plastic, so that, for instance, an association between the physiognomy of the signifier and the physiognomy of another signifier may contaminate the theme of the signified. Saussure gives the example of two French words: "décrépit" (weakened by age, dilapidated) and "décrépi" (the state of a wall for instance after the roughcast that covered it has been removed or crumbled down). These words have different origins and meanings, but most of the French speakers confuse them so that the two meanings contaminate each other. A sign is thus essentially dynamic: Its future avatars will be variations on an open-ended theme. The analogical regime is never overcome. The attempt to circumscribe and stabilize signs is not useless or devoid of value; however, this endeavor towards disambiguation becomes illusory when it forgets that it exists only as a surge of agency—certainly not mastery—grafted upon an oneiric field of proliferating meaning.

The origin of meaning: The oneiric institution of human culture

Language is what Merleau-Ponty calls an institution.[162] He defines this concept as a response to those who regard humans as Great Founders. This theory does not make the idea of creation and individual initiative irrelevant, quite the opposite; it states that our creations are meaningful and fruitful only as the taking up of already existing meanings.

Merleau-Ponty contrasts institution with constitution. A constitution—the reference here could be a political constitution—will remain inflexible and persist in its original form through time. Constitution implies the advent of mastery. By contrast, an institution can never be reduced to a founding act. It takes the form of a localizable event and a certain expression of a project or a tendency, but it also opens a tradition. An institution keeps resurfacing in new avatars and can always be taken up by others, through other forms/events/works. Good examples of institutions include the establishment of a law, the first book published by an author, or the first expression of the character of a child. An institution is fruitful precisely because it was never perfectly self-transparent. A lawmaker never fully masters the concept of absolute justice. She does not encompass all the consequences that the law she is creating implies, nor does she know what this law will become. The original estrangement of an institution begets its power of dissemination.

Further, every institution is a circular structure: An institution is both ahead of and lagging behind itself. Each institution draws from its ambiguity the boundless ability to inspire *new* re-institutions, through which it will live new unexpected lives: To take up is not to repeat. With the concept of institution, human culture ceases to appear as that which has *superseded* a natural existence. However, this theory of institution still recognizes human culture in its ability to create new forms and interpret imaginatively. Yet, as an institution, human culture appears as the response to an imaginative tradition that is older than the human realm.

According to Merleau-Ponty, we do not constitute the realm of culture. Cultural creations, the works of Great Founders, are in fact reinstitutions that draw on an oneiric field of meaning and that would never make sense otherwise.

As a consequence, dreams also play a crucial institutive role in human existence. In *Institution and Passivity*, Merleau-Ponty points out that, even after waking up, many of our dreams keep worrying, scaring, or pleasing us diffusely. Many relevant connections can be found between dreams and the preoccupations that are the core of our *real* life and our actual relation to the world.[163] Dreams and waking life are both shaped by lines of meaning such as the obsession, love, or disgust for this or that object, recurring questions, or strange remote connections between two persons or two situations. Dreams may even provide aha moments, when, for instance, the waking life is swamped with abuse and abasement. Thus Ida Bauer, also known as *Dora*, one of Freud's patients, struggled with existential tensions and somehow came to a crucial decision in the "burning house" dream.[164]

Precisely because the imagineareal unhinges every actual person and object, dream and reality cannot be heterogenous. In dreams, the existential themes—nascent essences, phantom-like individuals, questions, desires, haunting memories, recurring issues—that underpin our waking life come to the fore; the plasticity of such themes and the immense horizon of possibilities that *institute*—rather than constitute—each thing and person in our world become conspicuous. The insistent presence of actual perspectives and appearances, the obdurate "reality," are placed at a distance, but we still deal with the difficulty of making our own way through anonymous images and symbols.

Animal institution

Merleau-Ponty's comparison between animal communication and dream consequently points to a circular relation of institution between non-human and human animals. On this basis, the idea of a closed humanity or a closed culture can be overcome. As Merleau-Ponty suggests in a remark that remains to be fleshed out: "Human desire emerges from animal desire."[165]

Desire is classically distinguished from need and regarded as specifically human. In line with Plato's *Symposium*, desire is considered by the Western philosophical tradition as the sign of the relationship between human beings and the infinite. Desire is essentially beyond itself. It is defined by its indetermination and its contingency. Claiming that animals desire amounts to asserting that they institute an indefinite quest that opens them to the consideration of *all* the responses and new questions that emerge in the wake of this institution.

And indeed, human culture is obsessed with animals. Humans cannot but always seek to define themselves by referring to their allegedly lost or deeply subsisting animality. Moreover, sedimented images and symbols coming from animal expressiveness are still operating as the necessary lining of human symbolic language. Our language began to live its existence in non-human animals. Non-human animals' images and symbols continue to develop *their* potentialities in our world. To the indefinite open quest downstream, in the realm of human culture, must correspond, upstream, an animal institution. The openness to indefinite questioning was classically defined as a distinctive feature of mankind, but it must, in fact, be the taking up of an animal institution. Humans try to answer open questions that have been phrased in the realm of animal expressiveness.

Human culture creates an inexhaustible imagery of animals and, somehow, institutes animals as mythic figures and talking characters. But our access to meaning always unfolds as the taking up of animal meaning. In other words, in turn, the animal imaginary also institutes the human imaginary. On this basis, we can better understand the concepts of the "lateral universal"[166] and of a *lateral* kinship between human and non-human animals that Merleau-Ponty ventures in the *Nature* lectures without much explanation. The fundamental structure of human-animal co-institution makes it decisive for us to embrace the decentering process whereby we imagine non-human animals' perspectives: No truth, no reality, no knowledge should ever be claimed outside of such a dialogue.

Merleau-Ponty, like Portmann, stresses that mimicry and more generally animal appearances have consistently inspired human art and myths: "The question of mimicry is not yet regulated, to the extent that there is a good part of myth in the reported facts. But that such myths could have been created and have a long life is precisely what makes the fact interesting."[167] Likewise, it is an interesting *phenomenon* that the traditionally objectivist scientific approach to animals was somehow forced by its object to challenge its classical objectivist method. Biology was compelled to integrate a problematic reference to subjectivity, ambiguity, interpretation, images, imagination, and dream, in order to do justice to the very structure of the phenomena it examines. Instead of wondering whether this move toward a science that reckons with animal subjectivity is objectively legitimate or not, we should account for the stubbornness of what I have called the human-animal metaphor. Thus, how exactly does the animals' imaginary institute our human thoughts?

The animal roots of human imagination

In "Mythisches in der Naturforschung," Portmann contends that the human imaginary may very well stem from search images that are instinctive forms of subjective interiority.

Yet, Portmann insists on the rigid nature of animal search images. He gives the example of the cuckoo, who is able to *a priori* recognize conspecifics, and of the male stickleback, who attacks every crude dummy whose underside is brilliant red. Portmann accordingly contrasts the stock of animal archetypes that haunt us with the human manner of playing freely with them.

There is some oddity, however, in explaining human imagination, whose inexhaustible fertility was emphasized at the beginning of Portmann's essay, by describing it as hinging upon a stock of fixed clichés. If non-human animal images lend themselves to endless variations and parodies, they cannot be lacking the dynamism and the inner *écart* to themselves that characterize human images. And indeed, as the previous sections explored, animal images already enter into a playful process within the non-human animal realm.

The concept of institution helps us rephrase Portmann's intuition in a more accurate way. We can understand animal self-depiction through our experience of narcissism, fashion, or heraldic practices. We can re-enact the puppy look and turn it into a meme. The images of eyes, jaws, shadows from above, and so many other non-human animal archetypes are used to deepen and enliven the expressiveness of our pictorial works. In other words, the same living meaning thrives uninterruptedly through animal images and the human imaginary. Human and non-human animal imaginaries institute each other.

Merleau-Ponty thus consistently highlights that, at the level of non-human animal expressiveness, an *open* process of symbolization occurs.[168]

This process already includes a radical dimension of contingency and arbitrariness, which is a form of self-distance. Animal appearance such as studied by Portmann "manifests something that resembles our oneiric life" for the following reason: "In a certain sense, the sexual ceremony is probably useful, but it is useful only because the animal is what it is. Once they are there, these manifestations have a meaning, but the fact that they are this or that has no meaning. Just as we can say of every culture that it is both absurd and the cradle of meaning, so too does every structure rest on a gratuitous value, on a useless complication."[169] Merleau-Ponty stresses that, in different species—and possibly in different individuals—the same behavior can become the medium of completely different expressions: "Behavior take[s] on different significations: in one species of Cyclades fish, the behavior that originally indicates inferiority assumes the meaning of a threatening behavior for the dwarf Cyclades."[170] This contingency and these whimsical variations show that some subjective initiatives are at work in animal existence, as a form of undetermined intentionality. The morphology of the body is not utterly shaped by a teleology of expression; it has many other functions and evolves through a lot of accidental circumstances. Hence, expression emerges as a tendency grafted upon the life of this body, and, as such, it is intrinsically out of balance. Crystallized images and symbols are instituted that are also liable to be hijacked. Thus, animal expression, like dreams, governed by desire or a Platonic love rather than need, gropes for its way without ever knowing in advance what it is heading for. Portmann's statement that animal appearances may be "sent 'into the blue' ['*ins Blaue*' *gesendet*]"[171] makes more sense in the light of the concept of human-animal co-institution: Being sent into the blue is the very essence of institution and fundamentally defines meaning in general. Meaning only exists as a question posed to others. Both dreams and animal expression take up vague already-existing themes in an inventive manner and tentatively institute new markers.

Therefore, it is no surprise that instinctive behaviors can be turned into symbols through ritualization and even into fantasies.[172] Merleau-Ponty points to the example of the starling hunting imaginary flies.[173] Indeed, Konrad Lorenz has shown that instinct can be "objectless [*objektlos*]," namely aim at a vaguely defined object, which Merleau-Ponty hears as echoing the Platonic definition of desire. Instinct firmly asserts a subjective but not fully predetermined requirement and *turns toward a world* that in return provides all sort of surprising objects (a human as a *Kumpan* for instance). "[Instinct] does not know what it is nor what it wants,"[174] but, for that reason, its capacity for institution is immense. "With empty activity, instinct is going to be capable of being derailed or is going to pass from instinctive activity to symbolic activity. Empty or outlined activities are going to become means of communicating for the animals."[175] What is striking here is that the connection between these different steps is presented as a series of derailments and diversions: Neither finality, nor mechanism, but the oneiric process of institution. We should not think any longer in terms of "archaic" versus "advanced" forms. "No pure exteriority of biological space, no pure succession/sequence of biological time. There is going to be Being of ubiquity and Being of anticipation."[176]

The lateral kinship between non-human and human animals is symmetrical instead of hierarchical. Non-human animals intentionally and actively project their *Umwelt* and expose themselves to the experience of the discrepancy between actual and virtual, for instance, between the quasi-hallucinated prey and its actual incarnation. Similarly, non-human animals sense the gap between interiority and exteriority, as well as between the hidden and the expressed.[177] At least a nascent reference to other perspectives arises, a reference that becomes central in intentionally deceitful behaviors as well as in phenomena of joint attention, when, for example, animals such as goats, horses, dogs, many birds, and cetaceans follow another individual's gaze instead of simply focusing on an object,[178] or when the killdeer who performs a broken wing display regularly checks whether her target actually has its eyes on her and visually follows her.[179] Even below the theory-of-mind issue, at the level of the formation of *Selbstdarstellung*, non-human animals essentially play with perspectives, which makes them subjects, eminently, and parts of the intersubjective analogical system of meaning and transposition.

We are haunted by non-human animals for they actively haunt us: They launch a process of active de-centering and a quest for meaning that we inherit. Human beings seek themselves in non-human animals and non-human animals do seek themselves in others, including in humans. In this intersubjective quest, each expression is an enrichment to all. This opens up an understanding of non-human animals in the first-person mode, or more exactly, in the second-person mode. A dialogue—and certainly not a purely anthropomorphic illusion of dialogue—can and must take place between non-human and human animals. It will be based on adventurous transpositions and resumptions, but such is the essence of genuine communication. Donna Haraway's description of her relation with her dog Cayenne provides, as the next chapter will explain, a perfect illustration of this genuine and imaginative dialogue.

Benumbed and dreamy beasts gazing into the open: A dangerous myth

My theory, inspired by Portmann, biosemiotics, and Merleau-Ponty, thus breaks with approaches built upon the Rilkean concept of the open, namely with the tradition to which, in different ways, Heidegger, Agamben or Lacan and more recently Jean-Christophe Bailly have contributed. I agree with Elisabeth de Fontenay[180] and Kelly Oliver[181] in this regard. The discussions about the access of non-human animals to the unarticulated and suspended "open" only explore the way some humans fantasize their alleged extraordinary nature and fail to simply pay attention to existing ongoing intertwinements between non-human and human animals.

As Agamben indicates,[182] Rilke's and Heidegger's perspectives mirror each other. In Rilke's world, only animals have access to the open, for they are not bound under the power of symbolic language and the sharp "no." Heidegger turns this theory on its head: Only humans open up to being. In fact, both authors build upon the idea of the abyss between man and animal. Rilke's hazy concept of an "open" reserved for animals is in fact the secret accomplice of the concept of animal stupor [*Benommenheit*] in Heidegger. Rilke and Heidegger both rely upon what Agamben calls the anthropological machine, which creates the empty concepts of "animal" and "man" by trying to deny animality within humans. The paradox is obvious in Rilke, who gives glowing descriptions of that which man, allegedly, cannot see. As I have argued, Heidegger's theory also hovers among tensions and dead ends.[183] Agamben subtly deconstructs the anthropological machine and, to be sure, in the last chapters of *The Open*, the perspective of a "serene" and "happier" letting go of being is seductive. Agamben looks forward to the "saved night," beyond the desire to see, when the tension and the struggle are over. It is the esoteric and radical choice. We certainly may vow for this hypothetical future of relaxed sensuality and pliant contemplation, but something is missing in this approach, where all boils down to two alternatives: Either the merciless anthropological machine or the perfectly blissful [*perfettamente beata*] natural life. Agamben overlooks the fact that the anthropological machine was never the hegemonic matrix of modern thought. To paraphrase Latour, we have never been entirely subjected to the anthropological machine. The latter was never perfectly effective, so that multifarious relations kept developing between non-human and human animals, beyond a purely oppressive/reductive model, in everyday life, art, philosophy, and even sciences. Agamben's *between* is mute and serene, while the dream such as I have defined it, following Portmann, biosemiotics, and Merleau-Ponty's concept of lateral kinship, is swarming with intertwined symbols, questions, desires, and endless re-institutions.

Lacan also develops the concept of a paradoxical open/closed animal access to the imaginary realm. His reference to an imaginary dimension of the animal realm[184] falls back in well-trodden paths and cuts this animal "imagination" from what he calls the symbolic. As a result, in Lacan's view, animals lack language, plain and simple. In Lacan's theory, non-human animals are in fact denied creativity, the ability to respond[185] as well as a symbolic use of images. "But what is new in man is that something is already sufficiently open, imperceptibly shifted within the imaginary coaptation, for the symbolic use of the image to be inserted into it."[186] Lacan builds in particular on a Portmannian example: Sticklebacks are captivated by a general image of a mate, and they are easily fooled and hallucinated by what they desire.[187] Obviously, Lacan was not particularly eager to provide a careful description of the complexity of animal behavior, and he barely scratches the surface of Portmann's theory. Moreover, in these analyses, *the imaginary* is cut off from its fundamental ubiquity and plasticity, to such an extent that it becomes problematic to justify the use of the term "imaginary." Lacan grants imagination to animals with one hand, while taking it away with the other.[188] Correlatively, Lacan defines the imaginary in humans in a questionable manner: According to Lacan, the human imaginary is subjected to the symbolic, and the latter is defined as a hole in Being. The symbolic constitutes signifiers that form an autonomous system and are utterly *substituted* for things. When they lie, humans allegedly perfectly master the relation to the other and the absent as such. Animals, on the other hand, allegedly totally miss it, even though Lacan acknowledges that they are capable of pretense.[189] With Merleau-Ponty—who implicitly converses with Lacan throughout his works—a different conception of the imaginary is at stake. Merleau-Ponty emphasizes the floating nature of the imaginary field and its original connection with a non-logical meaning, from a phenomenological perspective. Such a phenomenological approach fruitfully overcomes Lacan's ambivalence by refusing to oppose or to subject the imaginary to the *logos*. As a result, while Lacan claimed that only humans have an unhinged identity and thus endlessly construct this identity through images, I argue that this existential, fractured identity and this openness to the virtual and ideas more fundamentally characterize animality.

The dream comparison also interestingly recurs in Bailly's depiction of slow and placid farm animals. Here again this comparison gently and discreetly enforces a suspension of judgement regarding what non-human animals think. In *Le dépaysement*, Bailly describes the thick and solid presence of "beasts" in a stable. "I was awed by the extraordinary density of presence of these animals, by their way of being there, in the wholeness of their being [*dans l'intégralité de leur être*], incapable of affectation."[190] Bailly then wonders "but do they dream?" And, indeed, dreaming is a form of not-being-entirely-*here*. Bailly concedes that beasts "most likely" dream. "About distant pasture land? Probably. But we do not know, we do not penetrate the thoughts of these pensive and maybe asleep minds."[191] The dream comparison returns here, together with its ambiguities. A dreamy thought is easily mistaken with an asleep thought. Is it a form of thought at all, then, as Heidegger suspected?

But who exactly is asleep? In *Le dépaysement*, Bailly undertakes to depict a certain rural life in regions that regard and experience themselves as *la France profonde*. This existence is at least sometimes still ruled by traditions, regular rhythms, terseness, and an idiosyncrasy of what Bailly describes as a "benumbed and dreamy brutality" combined with "craftiness and mètis."[192] More generally humans who are fascinated by the myth of the dreamy "open," such Rilke describes,[193] will fancy the image of contemplative, "pensive" animals and forget about the war of wits. They overlook the animal endeavor to create less ambiguous symbols as well as the cooperation between human and non-human animals in instituting new signs.

Let the cows, chickens, pigs, and other companions be free to roam as they please: Much more unpredictable animals will appear. They will display particular personal preferences and develop inventive and rich modes of communication. The farm life described by Rosamund Young in *The Secret Life of Cows* is thus rather hectic and full of surprise.[194] Cows in Young's world even prove able to transmit secret grudges against particular human individuals to others.

Everything demonstrates that non-human animals are widely opened onto the outside world. If everything contemplates, as Plotinus claims, animals do more than that. The animal swarming and manifold expressiveness is "sent into the blue": It functions as an inexhaustibly fruitful institution and spurs our imagination. The silence of the beasts is a myth, a meaningful myth, but it loses its pertinence if it is taken too seriously. Non-human animals are incredibly talkative. Their images and symbols invade our dreams, our works of art, and our everyday metaphors: They meddle in our language and thoughts. Communication being a matter of interference, the best way to respond is consequently not to "respectfully" or cautiously suspend our judgment, but to meddle in return.

Notes

1. Berwick, R.C. et al. "Songs to Syntax: The Linguistics of Birdsong." In *Trends in Cognitive Sciences* Elsevier, 15(3), 2011. See also Ouattara, Karim, Lemasson, Alban, and Zuberbuhler, Klaus. "Generating meaning with finite means in Campbell's monkeys." In *Proceedings of the National Academy of Sciences,* Vol. 106, No. 48, December 7, 2009. And Schlenker, Philippe et al. "Monkey Semantics: Two 'Dialects' of Campbell's Monkey Alarm Calls." In *Linguistics and Philosophy,* Vol. 37, Issue 6, 2014.
2. Savage-Rumbaugh, Sue. *Ape Language: From Conditioned Response to Symbol.* New York: Columbia University Press, 1986. See also Savage-Rumbaugh, Sue, Rumbaugh, Duane M., and Boysen, Sarah. "Do Apes Use Language? One Research Group Considers the Evidence for Representational Ability in Apes." In *American Scientist,* Vol. 68, No. 1, January-February 1980.
3. See for instance Lyytinen, Anne, Brakefield, Paul M., and Mappes, Johanna. "Significance of Butterfly Eyespots as an Anti-Predator Device in Ground-Based and Aerial Attacks," In *Oikos* 100, 2003.
4. See for instance Howse, Philip and J.A. Allen, "Satyric Mimicry: The Evolution of Apparent Imperfection," In *Proceedings of the Royal Society B: Biological Sciences,* 257, August 1994.
5. See for instance Maran, Timo, *Mimicry and Meaning: Structure and Semiotics of Biological Mimicry,* Dordrecht: Springer, 2017, p.119: "There are many semiotic processes that rely heavily on the non-determinism or fuzziness of culture or nature. All truly creative behaviors and artistic usages of signs are, in principle, nondeterministic. (...) The same appears to be true for living biological systems—at a fundamental level, they are non-deterministic and their complexity is far beyond our capacity of measurement and computation."
6. Hoffmeyer, J. "Semiosis and Biohistory: A Reply. Semiotics in the Biosphere: Reviews and a Rejoinder." In *Semiotica* (Special Issue) Vol. 120, No. 3/4, 1998, p.460. See also Artmann, Stefan. "Computing Codes versus Interpreting Life.

Two Alternative Ways of Synthesizing Biological Knowledge through Semiotics," in Barbieri, Marcello (Ed.), *Introduction to Biosemiotics. The New Biological Synthesis*. Dordrecht, Springer, 2007, p.230.

7. Uexküll, *A Foray into the Worlds of Animals and Humans. With A Theory of Meaning*. Minneapolis, London: University of Minnesota Press, 2010, p.181–2.

8. Souriau, Etienne. *Le sens artistique des animaux*, Paris: Hachette, 1965.

9. Van Leeuwen, E.J.C. et al. "A Group-Specific Arbitrary Tradition in Chimpanzees (Pan Troglodytes)," in *Animal Cognition*, Volume 17, Issue 6, November 2014.

10. Souriau (*Le sens artistique des animaux*, p.99 and 101) also refers to Evreinov, Nikolai, *The Theatre in Life*. Trans. Alexander Nazaroff, New York: Brentano's, 1927.

11. Portmann, Adolf. *Die Tiergestalt*. Freiburg/Basel/Wien: Herder, 1965, p.240. Trans. Hella Czech. *Animal Forms and Patterns*. New York: Schocken Books, 1967, p.106.

12. Merleau-Ponty, *La Nature*, Paris: Seuil (collection Traces écrites), 1995, p.246. Trans. R. Vallier. Evanston: Northwestern University Press, 2003, p.188.

13. Bateson, Gregory. *Steps to an Ecology of Mind*. Northvale NJ: Jason Aronson, 1972, p.69.

14. Bateson's *Metalogues* are dialogues between a father and his daughter.

15. *Ibid.*, p.60

16. Maran, Timo. *Mimicry and Meaning: Structure and Semiotics of Biological Mimicry*, Dordrecht, Springer, 2017.

17. Bateson, *Steps to an Ecology of Mind*, p.65–6.

18. Rilke, Rainer Maria. *Duineser Elegien* (Insel Verlag Leipzig 1923) Kapitel 8, Translated by A. S. Kline, in *The Poetry of Rainer Maria Rilke*, Poetry in translation, CreateSpace Independent Publishing Platform (11 May 2015): "The creature gazes into openness with all its eyes. (…) We never have pure space in front of us, not for a single day, such as flowers open endlessly into. Always there is world, and never the Nowhere without the Not: the pure, unwatched-over, that one breathes and endlessly knows, without craving."

19. Portmann, *Die Tiergestalt*, p.17. Trans. p.17.

20. Portmann, *Die Tiergestalt*, p.19, Trans. p.18–9.

21. Portmann, Adolf. "Selbstdarstellung als Motiv der lebendigen Formbildung," in *Geist und Werk. Aus der Werkstatt unserer Autoren. Zum 75. Geburtstag von Dr. Daniel Brody*, Zurich: Rhein Verlag, 1958, p.148.

22. *Ibid.*, p.162.

23. "Offenbar gibt es "Erscheinungen," die zum "Erscheinen" bestimmt sind - neben andern, die das nicht sind" (Adolf Portmann, "Selbstdarstellung als Motiv der lebendigen Formbildung" p.146).

24. *Ibid.*

25. Portmann, Adolf, *Die Tiergestalt,* p.32, Trans. p.31–2.

26. Portmann, *Die Tiergestalt*, p.56.

27. Portmann, *Die Tiergestalt,* p.17–8 (Trans. p.17–8), 137–8 (Trans. p.127–8).

28. Portmann, *Die Tiergestalt*, p.138 (Trans. p.127)

29. Caro, Tim et al. "The Function of Zebra Stripes," in *Nature Communications,* Vol. 5, Article No. 3535, 2014. See also Gibson, Gabriella, "Do Tsetse Flies 'See' Zebras? A Field Study of the Visual Response of Tsetse to Striped Targets," in *Physiological Entomology,* Vol 17, No. 2 (141–7), June 1992.

30. Larison, Brenda et al. "How the Zebra Got its Stripes: A Problem with Too Many Solutions," in *Royal Society Open Science*, Vol. 2, No. 1, January 2015.

31. Portmann, *Die Tiergestalt*, p.190–1 (Trans. p.180–1).

32. For instance: Kendrick, Keith, et al. "Sheep Don't Forget a Face," *Nature,* Vol. 414 (165–6), 2001; Stephan, Claudia et al., "Have We Met Before? Pigeons Recognise Familiar Human Faces." In *Avian Biology Research* 5(2), 2012; and Newport, Cait et al. "Discrimination of Human Faces by Archerfish (Toxotes chatareus)." In *Scientific Reports.* Vol. 6, Article number: 27523, 2016 among many others.

33. Knolle, Franziska et al. "Sheep Recognize Familiar and Unfamiliar Human Faces from Two-Dimensional Images." *Royal Society Open Science,* 4(11) 8 Nov 2017; and Stone, Sherril M. "Human Facial Discrimination in Horses: Can They Tell Us Apart?" In *Animal Cognition* 13(1), July 2009.

34. Baracchi, D. et al., "Speed and Accuracy in Nest-Mate Recognition: A Hover Wasp Prioritizes Face Recognition Over Colony Odour Cues to Minimize Intrusion by Outsiders." In *Proceedings of the Royal Society of London B. Biological Sciences,* Vol. 282, No. 1802, 2015.

35. See for instance Sheehan, Michael J. and Tibbetts, Elizabeth A. "Evolution of Identity Signals: Frequency-Dependent Benefits of Distinctive Phenotypes Used for Individual Recognition," in *Evolution,* 63(12), 2009.

36. Portmann, Adolf. *Biologie und Geist,* Zürich: Rhein-Verlag, 1956. Reprint Frankfurt am Main, Suhrkamp taschenbuch: 1973, p.115–6.

37. Portmann, Adolf. *Das Tier als Soziales Wesen.* Zürich: Rhein-Verlag, 1953. Trans. Oliver Coburn. *Animals as Social Beings.* New York and Evanston: Harper Torchbooks, 1964, p.136. Portmann refers to J.J. Ter Pelkwijk N. Tinbergen, "Eine reizbiologische Analyse einiger Verhaltensweisen von Gasterosteus aculeatus L." In *Zeitschrift für Tierpsychologie,* 1, 1937.

38. Tedore, Cynthia and Johnsen, Sönke. "Pheromones Exert Top-Down Effects on Visual Recognition in the Jumping Spider Lyssomanes viridis," in *Journal of Experimental Biology,* 216, 2013. There are innumerable other examples with various species. See for instance Matyjasiak, Piotr, "Birds Associate Species-Specific Acoustic and Visual Cues: Recognition of Heterospecific Rivals by Male Blackcaps," in *Behavioral Ecology,* Vol. 16, No. 2, 1 March 2005.

39. Portmann, *Animals as Social Beings,* p.95–6.

40. Maran, *Mimicry and Meaning,* p.15: "Mimetic resemblances can occur as colors and forms in the visual medium, as imitations of hissing, buzzing and other sounds, or as similarities of chemical components in pheromones."

41. Lanzing, W.J.R., Bower, C.C., "Development of Colour Patterns in Relation to Behaviour in Tilapia mossambica (Peters)," *Journal of Fish Biology,* 6(1), 1974.

42. Matsumiya, Lynn C. et al. "Using the Mouse Grimace Scale to Reevaluate the Efficacy of Postoperative Analgesics in Laboratory Mice," in *Journal of the American Association for Laboratory Animal Science,* 51(1), 2012. See also Miller, Amy L. and Leach, Matthew C. "The Mouse Grimace Scale: A Clinically Useful Tool?" in PLOS One. 10(9), 2015

43. Portmann, *Animal Forms and Patterns,* p.190.

44. See for instance Wathan, Jennifer et al. "Horses Discriminate Between Facial Expressions of Conspecifics," in *Scientific Reports,* Vol. 6, 2016; and Tate, Andrew J. et al. "Behavioural and Neurophysiological Evidence for Face Identity and Face Emotion Processing in Animals," in *Philosophical Transactions of the Royal Society,* 361, 2006.

45. Fedigan, Linda. "Social and Solitary Play in a Colony of Vervet Monkeys (Cercopithecus aethiops)," in *Primates,* 13(4), 1972.

46. Kaminski, Juliane et al. "Human attention affects facial expressions in domestic dogs," in *Scientific Reports,* 7(1), 2017

47. Portmann, "Selbstdarstellung als Motiv der lebendigen Formbildung," p.158. See also Portmann, *Die Tiergestalt*, p.225

48. Portmann, Adolf and Carter, Richard. *Essays in Philosophical Zoology by Adolf Portmann. The Living Form and the Seeing Eye*, Lewiston: The Edwin Mellen Press, 1990, p.25.

49. Hoffmeyer, *Biosemiotics: An Examination into the Signs of Life and the Life of Signs*, p.18.

50. Portmann, *Die Tiergestalt*, p.233.

51. Portmann, *Die Tiergestalt*, p.122.

52. John Langerholc, "Facial mimicry in the animal kingdom," *Bolletino di zoologia* 58.3 (185–204), 1991, p.189.

53. Portmann, "Selbstdarstellung als Motiv der lebendigen Formbildung", p.140.

54. Portmann, *Die Tiergestalt*, p.25–6 (Trans. p.25).

55. *Ibid.*, p.180 (Trans. 170). See also Portmann's remarks about the formation of external gonads in mammals (Portmann, *Die Tiergestalt*, p.188–192, Trans.178–81).

56. Kistner, D.H. "Revision of the African species of the termitophilous tribe Corotocini (Coeloptera: Stapylinidae). I. A New Genus and Species from Ovamboland and its Zoogeographic Significance." *Journal of the New York Entomological Society* 76 (213–21), 1968. See also Kleisner Karel et Anton Markoš. "Semetic rings: towards the new concept of mimetic resemblances", *Theory in Biosciences* 123, 2005.

57. Portmann, *Die Tiergestalt*, p.93 (Trans. 86).

58. Grene, Marjorie. "Beyond Darwinism: Portmann's thought", in *Commentary* XL (31–8), 1965.

59. Portmann, *Die Tiergestalt*, p.220 (Trans. 208–9).

60. Life implies "building processes of great complexity that would be useless, in any usual sense, for mere conservation", Portmann, "Selbstdarstellung als Motiv der lebendigen Formbildung," p.158.

61. Portmann, *Die Tiergestalt*, chapters 1 and 2.

62. Portmann, *Die Tiergestalt*, p.234.

63. *Ibid.* p.35 (Trans. 34–5)

64. *Ibid.* p.234, my translation.

65. Portmann also ventures the hypothesis that *Selbstdarstellung* may be a sketch of reflexivity, or a sort of communication to oneself similar to the one that we experience when our emotional state is revealed to us by our racing heartbeat, the heat rising to our head, the flushing of our face, or a feeling of suffocation (*Biologie und Geist*, p.286)

66. Plato, *The Republic*, 595c–598c.

67. Karel, Kleisner. "The Semantic Morphology of Adolf Portmann: A Starting Point for the Biosemiotics of Organic Form?", *Biosemiotics* 1, 2008, p.212.

68. This idea was pointed out by Portmann as well: Portmann, *Die Tiergestalt*, p.236–9.

69. Maran, *Mimicry and Meaning*, p.27.

70. *Ibid.*

71. Müller, Fritz. "Über die Vortheile der Mimicry bei Schmetterlingen. " *Zoologischer Anzeiger*, 1, 1878.

72. Rowland, Hannah et al. "Mimicry Between Unequally Defended Prey can be Parasitic: Evidence for Quasi-Batesian Mimicry," in *Ecology Letters,* 13, 2010.

73. Speed, Michael. P. "Muellerian Mimicry and the Psychology of Predation." *Animal Behavior*, 45, 1993, p.579.

74. Deacon, Terrence. W. *The Symbolic Species: The Co-Evolution of Language and the Brain*. New York/London: W. W. Norton. 1997, p.75–6.

75. See supra, p.140, the example of spiders Lyssomanes viridis. See also Maran, *Mimicry and Meaning*, p.44.
76. Maran, *Mimicry and Meaning*, p.48.
77. Oliveira, Paulo S. "Ant-mimicry in Some Brazilian Salticid and Clubionid Spiders (Araneae: Salticidae, Clubionidae)", in *Biological Journal of the Linnean Society*, 33, 1988.
78. Ceccarelli, Fadia Sara. "Behavioral Mimicry in Myrmarachne Species (Araneae, Salticidae) from North Queensland, Australia," *Journal of Arachnology* 36(2), 2008. See also the queen of behavioral mimicry: the octopus *thaumoctopus mimicus,* by electively changing the color and the shape of its body, through its movements, can mimic several of its predators. See for instance Norman, M. D., Finn, J., & Tregenza, T. "Dynamic Mimicry in an Indo-Malayan Octopus", in *Proceedings of the Royal Society of London B,* 268(1478), 2001
79. Langmore et al. "Socially acquired host-specific mimicry and the evolution of host races in Horsfield's bronze-cuckoo Chalcites basalis", *Evolution*, 62, 2008.
80. See supra, p.63.
81. Lorenz, Konrad. "Der Kumpan in der Umwelt des Vogels. Der Artgenosse als auslösendes Moment sozialer Verhaltungsweisen", *Journal für Ornithologie.* Beiblatt, 83, 1935.
82. Uexküll, *A Foray into the Worlds of Animals and Humans*, p.117 (see supra, p.63). Humans may also be under the spell of such search images: when we fail to see the glass carafe on the table, because we are unconsciously expecting a familiar clay pitcher, we experience such a perception of the world through our search images (*Ibid.*, p.113).
83. Howse, Philip and Allen, J.A. "Satyric Mimicry: The Evolution of Apparent Imperfection", in *Proceedings of the Royal Society B: Biological Sciences*, 257(1349), 1994, p.111.
84. Howse, Philip, "Lepidopteran Wing Patterns and the Evolution of Satyric Mimicry." In *Biological Journal of the Linnean Society*, 109(1), 2013.
85. See Driver, Peter M. and Humphries, David Andrew. *Protean Behaviour: The Biology of Unpredictability.* Oxford: Clarendon Press, 1988; and Byrne, Richard W. and Whiten, Andrew (Eds.) *Machiavellian Intelligence: Social Expertise and the Evolution of Intellect in Monkeys, Apes, and Humans.* New York: Clarendon Press/Oxford University Press, 1988.
86. Nelson, Ximena J., Garnett, Daniel T., Evans, Christopher S. "Receiver Psychology and the Design of the Deceptive Caudal Luring Signal of the Death Adder Ximena," In *Animal Behaviour* 79(3), 2010.
87. Kjernsmo, Karin. *Anti-predator Adaptations in Aquatic Environments,* Turku: Åbo Akademi University Press, 2014.
88. Van Zandt Brower, Jane. "Experimental Studies of Mimicry in some North-American Butterflies," In *Foundations of Animal Behavior: Classic Papers with Commentaries*, Chicago: University of Chicago Press, 1996, p.366.
89. Maran, *Mimicry and Meaning*, p.80–1.
90. Maran, *Mimicry and Meaning*, p.67 and Dalziell, Anastasia. H., and Welbergen, J.A. "Mimicry for All Modalities." In *Ecology Letters,* 19(6), 2016. See also Howse and Allen (1994).
91. Caillois, Roger. *Le mythe et l'homme.* Paris: Gallimard, 1938, p.105; Langerholc, "Facial mimicry," p.190 and 199.
92. Roper., T.J. and Cook, S.E. "Responses of Chicks to Brightly Coloured Insect Prey." *Behaviour* 110 (276–93), 1989. See also Alatalo, Rauno V. and Mappes, J. "Tracking the Evolution of Warning Signals." In *Nature* 382, 1996.
93. Maran, *Mimicry and Meaning*, p.67.

94. The *seme* is a unit of imitation; it can be a morphology, more generally a certain form, "color patterns, but also odors and kinds of behavior" first developed by one species or group of organisms and "consequently extended to the other often unrelated groups that were able to receive (or imitate) and built it up on their bodies or environment". (see Karel Kleisner & Markoš Anton "Semetic rings: towards the new concept of mimetic resemblances". *Theory in Biosciences* 123, 2005, p.218.)

95. Maran, *Mimicry and Meaning*, p.31 and chapter 11.

96. Kleisner, Karel and Anton, Markoš. "Semetic Rings: Towards the New Concept of Mimetic Resemblances." *Theory in Biosciences* 123 (2005), p.219.

97. Kull, Kalevi. "Ecosystems are Made of Semiosic Bonds: Consortia, Umwelten, Biophony and Ecological Codes." In *Biosemiotics*, 3(3), 2010, p.354.

98. Maran, Timo. "Are Ecological Codes Archetypal Structures?" In Maran, Timo, Lindström, Kati, Magnus, Riin, & Tønnessen, Morten (Eds.), *Semiotics in the Wild: Essays in Honour of Kalevi Kull on the Occasion of his 60th Birthday*, Tartu: University of Tartu Press, 2012, p.149.

99. Burt, Jonathan. *Animals in Film*, London: Reaktion Books, 2002, p.38–63.

100. See Deleuze about the "reume" (namely the sign that cancels every center and makes the meaning fluid) and the role of water and rivers in cinema, in *Cinema 1 L'image-mouvement*, Paris: Minuit, 1983, p.116.

101. See for instance Savage-Rumbaugh, Sue. "Communication, Symbolic Communication, and Language: Reply to Seidenberg and Petitto", in *Journal of Experimental Psychology*, 116(3),1987. Bovet, Dalila and Vauclair, Jacques. "Picture Recognition in Animals and Humans." In *Behavioural Brain Research* 109, 2000. Truppa, Valentina et al. "Picture Processing in Tufted Capuchin Monkeys (Cebus apella)." In *Behavioural Processes,* 82, 2009. And Kaminski, Juliane et al. "Domestic Dogs Comprehend Human Communication with Iconic Signs." In *Developmental Science*, 12(6), 2009.

102. "The first is that whose being is simply in itself, not referring to anything nor lying behind anything" (Peirce, Charles. *Collected Papers*, 8 vols. Hartshorne, Charles, Weiss, Paul, and Burks, Arthur (Eds.). Cambridge: Belknap Press of Harvard University Press, 2, 356). See also "A Firstness is exemplified in every quality of a total feeling. It is perfectly simple and without parts; and everything has its quality" (*Ibid.*, 1, 531)

103. *Ibid.*, 3, 362. See also Pharies, David A. *Charles S. Peirce and The Linguistic Sign*. Amsterdam & Philadelphia: John Benjamins, 1985 p.35.

104. See supra, §15.

105. See in particular Haldane, John Burdon Sanderson. "Rituel humain et communication animale," In *Diogene* 4, 1953; Huxley, Julian. "A Discussion on Ritualization of Behaviour in Animals and Man," in *Philosophical Transactions of the Royal Society of London*, Series B, Biological Sciences, 251 (No. 772), 1966; and McFarland, David. *Animal Behaviour* (Third Edition), Edinburgh: Pearson Education, 1999. See also Bateson, *Steps to an Ecology of Mind*, p.187.

106. Bateson, *Steps to an Ecology of Mind*, p.183–7.

107. *Ibid.*, p.57.

108. *Ibid.*, p.67.

109. *Ibid.*, p.61.

110. *Ibid.*, p.65.

111. *Ibid.*, p.62.

112. *Ibid.*, p.63. Reciprocally, play may escalate into fight. See for instance Pellis, Sergio M. "Keeping in Touch: Play Fighting and Social Knowledge," in Bekoff, Marc, Allen, Colin, and Burghardt, Gordon M. (Eds.), *The Cognitive Animal: Empirical and Theoretical Perspectives on Animal Cognition*, Cambridge: The MIT Press, 2002, p.422.

113. Bateson, *Steps to an Ecology of Mind*, p.190.
114. *Ibid.*, p.66–7.
115. *Ibid.*, p.67.
116. *Ibid.*, p.64.
117. *Ibid.*, p.61 and 67.
118. *Ibid.*, p.62.
119. *Ibid.*, p.62–3.
120. *Ibid.*, p.150 and p.296.
121. *Ibid.*, p.150.
122. Freud, Sigmund, "The Interpretation of Dream." In *The Complete Psychological Works* New York: Norton & Company, 1976 p.608 sqq.
123. *Ibid.*, p.634.
124. This dimension of creativity and unpredictability in communication also justifies the reference to a genuine cooperation between human and non-human animals: without joint significant efforts, without the will of non-human animals to actively take part in working sessions, the trainers' "commands" remain meaningless. This goodwill sometimes peters out and turns into play solicitation, ill will, display of power, stress, or fatigue for instance. See also Carlos Pereira's work, chapter 2 of this book, supra, p.79–80, and Haraway's analyses of becoming-with (chapter 5, infra §35).
125. Hearne, Vicki. *Adam's Task: Calling Animals by Name*. New York: Knopf, 1987, p.61.
126. *Ibid.*, p.248.
127. The story of Nim Chimpsky provides a saddening example of such evil outcomes. See Hess, Elizabeth. *Nim Chimpsky: The Chimp Who Would Be Human*, New York: Bantam Books, 2009.
128. Bateson "Why a Swan?" (1953), in *Steps to an Ecology of Mind*, p.44.
129. Bateson, *Steps to an Ecology of Mind*, p.57.
130. *Ibid.*, p.60.
131. *Ibid.*, p.62.
132. *Ibid.*, p.66.
133. *Ibid.*, p.66, my emphasis.
134. *Ibid.*, p.64–5.
135. *Ibid.*, p.57.
136. *Ibid.*, p.60.
137. *Ibid.*, p.57.
138. *Ibid.*
139. *Ibid.*, p.57–8.
140. *Ibid.*, p.59.
141. *Ibid.*, p.148.
142. *Ibid.*
143. *Ibid.*, p.149.
144. See for instance Mullally, Sinéad L. and Maguire, Eleanor A. "Memory, Imagination, and Predicting the Future A Common Brain Mechanism?" In *The Neuroscientist* 20(3), 2013. Clayton, Nicola Susan, Bussey, T.J., and Dickinson, A. "Can Animals Recall the Past and Plan for the Future?", In *Nature Reviews Neuroscience*, 4 (685–91), 2003. Roberts, William A. "Mental Time Travel: Animals Anticipate the Future," in *Current Biology*, Vol. 17, No. 11, 2007. Kabadayi, Can, Osvath, Mathias. "Ravens Parallel Great Apes in Flexible Planning for Tool-Use and Bartering", in *Science*, Vol. 357, No. 6347, 2017. And Zentall, Thomas R. "Animals Represent the Past and the Future", in *Evolutionary Psychology* 11(3), 2013.
145. An important concept in "Why a Swan?"
146. Bateson, *Steps to an Ecology of Mind*, p.66.

147. Bateson, "A Theory of Play and Fantasy," In *Steps to an Ecology of Mind*, p.184–5.
148. *Ibid.*, p.184.
149. *Ibid.*
150. Bateson, *Steps to an Ecology of Mind*, p.48.
151. *Ibid.*, p.49–50.
152. *Ibid.*, p.48–9.
153. Derrida, *The Animal That Therefore I Am*, p.30.
154. Bateson, *Steps to an Ecology of Mind*, p.61.
155. See in particular *Phenomenology of Perception*, Part one, chap VI, "The Body as Expression, and Speech"
156. Bateson, *Steps to an Ecology of Mind*, p.432.
157. See also, infra, p.214, n.34.
158. *Ibid.*, p.169.
159. *Ibid.*
160. *Ibid.*
161. See for instance Merleau-Ponty, *La prose du monde*, Paris: Gallimard, p.35–7, *Signes,* Paris: Gallimard, 1960, p.49–50.
162. Merleau-Ponty, *L'institution. La passivité. Notes de cours au Collège de France* (1954–1955). Paris: Belin, 2003.
163. *Ibid.* p.184–213.
164. "We hurried downstairs, and as soon as I was outside I woke up." Freud, Sigmund, *Fragment of an Analysis of a Case of Hysteria*, In *The Complete Psychological Works*, p.1401. In fact, it was more than time for Dora to run away from her current situation and a family circle that is indeed hardly metaphorically on fire. Through what we recognize as our dreams, we speak to ourselves and discover that we stem from lines of meaning that haunt us and that we cannot royally decide to abolish or transform. But these motifs are also ambiguous and call for original re-institutions. Dora's decision is both hers and the result of a secret maturation that grew in the bowels of an intricate situation.
165. Merleau-Ponty, *Nature*, p.288 (Trans. 225). In connection with animal desire, see also the inspiring reflections developed by Florence Burgat in *Liberté et inquiétude de la vie animale*. Paris: Kimé, 2006, esp. p.191–3.
166. Merleau-Ponty, *Signes*, p.150 (in "De Mauss à Claude Levi-Strauss").
167. Merleau-Ponty, *Nature,* p.242 (Trans. p.185).
168. *Ibid.*, p.231 (Trans.175). Merleau-Ponty also uses the vocabulary of imagination and dream to describe, first, the relation between the animal and its Umwelt (*Ibid.*, 233, Trans. 178) and, second, the impossibility to fully predict the way different vital functions are achieved in animal behavior (*Ibid.*, 207, 251, 253 and 255, Trans 155–156, 193, 194, the sentence "Il y a un caractère onirique de l'instinct" was not translated in the English edition).
169. *Ibid.*, p.246 (Trans. p.188).
170. *Ibid.*, p.258 (Trans. p.198).
171. Portmann, *Die Tiergestalt*, p.234.
172. Merleau-Ponty, *Nature*, p.255 (Trans. p.196)
173. *Ibid.*, p.250–1 (Trans. p.192).
174. *Ibid.*, p.253 (Trans. p.193).
175. *Ibid.*, p.254 (Trans. p.195).
176. *Ibid.*, p.305 (Trans. p.240).
177. The representation of disgust is another good example of an image that circulates from humans to non-human animals, back and forth. Paul Rozin contends that disgust is most certainly proper to humans. He emphasizes that moral repugnance is specifically human and that, in humans, disgust is

especially elicited by animals or "anything that reminds us that we are animals" (Rozin, Haidt, & McCauley, "Disgust". In Lewis, M., Haviland, J., and Barrett, L. F. (Eds.). *Handbook of Emotions*. New York: Guilford Press, 2008, p.761). Nevertheless, on the one hand, the moral dimension of human disgust should not be understood through a hypostasis of reason: this emotion remains first and foremost visceral. On the other hand, as shown by Toronchuk, Judith A. and Ellis, George F. R. ("Disgust: Sensory Affect or Primary Emotional System?", in *Cognition and Emotion*, 21:8), disgust in animals is a general pattern of behavior that involves a complex set of organs and learning processes, as well as a theatrical dimension. Even pleasant tastes can become disgusting. Rats can learn to avoid some foods, but they demonstrate gaping faces only when such aliments are associated with illness. Non-human animals do not only vomit or, when they are non-emetic, gape, they may stage and overemphasize the reject. "Rats do not just avoid food items after taste-aversive conditioning—they gape, open their mouths, gag and retch, shake their heads, and wipe their chins on the floor. A coyote may retch, roll on the offensive food, and then kick dirt on it, while a cougar may shake each paw. Monkeys may react to offensive objects by excessive sniffing and manipulation often followed by breaking and squashing the item, then dropping or flinging it away and wiping their hands" (*Ibid.*, p.1803). Some rats "claw out the filter paper containing this apparently offensive odor [that they have learned to averse], push the paper through slots in the floor at the other side of the cage, and only then drink the water" (p.1804). Spatial symbolism seems to be emerging here. A dialectic between interiority and exteriority is instituted and dramatized. Through rituals and a form of representation, distance, identity, and rejection are played out. More than that, they are constantly negotiated and appear as intrinsically unstable. I consequently refuse the claim "à la Nagel" that phenomenological data are missing with regard to disgust in animals (Joshua Rottman "Evolution, Development, and the Emergence of Disgust", in *Evolutionary Psychology*, 12(2), 2014) or that such a disgust lacks an existential dimension. The phenomenon of nausea is literally that of inside-out guts. Through nausea, human and non-human animals feel and represent the tension and reversibility between interiority and exteriority. In effect, animals are essentially defined by heterotrophy and, as I have argued in "Is a World without animals possible?" (Dufourcq, 2014), the disintegration-assimilation activity goes together with a dimension of resistance and risk in feeding. Heterotrophy entails a relation with other living beings where the latter become both the flesh-relatives and the rival aliens of the individual that eats them. The drama of disgust in non-human animals shows that the fantastic terror of devouring and being devoured which recurs in so many human myths is a symbol constitutive of the living flesh [*Leib*] of animals. And although the human disgust for animals is a noteworthy institution, it is in fact a new fold in a process that already exists in non-human animals.

178. Shepherd, Stephen V. "Following Gaze: Gaze-Following Behavior as a Window into Social Cognition", *Frontiers in Integrative Neuroscience*, 4, 2010. See also Seeman, Alex (Ed.). *Joint Attention. New Developments in Psychology, Philosophy of Mind, and Social Science*, Ed., 43–72. Cambridge: MIT Press, 2011.
179. Deane, C. Douglas. "The Broken-Wing Behavior of the Killdeer". *The Auk*, 61, 1944.
180. Fontenay, Élisabeth de. *Le Silence des bêtes*, Paris: Fayard, 1998, p.664.
181. Oliver, Kelly. "Stopping the anthropological machine: Agamben with Heidegger and Merleau-Ponty", PhaenEx, vol. 2, 2007.

182. Agamben, Giorgio. *The Open; Man and Animal.* Trans. Kevin Attell. Stanford: Stanford, University Press, 2004.
183. See supra, §15.
184. Lacan, Jacques. *Le Séminaire. Les écrits techniques de Freud* (1953–1954), tome 1, p.159.
185. Lacan, Jacques. "Fonction et champ de la parole et du langage en psychanalyse", in *Ecrits*, Paris: Seuil, 1966, p.273–7 and 293–300.
186. Lacan, Jacques. *Le séminaire, Livre II, Le moi dans la théorie de Freud et dans la technique de la psychanalyse,* Paris: Seuil, 1977, p.371, Trans. J. Forrester, Cambridge: Cambridge University Press, 1988, p.322–3.
187. *Les écrits techniques de Freud*, p.142, Trans. p.122.
188. See Oliver, Kelly, *Animal Lessons: How They Teach Us to Be Human.* New York: Columbia University Press, 2009, p.175–89.
189. Derrida has famously deconstructed this putative heterogeneity between human responses and animal reactions: see *The Animal That Therefore I Am*, p.122–6.
190. Bailly, Jean-Christophe. *Le Dépaysement,* Paris: Seuil, 2011, reprint Points 2012, p.427, my translation.
191. *Ibid.*
192. *Ibid.*, p.422.
193. See also Jean-Christophe Bailly *Le versant animal*, Paris: Bayard, 2007, p.40, a book that certainly succeeds in developing an original and inspiring relation to animality.
194. Young, Rosamund. *The Secret Life of Cows,* London: Faber & Faber 2017.

5 Metamorphoses and corporeal imagination
The "second person" at stake

§32 Several takes on becomings-animal: The issue of imagination

In *A Thousand Plateaus*, Deleuze and Guattari argue that non-human and human individuals have no fixed essence and can ontologically contaminate each other. Thus, for instance, Captain Ahab could enter a becoming-whale, which also modified the "being" of whales altogether. This famous concept of becomings-animal [*devenirs-animaux*], built upon a non-positivist ontology, shares significant common features with my concept of the chiasm between non-human and human animals, inspired by Merleau-Ponty's phenomenology. Deleuze and Guattari's theory certainly helps us understand how the human imaginary of animals involves actual metamorphoses. With Deleuze and Guattari, we learn to focus less on the representation of what it is like to be an animal, and more on the actual metamorphoses that the communication with other animals implies. How exactly do our bodies and dynamic attitudes enable the "transposition" into other individuals? This is what Bachelard's theory of muscular imagination and Deleuze and Guattari's theory of affects help us find out. I will thus define dynamic imagination as the most fundamental form of imagination that fuels and shapes ossified myths and storybook images: Understanding images consists in enacting and experiencing this dynamism, in resonance with non-human animals' energy and potentialities, rather than in categorizing and ascribing a symbolic meaning to a set of clichés.

But a problem immediately arises. Whereas I have consistently put the concepts of images, the imaginary, and imagination forth in my analyses, Deleuze and Guattari firmly reject a description of becomings-animal as images or as involving our imagination in any way. "Becomings-animal are neither dreams nor phantasies [*fantasmes*].[1] They are perfectly real."[2] Likewise in *Kafka: Toward a Minor literature* metaphor is presented as the enemy:

> "'Grasp the world,' instead of extracting impressions from it; work with objects, characters, events, in reality, and not in impressions. Kill

metaphor. Aesthetic impressions, sensations, or imaginings still exist for themselves in Kafka's first essays where a certain influence of the Prague school is at work. But all of Kafka's evolution will consist in effacing them to the benefit of a sobriety, a hyper-realism, a machinism that no longer makes use of them. This is why subjective impressions are systematically replaced by points of connection that function objectively as so many signals in a segmentation."[3]

Therefore, should we give up the conceptuality of the imaginary because it weakens the power of actual metamorphoses? I will contend that, for several significant reasons, the theory developed in the present book does not coincide with Deleuze and Guattari's perspective.

I begin by summarizing the many points of tangency between the preceding chapters and Deleuze and Guattari's approach to animals.

I have defined, in line with Merleau-Ponty's philosophy, living beings as phantom-like. The living emerges through an unpredictable folding process immanent in matter, where no absolute essence, no plan, no fixed nature can be identified. Deleuze famously shattered the boundaries between species as well as between individual organisms and focused on what he calls the plan of immanence, namely a plan of multiplicities, ultimately fissured by and consisting in difference and repetition. This unilinear plan dooms every formation (*stratum*) to uncontrolled metamorphoses, and, overall, to a splintered and impossible identity. Thought along the model of the egg, this meta-stable matter is the "prebiotic soup"[4] where relatively stable individuals form, morph, and dissolve in an unpredictable manner. Deleuze has radically decentered anthropology as well as the philosophy of life into a philosophy of intensities and metamorphoses, making even the concept of "center" obsolete.

Furthermore, a return to what I have called an oneiric expressiveness is also at stake in *Proust and Signs, Kafka: Toward a Minor literature,* and "1837: Of the Refrain" in *A Thousand Plateaus.* How can sense [*sens*] arise from bodies independently of the norm imposed by a conventional system of symbols? "Of the Refrain" studies animal expression and, more particularly, the emission of signs whereby animal territories form, out of the blue. Deleuze and Guattari conclude that art is a vital process that creates its own meaning and its own world, while remaining connected to other possible meanings and worlds. "[Human] art is continually haunted by the animal."[5] More radically, in Deleuze and Guattari's view, art was not invented by humans and has nothing essential to do with traditional key concepts of the philosophy of art, such as representation, copy-images, or spiritual expression.

I also share with Deleuze and Guattari a criticism of the model of copy-images, reproductions, and re-presentation: We cannot start from alleged clear-cut entities and claim that they would be connected externally by the artificial device of the human talent for metaphors.

In truth, the concept of imagination and image were not banned from Deleuze's philosophy, and I will contend that many bridges can be built between Deleuze's concept of the virtual and my theory of the imagineareal that was drawn from a phenomenological influence. This is also clear from Deleuze's reference to imagination in Empiricism and his theory of images inspired by Bergson.[6]

Could an agreement be found between my approach to the imaginary of animals and Deleuze and Guattari's concept of becoming-animal if I simply emphasized that, in my conception, the imaginary is actually not opposite to the real and that "imagination" can never be defined as a specifically human capacity for arbitrary invention? In fact, the discrepancy between these two approaches has deeper roots than contrasting definitions of words. What is at stake fundamentally is the role that should be ascribed to the imagination of human and non-human animal individuals in the processes of becoming-animal. As I will argue, it is crucial to engage non-human animals not only in their virtual dimension, but also in their subjective, imaginative activity and phenomenological perspective. Through such consideration, I will also tackle the issue of the thorny relationship between Deleuze and phenomenology.

It will be illuminating to invite Bachelard and Haraway into the discussion. Bachelard's study of Lautréamont's *Songs of Maldoror* constitutes a reference and a major source of inspiration for Deleuze and Guattari in "1730: Becoming-Intense, Becoming-Animal, Becoming-Imperceptible...." However, in Bachelard's view, Lautréamont's metamorphoses are both real and essentially connected to our imagination. Seventy years later, in *When Species Meet,* Haraway resumes this issue and rethinks becomings-animal against Deleuze and Guattari. She firmly reasserts the necessity of acknowledging the effective and operational nature of such becomings, but she also wants to stress that these becomings are *both* real and imaginary. Again, interestingly enough, "imaginary" and "real" are not regarded as opposites. There is no disagreement among all the authors involved regarding the fact that becomings-animal are real. The question is: Should we still give some significance to human and non-human individuals' imagination in these becomings? I see a cohesive bond between the role played by the concept of "figures" in Haraway's theory of becoming-with-animals and her critique of Deleuze and Guattari's violent scorn for companionship between non-human and human individuals. There is, in principle, still room in Deleuze's philosophy for individual subjects and personal agency. But in fact, in the theory of becoming-animal, as well as on several other occasions, Deleuze and Guattari express a harsh contempt for a relation that would enhance such individual subjectivity and agency in animals. When ideologies threaten to take over, an excursion through the mediation zone constituted by our imagination and its play with possible realities can prove salutary. At stake is a broader space for *all* the complex options offered by becomings-animal, including becomings-*with*-animals.

I will first analyze Bachelard's concept of motor imagination in *Lautréamont* and show how studying processes of becoming-animal can help us unveil the corporeal roots of the imaginary of animals (§33). Starting from Deleuze and Guattari's critique of Bachelard's approach, I will show that an ontology of the virtual is substituted for an ontology of the imaginary in *A Thousand Plateaus*. From this substitution, Deleuze and Guattari conclude that becomings-animal should be considered as real and *in no way* metaphorical, although in a deeply subverted sense of the word "real" (§34). The next paragraph will concern itself with the analysis of the concepts of figures and fact-fiction in Haraway's philosophy of animal becomings. Of particular interest here will be the concepts of becoming-with, inter-subjectivity, other-worlding, and dance, which reflect Haraway's concern for individual perspectives and agency, namely, so to say, a phenomeno-logical perspective. We will see how the reference to figures and fiction in Haraway is closely tied to a concern for individual imagination (§35). I eventually contend that Deleuze and Guattari's use of the concepts of real and imaginary in "1730: Becoming-Intense, Becoming-Animal, Becoming-Imperceptible..." is Janus-faced and ideological. Their con-tingent choices—their own imagination—put the mask of an ontological discarding of individual human and non-human imaginations. I argue in favor of the maintenance of the conceptuality of imagination and the imag-inary in order to ensure a better balance in the description of two aspects that are both integral parts of becomings, namely, on the one hand, deep anonymous processes, and, on the other hand, the personal creative way of engaging such processes. Finally, I show that other texts in Deleuze's and Deleuze and Guattari's works allow for a more nuanced approach (§36).

§33   Bachelard and Lautréamont: Motor imagination and animal dynamogeny

Lautréamont

Animals thrive in Lautréamont's *Songs of Maldoror*. The poem contains a wealth of—sometimes anthropomorphic—descriptions of animal behav-iors. "A sperm-whale raises itself gradually out of the sea's depths and shows its head above the water in order to see the vessel that passes by this sol-itary place. Curiosity was born with the universe."[7] Lautréamont's poem also consistently develops animal metaphors to describe human behaviors. Lautréamont explicitly compares humans with non-human animals: "Toad-faced man no longer recognizes himself and is continually lashing himself into fits of rage that make him look like a wood beast."[8] Lautréamont refers to specific animal parts (claws, fangs, tentacles, etc.) to describe the atti-tudes, moods, or actions of human characters. "Know that (...) the fever that fingers my face with its stump, every unclean beast that brandishes its claws, well it is my will that keeps them going round and round in order to provide solid nourishment for its perpetual activity."[9] More radically, in *The Songs of Maldoror*, living beings unrelentingly morph into each other, and hybrids multiply. "How astounded he was when he saw Maldoror, changed into an octopus, bear down upon his body with eight enormous arms."[10] "I applied my four hundred suckers to his armpits making him cry out horribly. The cries changed into serpents as they left his mouth...."[11] Maldoror, in particular, takes on multifarious forms. The reader barely manages to circumscribe them and to identify the subject that connects them together: Here "I," there "he," most of the time an invisible narrator. Maldoror also becomes a pig, an octopus, a dog, and, quite often, a blend of all of these guises. More consistently, so to speak, Maldoror comes across as a flood of hatred, anger, and aggressive energy that invades and torments words and images, governing the series of metamorphoses. The reader herself is transported by these transformations: "Reader, it is perchance hatred that you would have me invoke at the beginning of this work! How do you know that you would not snuff it up (...) through your wide, thin, prideful nostrils, turning up your stomach like a shark in the fine dark air (...) I assure you the savor will rejoice those two malformed holes in your hideous snout, O monster."[12]

In *Lautréamont*, Bachelard emphasizes the hypnotic and hallucinatory power of *The Songs of Maldoror* and contends that this poem draws on and heightens a deep form of imagination that implies a genuine becoming. An effective transformation occurs that shapes the sentences and carries the reader away: One cannot imagine with Lautréamont without *becoming*. "Animal life is no empty metaphor in Ducasse's work; he presents not symbols of passions but veritable instruments of attack."[13]

Lautréamont does not simply describe animals' outward appearance; nor does he tack human speech and social behaviors onto non-human animal bodies. Bachelard thus rightly contrasts *The Songs of Maldoror* with the *Fables* of La Fontaine.[14] Animals are invoked by Lautréamont as a certain style of being. Animals display a muscular dynamic that has the power to communicate new attitudes to our body and to infest it through contagion. It is worth noting, in passing, that Lautréamont's selection is consistently oriented toward aggression, attack, suffering and desire: We will return to that interesting bias later in this chapter (§35). As a result, to imagine consists in incorporating a new interiority. The latter stems from "subjective impulses"[15] and consists in a way of being whereby the entire world is remolded, a new *Umwelt* arises. Each animal, each organ, each gesture given in a first image becomes the living theme that animates a sequence of sentences and new images, orchestrating their rhythms and their tonalities. The poem thus becomes running, biting, attacking, crying *in energeia* and resonates in the reader's body. Life is then "polarized" by "animal speed and energy."[16] Each non-human animal conjured by Lautréamont brings a certain orientation. She has her idiosyncratic impulses, her own tempo, her accelerations and decelerations.[17] In Kafka's *Metamorphosis*, for instance, we experience a general slowing down and get to languish at a speed below which life disintegrates.[18] "Lautréamont's is a poetry of provocation [*de l'excitation*], of muscular impulse, and in no way a visual poetry of shapes and colors. Animal forms are poorly delineated in his work. In fact, they are not reproduced, but truly *produced*. They are induced by movements (...) Animality is seized from within."[19]

Material and dynamic imagination:
The connection to nature's imagination

In relation to the conceptual framework that Bachelard more systematically draws up in his research devoted to the imagination of elements,[20] it appears that the imagination at work in *The Songs of Maldoror* is what Bachelard calls "material imagination" and "dynamic imagination," in contrast to "formal imagination." The latter "takes pleasure in the picturesque;"[21] it delineates clear-cut images, and whimsically jumps from one image to the other. Formal imagination does not fathom the deep themes that connect images together in a series and give them the power to viscerally haunt us.[22] Through formal imagination, consequently, I superficially mimic that which remains foreign to me. By contrast, material imagination does not confine itself to any fixed image: It takes up a principle of variation and lets it inspire endless metamorphoses. Bachelard defined the concept of material imagination to insist on the possibility for the daydreamer to penetrate the sensible matter of elements, from which beings actually emerge. This matter secretes its own diffuse forms. From the contemplation of fire, the kneading of dough, or the walk against the wind, a world with a certain tone, style, and vivacity arises. These elemental reveries even engendered some pre-Socratic or Nietzschean philosophies.

Material imagination is defined as essentially dynamic in *Water and Dreams*, for it never sticks to one fixed image. Bachelard focuses more specifically on dynamic imagination in *Air and Dreams* and highlights that imagination overall is less the ability to shape images [*former des images*] than the ability to distort [*déformer*] them.[23] Such a distortion yet retains a certain consistency, for it can be achieved by following the inspiration of an immemorial imaginary that transcends human faculties and underlies the realm of clear-cut individuals and species.[24]

"The fundamental term that corresponds to imagination is not 'image', but 'the imaginary.'"[25] Imagination gets in contact with a principle of deep transformation of our affectivity, our attitude, and the rhythms of our existence. Correlatively, imagination does not first and foremost arbitrarily entertain fantasies, but rather detects, in things and living beings, stubborn open-ended themes from which countless avatars derive. In this way, imagination can endlessly move from one form to the other, from one animal to the other. This imagination in us is akin to *imagination of nature*, as Bachelard claims in *Lautréamont*.[26] Indeed, for instance, Lautréamont intertwines the image of a fish and the image of a bird through a sequence of metamorphoses, with a view to desegregating the well-delineated form of recognizable species or individuals. Lautréamont thus releases a general, dynamic scheme of flying-swimming that can develop beyond the limit of existing bodies and carry us into unexplored territories. In doing so, he also echoes the chimerical, natural creation of flying fish.

As a result, the reader is *really* transformed by the poem, insofar as the latter stems from material and dynamic imagination. The poem "creates its reader."[27] The howling dogs that haunt Maldoror, Lautréamont, and the reader are consequently neither arbitrary inventions, nor the essence of "the dog" in itself. Through the poem, human beings are transformed, and so are dogs since they inspire our becoming and lend themselves to these variations in our imagination. The poem creates new beings by bringing different animal corporeities into resonance.

Bachelard and the psychoanalysis of human imagination

Aren't Lautréamont's animals only figures in the human psyche? Aren't Maldoror's immense anger, hatred, and aggressiveness linked to what Bachelard calls the *Lautréamont complex* rather than to the lives of animals? Bachelard's approach is ambiguous in this respect and clarification is needed to eschew a psychologizing and anthropocentric interpretation of this theory.

Bachelard initially conceived of the imaginary field as a set of complexes that are epistemological obstacles. Bachelard called upon philosophers and scientists to challenge and overcome such obstacles. This position is central in *The Psychoanalysis of Fire*. But, in later works on water, air, and earth, Bachelard insists on the aesthetical and ethical potential fecundity of complexes, eventually regarding the latter as fundamental structures of our relation to the world.

The so-called "Lautréamont complex" is not any more the complex of Isidore Ducasse than the Oedipus complex is the complex of a singular Greek mythical character. A complex, in Bachelard's philosophy, following a psychoanalytical and particularly a Jungian inspiration, is a set of conflictual tendencies, emotions, and representations that irresistibly develop in connection with one specific aspect of our existence. A complex possesses a relatively identifiable core but is also full of potentialities and functions as a set of critical variations on an existential theme. And indeed, fundamental existential problems are raised by the interlacing between human and non-human animals, or by the conflict between, on the one hand, social norms, disciplining processes, and, on the other hand, the unlimited possibilities offered to us by animal modes.

To be sure, each complex is problematically connected to particular circumstances, idiosyncratic moods, and personal interpretations. The idiosyncrasy of Bachelard's approach is unveiled by the reductive and systematic selection of attack and brutality in his description of non-human animals, as well as by the identification between animality and aggressiveness or bestiality. Bachelard's reading of Lautréamont is in fact quite selective, therefore symptomatic.

However, Bachelard claims that complexes are challenges to which individuals have to respond. Complexes are open questions that invite us to learn how to better imagine.[28] And Lautréamont is a guide to every reader, in that he succeeded in bringing this issue of *animals within us* to expression, in all its complexity, via the creation of lively and inspiring figures.[29] Imagination takes up an existential complex, brings mediations, and set picturesque screens through which this complex can be channeled and turned into more creative forces. Lautréamont helps us better imagine for two reasons. First, according to Bachelard, a successful imagination must be faithful to its inspiration. Meaningful fantasies open themselves to the singular style of this or that animal movement or behavior; thus, avoiding arbitrariness, they engage in the unique manner these movements and behaviors reveal new aspects of the world. The daydreamer learns to act with more precision and through a wider range of motion and behavioral patterns. Second, a successful imagination raises itself to a cosmic level: It lets itself be animated by original ways of being, but, in return, it gives the latter a tenfold increased intensity. It multiplies and playfully invigorates variations on these themes. Imagination is the way toward "superexistence."[30] Further, through the mediation of images and variations, the daydreamer turns to meaning and meaning-giving processes instead of merely undergoing the weight of existential tensions and deadlocks. Lautréamont thus overcomes the great divide between humans and animals: With his work, we learn to harbor, in our body, specific animal behaviors, which, thereby, take on new forms and unfold in a new fashion.

The circular concept of co-institution, that was defined in the previous chapter, is the most adequate to describe the relation between this creative cosmic imagination and the inspiration it receives from actual animals. One cannot easily draw a boundary between the personal dimension of the complex and the very being of animals addressed through Lautréamont's images. What inspires and provokes our imagination, even if it is a psychological complex *via* the reference to animals, must find motifs that resonate with human concerns *in animals themselves*. Indeed, with Bachelard, in a way that foreshadows Deleuze's approach, we stand at the level where complexes take the form of rhythms, trajectories, orientations, and impulses. Fantasies are not to be reduced to psychological facts. Accordingly, imaginative transposition processes are now deemed legitimate, without there being any need to prove that animals consciously thematize existential problems. Living bodies vibrate in resonance with each other and subjectivity—as a certain way of engaging the world and others—ensues from this set of body attitudes. The dynamic of becoming cannot begin with human beings exclusively.

The emergence of thought from a
human-animal muscular imagination

How does Bachelard theorize the rooting of Lautréamont's imagery in a deep kinship between human bodies and non-human animal bodies? Bachelard particularly underscores the role played by "muscular lyricism" and "muscular syntax"[31] in human thought, which allows him to essentially link non-human animals and *every* human thought.

Bachelard draws to our attention the fact that our imagination essentially tends to animalize everything:[32] It finds some ideas fishy; it feels butterflies in our stomach, creepy-crawly tingling in our spine, ideas and questions buzzing around our head; it places elephants in rooms or, in French, in China shops, and makes time fly.[33] In fact, imagination cannot understand anything without actively adumbrating gestures, virtually drawing lines, and moving in virtual space. Indeed, imagination deals with the ubiquitous dimension of every being and must therefore play out the imaginareal themes through space and time. More fundamentally, understanding in general necessitates the same dynamic imagination, for it must connect the multifarious realm of particular individuals and general schemas/ concepts.[34] There is accordingly an essential link between thought, imagination, and the motricity of living bodies.

Consider the familiarity that is achieved in our body through the incorporation of a certain dexterity in the handling of a car, for instance, or when we acquire a sense of direction in a city. These gestures are much more than actual means to produce a specific useful outcome, they reach the virtual field by linking various experiences and they can adjust to an infinity of concrete situations to the point that, on the basis of a certain attitude and a certain state of tension in my body, I can experience concrete metaphors. I viscerally feel the viscosity or the acidity in a voice or, when a music takes possession of my muscles and limbs, simultaneously new visual, olfactory, tactile, or gustatory sensations arise.[35] Body attitudes and motion patterns can function as Gestalten that bear endless possibilities and, thus, as symbolical matrices. I recognize a sense of familiarity in my hands and my legs, in the tingling and twitching, the restlessness (what is called "impatiences" in French), the quivering in my muscles even before I achieve fullfledged movements. Bergson coined the concept of motor schema [*schème moteur*] in *Matter and Memory* to designate the rehearsal of a certain type of action through nascent micro-movements that accompany our conscious or half-conscious recognition of things or our understanding of a speech we are listening to.[36] Against this backdrop, Bachelard's claim that "there is nothing in the understanding [*l'intelligence*] that was not first in the muscles"[37] can become more intelligible.

In *Les Chants de Maldoror*, it is precisely this *muscular imagination* that is invigorated and thus proves to be essentially tied to animal bodies. For instance, Bachelard describes how, guided by Maldoror, we can discover the common muscular origin of human and leonine pride. Lautréamont prompts us to focus on the movements of our neck: "Like a condemned man tries his muscles while reflecting upon their destiny, knowing that he is about to mount the scaffold, I stand upright on my straw pallet, my eyes closed and I turn my head slowly from right to left, from left to right, for whole hours on end."[38] We can feel how lifting up our neck and acting out surveying attitudes give us a sense of physical self-assertion and of a more dominant presence. Hence, Bachelard argues, the importance of collars, ruffs, and ties in etiquette, in contexts where power relations prevail, and, overall, in the "psychology of majesty."[39] As Bachelard states it, "A muscular syntax"[40] is at play far from the focal point of our common consciousness. Bachelard also calls this muscular syntax "muscular lyricism," since it develops metaphors that move, for instance, from muscles to lions, pride, and ties. Through muscular lyricism, the meaning of body attitudes, animal behavior, and cultural artifacts is *simultaneously understood and artistically created*. In other words, we think less through the rigorous subsumption of individuals under fixed concepts, than through a corporeal matrix that uses virtual movements to endlessly create variations on protean themes.

Bachelard suggests that muscular imagination essentially draws inspiration from the animal "dynamogeny." In effect, embodied imagination possesses its own inertia. In other words, the medium of our imagining body, in its actuality, is limited, in that it has crystallized into a certain center of gravity and a set of ossified habits. For instance, human thought obviously has a predilection for visual forms. Admittedly, such habits are not absolutely rigid properties of our body and organisms have a profound virtual dimension, but the move toward a radically original and decentered imagination requires some effort. The imagining body therefore benefits greatly from a stimulation that comes from other habits and other patterns of behavior that have crystallized into radically different morphologies and *Umwelten*.

Intercorporeity with animals spurs our imagination: Non-human animals, for instance, propose countless "patterns of swimming [*formules de nage*]."[41] Plants, and, more broadly, elements, certainly also inspire our imagination and expand the range of motion patterns. But, in comparison to elements and plants, animals are characterized by a greater number of structures that store energy and release it discontinuously. The animal realm is an exuberant demonstration of breaks in rhythms, multiple phases of activity and passivity, movement and rest, observation and attack. With Lautréamont and through the animal dynamogeny, we can learn to feel, play with, and intensify the nascent characters and habits (ἤθη [ethe]) that hatch in our muscles.[42] We get to discover the meaning of such nascent characters as well as their potential further developments. "[The reader] can feel the myopsyche reviving within him almost fiber by fiber. An animalized imagery helps him reach that curious state of muscular analysis."[43]

Fantastic creatures: Symbols and motor imagination. The example of the unicorn

Bachelard's concepts of motor imagination[44] and muscular lyricism are crucial for analyzing fantastic creatures. Of course, dragons, phoenixes, mermaids, unicorns, and griffins are the bearers of complex symbols. But a cultural description of the symbols traditionally attached to these imaginary animals should not be an exclusive starting point for their study. The interpretation will remain cold and dry as well as too univocal, if not sustained by muscular lyricism. Hence, we cannot simply rely upon a dictionary of symbols or a structuralist analysis of the system of values where symbols acquire their meaning through formal relations. The approach through muscular imagination presents the significant advantage of rooting symbols in materials that bridge the human-animal divide.

Unicorns, for instance, first described as real exotic animals, have been defined as the symbol of innocence, chastity, or as the symbol of Christ. Today, they also symbolize the creative and imaginative relation to sex and gender claimed through a queer approach. As unreal, deprived of any obvious gender,[45] and fiercely resisting human understanding and domination, unicorns thus have become the symbol of the social invisibility and the noncategorizability of the LGBTQI+ community. Symbols branch out exponentially, also through more formal and accidental associations. The myth of unicorns made its way through a bumpy road. For instance, in Greek versions of the Old Testament, the Hebrew "re êm" was sometimes translated into "unicorn," sometimes into "wild ox" or "rhinoceros."[46] Saint Basil provides a good example of a formal symbolic root: He explains the link between Christ and the unicorn as follows: "Christ is the power of God, therefore he is called the unicorn on the ground that He has one horn, that is one common Power with the Father."[47] Symbols often freeze into clichés or even commodities, as also shown by the recent unicorn trend. A purely symbolic approach is thus doomed to either superficial univocity or motley multivocity.

However, Cassandra Eason, in *Fabulous creature, Mythical Monsters, and Animal Power Symbol,* importantly reminds us that non-human animals provide exceptional and dynamic inspiration to human imagination: "The human imagination has run riot regarding the magical equivalents of powerful birds such as the eagle. If such creatures could exist in the everyday world, then their spiritual equivalents must be much more incredible."[48] And, in effect, through history, the unicorn is depicted as the most relentless and belligerent animal, a powerfully-armed goat/horse that, nevertheless, lets itself be tamed by sweetness and graciousness. In medieval art and legends, only virgin maidens can approach unicorns. Once it has surrendered, the unicorn is entirely vulnerable, but also, still, the blatant figure of fierceness. "He is alert and could spring away at any moment on his coltish legs."[49] Half-coaxed, half-consenting, it remains awe-inspiring and dangerous, although turned into the most gentle and faithful creature. The unicorn is the combination of extreme shyness, gentleness, and potential fury—such as can be experienced in many hunting/taming interactions with non-human animals—taken to a paroxysm and cast into the adequate fantastic body. This complex dynamism is an integral dimension of the myth of the unicorn. It can thus sustain and be sustained by correspondences with Christianity, courtly love, and, in more recent revivals, queerness, or even, according to the charming interpretation ventured by Alice Fischer—who tries to account for the half-appalling, half-enticing recent proliferation of unicorns in girls' toys—a wild leap of hope in the midst of dark times.[50]

Another example of human-animal corporeal imaginary: Becoming animal through dreams in Ojibwa culture

The essential connection between animals and an effective, dynamic imagination has also been highlighted in a non-Western context, which tends to substantiate Bachelard's theory of a stubborn imaginary field stemming from the fundamental structures of animal existence. I am here referring to Hallowell's work on the Ojibwa ontology and on a form of becoming-animal through dreams.[51]

An Ojibwa tale narrates the story of a man called Iron Maker who "sank to the bottom of a lake after his boat has capsized."[52] He was drowning, but then "he thought of the beaver, whereupon the beaver came to him and gave him his body. He swam towards the shore, but before he could reach it, he felt himself losing the power to keep the shape of the beaver. So he thought of the otter. Then the otter gave him his body, and in that form he reached land."[53]

This story interestingly intertwines real metamorphoses and the mediation of thought ("he thought of the beaver," "he thought of the otter"). Yet, the terms "imagination" and "the imaginary" must be applied to this story only with great caution, for they are deeply imbued with Western ontologies. Nevertheless, the emphasis placed by Ojibwe on the significance of dreams in their relation with non-human persons is striking. According to their ontology, other-than-human persons can be encountered in waking life, but mostly by humans who have been made especially receptive and sensitive to them through dreams and, in particular, by undergoing a dream fast. Every boy, on reaching puberty, embarks upon a prolonged period of fasting, during which what matters most is to dream a lot and to dream well, namely to make significant encounters in dreams. But why are dreams so effective?

According to the Ojibwe people, a subject is a soul contained in an outward appearance that is susceptible to transformation.[54] More capacities and broader experience go together with more metamorphoses. Also, "metamorphosis" equals "power." Such metamorphoses are neither real nor imaginary. They are certainly not regarded as unreal or purely fictitious by Ojibwe, even when they are experienced in dreams. But it is still remark-able that Ojibwe regard dreams as the place where such metamorphoses bloom and ultimately take roots. "This is not to say that Ojibwa confuse dream experiences with those they have while wide awake. The difference is that in dreams, the vital essence of the person—the self—is afforded a degree of mobility, not only in space but also in time, normally denied in waking life."[55] Ojibwe thus distinguish between a surface of relatively stabi-lized bodies—which defines the experience of the waking life—and a plane of ontological fluidity, where metamorphoses intensify and the boundaries between species and individuals are porous. Likewise, in our account of Bachelard's approach, dynamic imagination is much less a mental faculty than a realm, a level of existence, where bodies enter into resonance with each other, increase their power, and invent new movements.

The ambiguities of Bachelard's theory

Are the animals mentioned by Lautréamont in any way modified by the metamorphoses described in *The Songs of Maldoror*? Deleuze and Guattari remarkably take the decisive step of describing human zoomorphic met-amorphoses as an "aparallel evolution"[56] that entails a becoming exactly to the same extent on the part of the animals involved. Bachelard does not. Yet, I have argued that Bachelard's concept of material imagination allows, in fact, for the overcoming of an anthropocentric concept of imag-ination. Moreover, in *Lautréamont,* Bachelard indeed unveils a field of vital orientations and bodily rhythms that lies below what Deleuze and Guattari call the striated space, namely the realm of apparent clear indi-vidual and specific boundaries. The fact remains that Bachelard did not explicitly put forward such a claim in *Lautréamont.* Moreover, consist-ently emphasizing the human specificity of imagination, Bachelard casts intensification of schemes and multiplication of possibilities as the pre-rogative of humankind. "Man appears as the sum of vital possibilities, as a superanimal."[57] Bachelard furthermore closes his book with a plea for the rehumanization of an imagination that has been, in his view, overly animalized by Lautréamont. Ducasse's imagination, being too brutal, not spiritual enough, has lost a great part of its fecundity.[58]

The issue is still pending: Are such processes "imaginative"? Do Lautréamont and Bachelard describe actual transformations? Bachelard himself occasionally contrasts "metamorphoses" with mere "metaphors"[59] but continues using the concept of metaphor in his book.[60] Do we imagine and/or undergo metamorphoses? Bachelard's commitment to the classic idea that only humans thematize the virtual and can *play* with a variety of animal behaviors may be the reason why he stubbornly maintains the concept of imagination in his account of zoo-anthropic metamorphoses. Deleuze and Guattari clearly shift the analysis to the ontological level and dispel the ambiguity regarding the possible purely psychological nature of such metamorphoses. Should we conclude that the conceptual apparatus pertaining to "images," "fantasies, "imagination," and even "the imaginary" in the account for becomings-animal is obsolete?

§34 Becoming-animal in *A Thousand Plateaus*: "Perfectly real"

Bachelard's *Lautréamont* is a key reference in Deleuze and Guattari's "1730: Becoming-Intense, Becoming-Animal, Becoming-Imperceptible...," beside a wealth of other examples, borrowed from literature (Kafka, Melville, D.H. Lawrence ...), science fiction B movies (*Willard*), social sciences (secret societies of leopard-men, shamanism, witchcraft practices), legends and popular beliefs (werewolves, vampires), and psychoanalysis (Freud's little Hans and the wolfman). No doubt the majority of these examples are commonly regarded as pertaining to the realm of the imaginary. But Deleuze and Guattari carefully deflect us from this spontaneous categorization. First, they systematically use the vocabulary of memory (the "plateau" is divided into ten sections of "memories": Memories of a moviegoer, memories of a naturalist etc.), which already grants a step closer to reality than the vocabulary of fantasies. Second, just when the reader starts getting used to this strange concept of memory, Deleuze and Guattari explode this psychological guise. The "subjects" of these "memories" become less and less subject-like ("memories of a haecceity," "memory of a molecule" ...), and eventually the reference to "memory" becomes irrelevant: "Wherever we used the word 'memories' in the preceding pages, we were wrong to do so; we meant to say 'becoming.'"[61] Disconcerting and humbling the too conceited subject and dismissing a psychological approach are essential steps in Deleuze and Guattari's strategy for escaping the human-animal dichotomy. The shift from the alleged realm of fantasies to the sheer reality of becomings is an integral part of the radical transformation Deleuze and Guattari want to induce in us.

Let us examine Deleuze and Guattari's explicit criticism of Bachelard and their new conceptuality. The theory of becoming-animal has been extensively commented upon and abundantly explained: I will focus on the relation between becomings-animal, reality, the virtual, imagination, and what I have called the imaginareal.

Neither Lévi-Strauss, nor Bachelard

Deleuze and Guattari's first move is to play Lévi-Strauss' structuralism against Jung and Bachelard's psychology of imaginary metamorphoses and conversely.

The Structuralist approach to totemism intends to replace analogies of proportion (Maldoror is similar to an octopus), which it judges to be irrational and muddled, with analogies of proportionality. Such analogies of proportionality rigorously identify a specific relation at work in two pairs of beings. For instance, a man can never say, "I am a bull, a wolf...," but he can say: "I am to a woman what the bull is to a cow."[62] Analogies of proportionality claim to be logically legitimate: "The whole world becomes more rational,"[63] but they *reduce* the phenomena they attempt to account for, instead of doing justice to the complexity of reality.[64] They gloss over the fact that relations between humans and totems imply a dimension of identification and deep transformation. Lévi-Strauss acknowledges that the individual may cease to focus on the relation itself and dive into a feeling of concrete affinity with the animal term of the relation, but, in the framework of his theory, he can only describe these affinities as "complications or degradations"[65] of the *right* form of analogy.[66]

Such affinities between human and non-human animals, as well as the concrete metamorphoses that they enable, reside at the center of Jung and Bachelard's theories. But, according to Deleuze and Guattari, the conceptuality of resemblance, analogy, and imagination hinders the understanding of the phenomena at issue. An approach in terms of images and imagination strongly suggests the reference to beings that are allegedly, in the first place, *really* separate. Conceiving of becomings as imaginative processes also strongly encourages the search for bridges in the ideal realm. Bachelard and Jung do indeed seek, beyond the terms of a sequence (the claw, the fang, the horn, the tentacle etc.), the principle that supports the resemblance between such terms. It is assumed by Jung and, to a certain extent, by Bachelard as well, that a certain "essence" constitutes, as it were, the reason of the series.[67] Although, Jung emphatically contrasts his empirical method with the Platonic metaphysical theory of "supracelestial ideas,"[68] he still hypostasizes *archetypes*. The concept is true to the etymology of the term: Jungian archetypes are supra-personal, pre-existent, instinctual vectors of existential developments that provide a determining framework for every individual's psyche. "[Archetypes] are forms existing a priori, or biological norms of psychic activity."[69] In a Jungian fashion, Bachelard strives to find dominant themes behind the series of images unfolded in *The Songs of Maldoror,* and he finally negatively *evaluates* Lautréamont's poetry according to its *failure* to humanize wild forces in us, thus promoting a questionable ideal of enculturation. Bachelard contends that the key to the series of images in Lautréamont's poetry consists in *pure* aggressiveness and *pure* animality, which are more or less intensely participated by—in a platonic sense of the word "participated"—and expressed through various images. This reading of Bachelard is reductive,[70] but it must be conceded that Bachelard follows several trails at once and never explicitly chooses against essentialism. Bachelard's reference to *pure* animality and to animality *itself* indeed offers an impoverished perspective on the *becomings* he wants to account for. How could animals be characterized by a set of stable and well-defined features when, precisely, the most surprising metamorphoses are possible?

Individuals, species, and individuation: The virtual

Deleuze wants to start from individuals, which entails the overturning of Platonism.[71] He resumes a classic philosophical problem and draws in particular on Schelling and Simondon. An individual is that which cannot be logically analyzed anymore; it is not a class, but a particular being, a "this" or "this-here" (τόδε τι [tode ti][72]). The challenge with individuals is consequently to think their dynamism and their precariousness. The individual is characterized by its ability to drift, wander, and err, as well as by its unachieved tendency to detach itself from the world as a systematic and readable whole. Schelling thus emphasizes the creatures' independence from God[73] and their capacity to become out of phase with the global harmonious divine order. Likewise, "the human" takes on multiple singular forms in human individuals and becomes, as already acknowledged by Plato in *The Sophist*, an ontological chasm, for it is simultaneously, on the one hand, self-identical, eternal, and ever-recurring (we do *recognize* at least vaguely human beings), and, on the other hand, other-than-itself and sheer becoming.

Deleuze's concept of repetition also directly relates to this issue of individuation. How can the same event occur repeatedly? I must be able to identify the first term, the first stroke of the clock, for instance, which will then be recognizable. I also must grasp a common nature shared by these different terms and without which I could not speak of repetition, that is, of *the same* A. But how can there be several *different and same* terms? It does not suffice to claim that they are simply differentiated by various locations in space and time. Their raw factual presence combined with their relative stability and recognizability evinces the ultimate intrinsic fission and self-alienation of the essence. Hence repetition is "difference without a concept."[74]

The ontology whose ultimate principle resides in eternal essences is therefore to be overcome. There is nothing but singularities. In other words, everything boils down to a plane of singular, sensible "τάδε τινά [tade tina]." Being immanent in this plane of differences,[75] repetition processes happen to take place without *reason*. Such processes are neither steered nor guaranteed by pre-existing eternal essences. They nevertheless sketch elusive, relatively stable and recognizable individuals and species. "Alogical essences"[76] emerge, although only in the form of problems that launch a quest for understanding, rather than providing fully recognizable identities.

The τάδε τινά are manifold. They are both material and ideal, or, to put it more accurately, they make this distinction artificial. These singular beings give birth to a thinkable and recognizable reality. They are ontological and transcendental. Deleuze describes them by referring to sensations and their syntheses,[77] appearing,[78] contemplation ("all is contemplation"[79]), and little selves conspiring in all the recesses of being.[80] But he also uses terms borrowed from physics: These τάδε τινά are intensities,[81] forces,[82] and molecular processes.[83] In fact, no "ideas" in the psychological or platonic sense of the term are involved in individuation processes, so that Deleuze has tended to emphasize more and more a machinic terminology that refers as little as possible to thoughts, subjects, meaning in the world, and meaning for us. This is a major difference with phenomenology and, actually, using this machinic terminology is a manner for Deleuze to conspicuously differentiating himself from phenomenologists; we will return to that in §35. But the dimension of contemplation and imagination at play in "machines" should not be overlooked. Differences and repetitions make the phenomenal world and our thoughts possible. To put it in phenomenological terms, as Husserl highlights: A nascent and scattered *meaning* is older than us, but confusingly *speaks to us*.[84] In that sense, Deleuze's thought is, to a certain extent, aligned with phenomenology, although Deleuze dismisses the deceitful concept of *the* transcendental subject (that indeed regrettably suggests a unique origin of everything), a move that comes, as I will contend, with advantages and drawbacks.

Individuals are thus more exactly individuation processes. Deleuze draws on Simondon's *L'individu et sa genèse physico-biologique*.[85] An individual is essentially a process that propagates through the molecular renewal of a body, as well as through various behaviors and a territory. Individuals change and vanish without our being able to identify turning points and clear boundaries. An individual therefore consists in the problematic emergence of an unfinished dynamic form within a meta-stable, pre-individual medium. Deleuze also refers to the highly meta-stable matter of embryos and eggs to define such a medium.[86] Like Merleau-Ponty, like Simondon, Deleuze insists on the idea that emerging individuals do not result from a continuous and necessary causal or teleological chain; otherwise, indeed, the ontological anomality of individuals would never occur. "Events explode, phenomena flash, like thunder and lightning."[87] "This body without organs is permeated by unformed, unstable matters, by flows in all directions, by free intensities or nomadic singularities, by mad or transitory particles."[88]

Imagination still has a key role to play in Deleuze's philosophy, but merely as an anonymous imaginary that precedes *my* imagination. Deleuze continues a long tradition that steadily led to the conception of imagination as an anonymous ontological principle. Imagination was never tamed, even by Kant—or especially *not* by Kant who tried to reduce imagination to a rationalist approach, but somehow beautifully surrendered in the chapter about schematism in the *Critique of Pure Reason* as well as in the *Critique of Judgement*. Imagination always returned as the troublemaker in rational philosophy. The famous Kantian reference to imagination as an art hidden in the depths of nature[89] has also deeply troubled and inspired Fichte, Husserl, Heidegger, or Merleau-Ponty among many others. Ultimately, a world cannot emerge from purely subjective syntheses, otherwise it would appear as an arbitrary invention. Deleuze thus defines the plane of immanence as animated by imaginative *passive* syntheses[90]—a Husserlian concept—that precedes subjective acts of interpretation. In other words, the syntheses described by Hume and Kant and through which the world forms must be ascribed ultimately to an anonymous and ontological "imagination": Forms arise, although in a way that is irreducible to a set of clear-cut essences. Accordingly, the concept of "ultimate" principle takes an ironic twist. Indeed, the anonymous imagination is immanent and scattered in the sensible flux and is in no way subjected to the supervision of transcendental rules, concepts, or categories. Before God, before reason, before us: Imagination. The plane of immanence is thus also defined by Deleuze as the virtual, a term that has the advantage of not misleadingly pointing to a mental faculty. "The virtual is opposed not to the real but to the actual."[91] The real is made up of actual individuals and of the swarming of differences that are also an integral part of their reality.

Reality of becomings-animal: From Bachelard's muscular lyricism to the concept of molecular dog in Deleuze and Guattari

When specific and individual boundaries cease to be regarded as the key to beings, human and non-human animals are then to be defined by *what* they do, *how* they do it, and what they *can* do: They bite, run, creep, or fly at a pace and with a style of their own. Deleuze and Guattari draw on various references, among which are Spinoza's concept of affects, Uexküll's description of animals in terms of selective perception/action patterns, and Bachelard's study of muscular lyricism. Individuals and species form the *molar* scale. Becomings occur as *"molecular"* processes. While, on the molar scale, Maldoror and the octopus are two separate individuals, on the molecular scale they develop conjointly as variations on a certain rhythm of action, for instance. At the molecular level, a tension in a muscle can develop in many possible individuals and forms of action; or a certain melodic vibration, a jazz theme, for instance, gives birth to a pink panther. In this framework, it becomes relevant to claim that "there are more differences between a racehorse and a workhorse than between a workhorse and an ox."[92] Any individual—more exactly: Individuation process—may enter in resonance with the singular way of being that defines another individuation. Barking is not sufficient to enter a becoming-dog[93] and to "emit a molecular dog:"[94] One has to arise from a unique tonality that pervades everything as a dog-atmosphere. Likewise, the pink panther, that was born from a fantastic and improbable resonance between sneaky, powerful, pugnacious animals, an elusive diamond, and a haunting jazz theme, "paints the world its color, pink on pink."[95] Becomings-animal hatch in and emerge from a certain tension in the jaws, a way of walking, a manner of frolicking around, or of "scrap[ing] at one's bread like a rodent."[96] "My jaw aches like a young dog's that craves to chew a bone."[97] "I must succeed in endowing the parts of my body with relations of speed and slowness that will make it become dog, in an original assemblage proceeding neither by resemblance nor by analogy."[98] Bachelard already showed how orientations, rhythms, and trajectories, which are nascent in the body, bear endless potentialities and secretly sustain human thought. Interestingly, Bachelard insisted on the muscles, which was already a way of atomizing the organic whole and moving toward a molecular level.

Spinoza's concept of affect also overcomes the duality between the psychical and the physical. Moreover, it ultimately defines individuals as unstable relational complexes. "Little Hans's horse is not representative but affective. It is not a member of a species but an element in a machinic assemblage [*agencement*]: Draft horse-omnibus-street. It is defined by a list of active and passive affects in the context of the assemblage it is part of: Having eyes blocked by blinders, having a bit and a bridle, being proud, having a big peepee-maker, pulling heavy loads, being whipped, falling, making a din with its legs, biting, etc. (...) What a horse 'can do.'"[99] Modifications in the unstable assemblage that defines an individual are described by Deleuze as "affects." An affect is simultaneously something felt, the ability of being affected, and the ability to affect. The individual always consists in a relational being. It emerges within a fluid pre-individual plane where only fuzzy and moving boundaries arise and resorb.[100] At a given moment, the matter folds in such a manner that some forces can build up. Potential energy is stocked, a certain relation to an "external" environment can emerge. An individual is at once a physical structure, a relation to an environment, a state, an emotion, and a set of abilities/potentialities. Thus defined, individuation appears as that which is especially achieved in living beings, and even more strongly in animals. Affects do belong to the emotional and psychological realm, but not less to the field of bodily capacities, states, and tempers. Affects depend on the way my body can receive the influence of other bodies and respond through specific organs, blood pressure, respiratory rhythm, and muscular tension, for instance. Bachelard also shows how meaning and thought are born from rhythms, trajectories, and motor patterns. However, Deleuze's philosophy of differences and individuation processes offers an ontological framework where the muscular lyricism tentatively described by Bachelard can appear as the core of beings, somehow the most real aspect of reality.

Reality of becomings-animal: The orchid does not form a wasp image

"Little Hans's horse is not *representative* but affective."[101] Although, in Deleuze and Guattari's descriptions, becomings-animal involve two relatively identifiable terms that enter in resonance, it is inadequate to understand them through the paradigm of the copy-image. Nor, for that matter, are becomings-animal subjective mental representations of an original object. A becoming does not copy a model: It takes up the musical theme[102] of an individuation process—the style of this dog or of workhorses/

oxes—and interprets it in its own unique and unpredictable manner. There is indeed no stable model that may be reproduced. Moreover, the individual that is inspired by the theme develops it in a totally original way. Lastly, the open nature of the theme dooms it to be essentially modified by the new variations, so much that the "original" is deeply affected by the "copy." The "essence" of dog is also affected by the becoming-dog of some humans. I am not becoming dog "by resemblance nor by analogy. For I cannot become dog without the dog itself becoming something else."[103]

For all that, however, becoming processes do not amount to the advent of a fusion, hence the concept of "aparallel evolution."[104] Several becomings, each absolutely unique, occur together, in resonance with each other. Deleuze and Guattari put forward two models: Viruses and the symbioses or the wasp and the orchid. "The wasp and the orchid provide the example. *The orchid seems to form a wasp image, but in fact there is a wasp-becoming of the orchid, an orchid-becoming of the wasp*, a double capture since 'what' each becomes changes no less than that which becomes. (...) One and the same becoming, a single block of becoming, or, as Rémi Chauvin says, an 'a-parallel' evolution of *two beings who have nothing whatsoever to do with one another*."[105] The latter assertion is polemical. Deleuze and Guattari choose to push and so to say beef up an idea that was also put forward by Uexküll: An *Umwelt* cannot be reduced or entered. The wasp has its own unique way of becoming and of performing the orchid-style. The orchid likewise. There is neither an appropriation of what would allegedly be the "being" of the orchid by the wasp, nor any encounter, nor participation in *one and the same* common essence. Deleuze and Guattari seek to avoid a phenomenological conceptualization of becomings in terms of encounter, world-sharing, and *Einfühlung* at all costs. Such a phenomenological frame would point to the perspective of individuals, while the priority is to challenge it. A phenomenological approach may also give the impression of naïvely presupposing a common ground or world that makes differences secondary and inessential. Yet, the becoming-orchid of the wasp and the becoming-wasp of the orchid consist in a form of remote resonance or counterpoint[106] that would never occur without a *communication* between them. Deeply-reworked concepts of *communication* and *encounter* are at stake here. It is regrettable that Deleuze and Guattari overlook this dimension of communication in becomings-animal, while such a dimension could be analyzed on the basis of a conceptuality that was developed in other books (*Proust and Signs* in particular). This would bring their theory much closer to a phenomenological approach: In fact, as I suspect, closer than they wish.[107]

In what sense exactly are becomings "perfectly real"?

As I will argue, this claim is questionable, but certainly not in the sense that it may be contended that becomings are fictitious.

"Becoming does not occur in the imagination, even when the imagination reaches the highest cosmic or dynamic level, as in Jung or Bachelard.

Becomings-animal are neither dreams nor phantasies. They are perfectly real."[108] "Perfectly real" sounds unduly positive given the fact that "the real" is floating, open, and indeterminate. But Deleuze and Guattari specify that "What is real is the becoming itself, the block of becoming, not the supposedly fixed terms through which that which becomes passes.[109]" The terms of a becoming-animal are phantom-like, but there is no "reality" that would be more stable and more original than the becomings. There is no genuine reality of which these becomings could be copies, or, in other words, beside which they could appear as less rich and less fruitful. Moreover, becomings are not pure constructs of the mind: Whether I want it or not, everything always becomes. Becoming is thus as real as it gets.

However, do individuals simply *undergo* becomings? Is this fluid reality opened to the individuals' creative interpretation? What can be the role of imagination apart from seeing itself as a subjective arbitrary fantasy that simply entertains mental pictures within the limits of its vain representations? If the real includes the virtual, is there not a leeway for our imagination to play with becomings? I have shown that the phrase "*in* the imagination" does not make sense in Bachelard's approach. Imagination cannot be reduced to a well-delimited, purely mental field that represents the world instead of actually changing it. Therefore, I find the leap from "becoming does not occur in imagination," to "it is perfectly real" rather harsh. I take Deleuze and Guattari's assertion to hastily imply that, if becomings are ontological, they have nothing to do with personal imagination, personal agency, and creativity. Machines are *compatible* with subjects, but Deleuze and Guattari want the former to take over. The reference to anonymous imagination, such as it was developed in *Difference and Repetition*, for instance, vanishes in *A Thousand Plateaus*, perhaps because this concept is already too psychologizing. But this turn of event is unfortunate. Indeed, a reference to imagination has the potential benefit of raising questions regarding the kinship between two "imaginations" that cannot be mere homonyms: Passive syntheses and active syntheses, or, put otherwise, the imaginareal and the active imaginations of human and nonhuman individuals.

§35 Figures, fiction, and intersubjectivity in Haraway

Haraway's approach to becomings and Deleuze and Guattari's philosophy have a lot in common. "I am indebted to Deleuze and Guattari, among others, for the ability to think in 'assemblages,'"[110] Haraway concedes in *When Species Meet*. But, above all else, she highlights that such an apparent kinship is misleading: There is no possible resonance between "the dance of becoming with" that she wants to describe and the "fantasy wolf-pack version of 'becoming-animal' figured in Gilles Deleuzse and Félix Guattari's famous section of *A Thousand Plateaus*."[111] Haraway is even tempted to refer to Deleuze and Guattari as "the enemy."[112] The disagreement between

Haraway and Deleuze and Guattari has two major components. First, Haraway criticizes Deleuze and Guattari for despising and discarding the relation of cooperation between individuals.[113] Second, Haraway explicitly connects becomings to fiction,[114] figures,[115] and uncertain transpositions into the other's perspective.[116] I contend that these two aspects are interlinked. Let us first examine Haraway's concept of becoming-with and its essential connection with what she calls "figures."

Molecular becomings

Haraway's becomings-with have a lot in common with Deleuze and Guattari's becomings-animal. At stake for her, as it was for Bachelard, Deleuze and Guattari, is to describe relations based on real transformations. Moreover, like these authors, she refers to a fluid reality beneath the individual forms. Haraway emphasizes, in the very first pages of *When Species Meet*, that "human genomes can be found in only about 10 percent of all the cells that occupy the mundane space I call my body; the other 90 percent of the cells are filled with the genomes of bacteria, fungi, protists, and such, some of which play in a symphony necessary to my being alive at all, and some of which are hitching a ride and doing the rest of me, of us, no harm."[117] Haraway also draws inspiration from Lynn Margulis' works on symbiogenesis,[118] thus promoting, in line with Deleuze and Guattari, the paradigm of "infectious exchanges" and symbioses (horizontal open-ended relations), over the model of the tree (fixed hierarchy between species).[119]

According to Margulis, changes in morphology or behavior as well as the development of new hereditary features are not first and foremost linked to hypothetical random mutations in the genome of an individual, but rather to the circulation of bacteria from one organism to another. For instance, when an animal ingests another organism, she also absorbs all the bacteria it contains and may fail to digest them entirely.[120] The bacterial population then prospers in her and "the inheritance of trapped populations, especially in the form of microbial genomes, creates novel evolutionary lineages."[121] Haraway consequently rejects the modern paradigm of well-circumscribed individuals, defined by a fixed identity and belonging to clear-cut species.

In Haraway's descriptions of her becoming with her dog Cayenne-Pepper, the molecular level, both in a classical and a Deleuzian sense, plays a key role. "Ms Cayenne Pepper continues to colonize all my cells–a sure case of what the biologist Lynn Margulis calls symbiogenesis. I bet if you were to check our DNA, you'd find some potent transfections between us. Her saliva must have the viral vectors. Surely, her darter-tongue kisses have been irresistible."[122] Communication between Haraway and Cayenne is based on "unintentional" micro-movements and the unconscious attunement to

them. In the same way, for instance, the rider just thinks of a possible movement and, without knowing it, because her body infinitesimally sketches
this movement, communicates her intention to the horse, who, then, actually performs it.[123]

More generally, through the figure of the cyborg—"the cyborg is our
ontology"[124]— Haraway challenges any classical dualities and boundaries.
Inert matter, machines, non-human animals, and human animals are only
fragile virtual poles in a set of biotechnological processes from which literally anything could emerge in the future.[125] This theory goes together with a
strong suspicion toward the concepts of subject and choice: "Through patient
practices in biology, psychology, and the human sciences, we have learned
that we are not the 'self' or 'transparently present to the self' either."[126]

The importance of intersubjectivity

Nevertheless, in *The Companion Species Manifesto* (2003) and *When Species
Meet* (2008), Haraway starts paying more attention to a different and apparently competing conceptuality: That of intersubjectivity. Becomings-with
are not only changes in rhythms, modes of being, or even modes of feeling, that transform individuals and from which subjects stem. They cannot
be reduced to molecular processes through which subjects are "formed"
or "made"[127] in new fashions. They are relations that require the attention,
efforts, and initiatives of subjects.

When Marco, Haraway's six-year-old godson, learned to train Cayenne,
he quickly got to master some basic orders such as "sit" or "fetch." He thus
first "treat[ed] her like a microchip-implanted truck for which he held the
remote controls."[128] But Haraway then felt her "ideals of intersubjectivity"[129] to be thwarted: The relation between Marco and Cayenne would be
brought to a whole new level if it could stem from mutual attention and
respect. Haraway describes enthusiastically how this actually, eventually
happened. The relation of becoming-with also implies wondering about
what the other thinks, feels, and wants. Haraway therefore takes the lib-
erty of using anthropomorphizing perspectives, claiming for instance that
Cayenne forgives her when, during the training session, Haraway "misin-
terpret[s] her invitations, preferences, or alarms."[130] "I know perfectly well
that I am 'anthropomorphizing' (as well as theriomorphizing) in this way
of saying things, but not to say them in this manner seems worse, in the
sense of being both inaccurate and impolite."[131] The question "what does
Cayenne think?" is thus accepted and embraced by Haraway as problematic. She regards the transposition as imaginary and risky. However, she
elects such an imaginative transposition as the only way towards both
knowledge and politeness. "Knowledge and politeness" make up a rather
surprising and unusual pairing. If imagination is used out of generosity,
how does it provide a solid foundation for accurate knowledge?

Fiction and the practice of agility

One of the key figures of *When Species Meet* is the hybrid that Haraway and Cayenne Pepper form through their practice of agility. This sport and its rules are human creations, but without the personal commitment of the two partners, without desire, good will, and attention also on the dog's part, the trainings fail to make any sense.

Haraway insists on Cayenne's personal taste for agility: "On the mornings when we are driving to a trial, she tracks the gear and stays by the car with command in her eye. It's not just the pleasure of an excursion or access to a play space. We do nothing else in the agility yard but work on the obstacle patterns; that is the yard she wants access to."[132] Further, training is based on a method of positive reinforcement that fosters the non-human partner's initiative and creativity. A certain behavior is circumscribed, marked, and carved by a click or a reward. This also necessitates intense focus and significant inventiveness on the human partner's part. The trainer must judiciously make numerous decisions regarding the timing for clicking, the progression towards more and more precise delineation, and the number of repetitions that are effective just before they become stressful.[133] It would be unfair to understand this method as essentially Pavlovian. Positive reinforcement chooses to never chasten the dog for venturing all sorts of behaviors, including ill-suited or poorly effective ones, or simply behaviors that do not match the trainer's expectations. It is a manner of leaving the non-human partner free to invent and propose new solutions. Thus, for instance, positive reinforcement is to be used in playful situations, where non-human partners have the opportunity to offer new material that can be reinforced.[134] Positive reinforcement proved pivotal in the research about animal creativity.[135] Haraway consequently describes marked behaviors as "inventive construction[s]" and "generative fact[s]–fiction[s]."[136]

Each agility trial offers new situations to prevent the participants from relying on sheer reflexes. Human partners discover the course before the teams actually run it, and they have the responsibility to guide their partners on the spot. But, Haraway insists, overhandling is disastrous, unless the dog takes it upon herself to bypass the awkward directives given by her human partner: "Cayenne saw me coming, clipped her smoothly curving stride slightly, and dodged around me, all but shouting, "Get out of my way!" while she slipped magically between poles one and two and wove very fast without break in rhythm through the twelve poles. In my mind's ear, I heard my agility teacher Gail Frazier telling me over and over, 'Trust your dog!'"[137]

Invitation to dance: Making space for other subjects. The birth of subjects from subjectification processes and second-person solicitations

To be sure, through training, non-human animals do learn some specific behaviors. But the crux of the matter resides elsewhere, in a, so to say, deeply existential dimension of behavior. In order to describe what is at stake in becoming-with, Haraway borrows the dance paradigm from ethologists, such as Barbara Smuts or Sue Savage-Rumbaugh, who research operations of negotiation and proactive attunement in play, greetings, and education.

Synchronization processes rely on the active solicitation of the other. They involve questions and invitations and necessitate the opening of a personal space for the other. Barbara Smuts compares the subtle sequence of gestures, glances, and adjustments in greetings behaviors to dance. "In slow motion, some greetings look like the awkward steps of two people first learning to dance together, whereas others look like Fred Astaire and Ginger Rogers."[138] Smuts does not only find the similarity striking, but also would like to understand what such a comparison means exactly and to what extent it unveils the actual nature of animal behaviors. She refers to a decisive monograph by Sue Savage-Rumbaugh: *Language comprehension in Ape and Child*.[139] In this text, Rumbaugh and her co-authors highlight in what decisive manner the formation of interindividual routines scaffolds language learning. Again, such routines would be inaccurately described as automatisms. They are rather comparable to dances, which Savage-Rumbaugh explains in a detailed fashion.

At stake in the relations between the caretaker and the child or the young ape is that they all acquire the ability "to join behaviors in a smooth manner and to pass each change of interacter-cum-interactant seamlessly back and forth."[140] The acquisition of this skill requires the implementation of a specific method on the part of caretakers, a process in which some individuals are more talented than others. This method is twofold.

1   The caretakers create "markers," which needs not to be conscious, but sure must be intentional. Such a creation of markers is indeed an effort toward the other, an endeavor to make oneself more readable for the other. Caretakers "utilize frequent and sometimes exaggerated postural, gestural, and verbal markers when engaging in interactions with very young children or apes."[141] They, for instance, point to an object, describe what they are doing, and repeat the same description each time they start a routine ("Let's play with the bubbles," "Let's find the bubbles," "*and now* I am going to" ...). Non-human caretakers also use this method: A chimpanzee mother may, for instance, sacrifice the effectiveness of her gestures, slow down here movements, and exaggerate some key gestures in order to patiently teach her cubs how to crack nuts.[142] The clicker or the use of particular signals during training sessions with dogs or horses, for instance, can play a similar role. These markers certainly could be intended as a mere description of obdurate facts or as Pavlovian triggers, but this would already be slightly paradoxical since the process actually implies that the reality is lined with signs, saturated with intentionality, and

that the caretaker undertakes a form of communication and tries to reach out for the fleeting attention and the understanding of another animal. And indeed, the method proves fruitful only under a second condition.

2 Rumbaugh points out that these markers achieve full effectiveness when they are intended as invitations and questions. The best care-takers, who obtain the most impressive results, do not only comment upon what they are doing or what was just done, but also emphatically announce significant actions just before they initiate them.[143] They solicit the ape or the child and ask them questions about the situation. "Oh look, what is this in the bottle?"[144] These successful caretakers pay heed to the reaction of their partners and adjust their behavior until they catch their attention. More than that, they pause awhile before achieving an announced action, and they watch out for possible signs of "agreement" or synchronized impulse in their partner. Even if such an "I'm in" sign was in fact an accidental gesture or sound, this strategy makes room for such signs to develop, so that a possibly acci-dental sign may be reinforced and become a more personally endorsed sign in future similar situations. Babies "whose caretakers wait for the emergence of a contingent signal are the ones who quickly learn to use that signal to indicate to the mother that they anticipate her intent. Similarly, when such babies do not want to be picked up, they can withhold the 'I'm ready' signal. (...) Babies who have not experi-enced such contingent 'picking-up' behavior do not develop a signal between themselves and the caretaker that indicates their readiness to be picked-up."[145]

Such nursing processes play a crucial role in the birth of free subjects. Freedom and subjectivity are not metaphysical entities or heaven-sent substances; they are, in fact, absolutely dependent on the way generous indi-viduals actively create new folds and spaces of pure contingency that inspire the initiatives of others. Without support, without proper impulses, without invitation, without the active intentionality to draw a virtual space where nascent intentions are welcome to thrive, freedom and subjectivity remain empty words.[146] In a carefully folded flesh, in a carefully folded world, a recursive, mirroring process of subjectivation can occur. Interindividual routines constitute an invitation to dance. They *solicit* the intention to, in response, willingly and carefully pay attention to and make room for the other's intentions.

The result of such invitations to dance is a fragile and dynamic co-fiction that requires not only molecular processes and a certain assemblage, but also the active, wishful, *and* fruitful reference to another *subject* over there, as well as the active response and engagement of the subject thus conjured up. Hence, synchrony can never be taken for granted and participants must constantly fine-tune their interactions, applying themselves to attending to each other, using existing markers, and producing more and more fine-grained markers. Routines operate as scaffoldings that open the possibility for more graceful interactions, although the latter never lose their dimension of improvisation and creativity.

The "significant other"[147] in Haraway's theory of becoming-with thus emerges as a pole in a dynamic relation. Intention and commitment make space for a companion, while nonetheless emphasizing its looming at a distance, on the horizon, as a virtual center of a unique *Umwelt*. The other is thus another "I" that must be *solicited* as a "you." This idea constitutes the core of a subject-centered, second-person approach to non-human animals. To be sure, the subject arises from a favorable medium and opportune folds in matter.[148] Yet, it also *consists in*—and, in the invitation to dance, *is invoked as*—that which takes up and endorses the process of emergence as *its own* movement. It can take the initiative, interpret, and create unexpected behaviors, with the result that an interiority is carved and carves itself in the flesh. We may *or may not* choose to make room for the other's interiority, which is the opportunity for us to try and experience our ability to make decisions that matter. Without our commitment, the nascent other will likely wither away. However, our engagement does not amount to entertaining a *pure* fantasy. The other, in the process of becoming-with, necessarily takes the form of a foreign perspective that cannot be referred to without recourse to imagination, also because what is going on "over there"—"in" the other's mind—is also a form of imagination. Dance is consequently, and without contradiction, a combination of imagination, politeness, and accurate knowledge.

Is the invitation to dance purely human? Although the agility sport is a weird human invention, there is no reason to presuppose, on principle, that non-human animals cannot invite partners to dance. Dolly's milk-smoking is doubtless a compelling invitation to dance. I contend that images produced in animal morphology are invitations to dance. We have also explained in Chapter 4 that the tendency to intentionally create markers and signs for the sake of disambiguation is present in the non-human animal realm. It is thus not surprising that so many non-human animals may actively, even sometimes joyfully and consistently, engage in training processes of work/play mainly fashioned by humans. Bekoff and Allen's article about play in *The Cognitive Animal* also gives excellent examples of becomings-with. Non-human animals can enter playful interactions where hierarchical relations become flexible and transgressions are allowed. They often use signs that indicate the playful nature of the interaction and have to "finetune ongoing play sequences to maintain a play mood and to prevent play from escalating into real aggression." "Detailed analyses of films show that there are subtle and fleeting movements and rapid exchanges of eye contact that suggest that players are exchanging information on the run, from moment to moment, to make certain everything is all right—that this is still play."[149]

Figures

Haraway argues that the proper concept for understanding becomings-with is that of "figures," which was already central to her previous works. This concept has the same structure as the invitation to dance. More than that: Precisely because of its conspicuous connection to imagination and the imaginary, it constitutes in itself a perfect invitation to dance.

The Latin term "figura" designates a drawing, a form, or a silhouette and insists on the particular way in which a matter was shaped. *Figurae* are thus essentially connected to the realm of appearances and the faculty of imagination. "Figura" may also mean "image," "apparition," "specter." It is etymologically linked to the verb *fingere* (to create, to invent, to mold, to fake[150]), which is also the root of the word "fiction."

"The figures are at the same time creatures of imagined possibility and creatures of fierce and ordinary reality."[151] My body, for instance, is such a figure: It has been shaped by biopolitical processes, through certain representations of appropriate or inappropriate behaviors, shameful or worthy-of-attention body parts, aesthetical codes, symbolism of body parts, the air I breathe, the industrial food I eat, the pills I was told to take. My body is also molded by my body image, which reflects processes of introjections and projections connected to four factors: Archaic urges, social structures and imageries, the accidents of my personal history, and more thoughtful personal decisions. And finally, my body is a figure in that it constitutes a surface that lets these facets and influences manifest and leads us into imagining other possibilities.

But a body becomes a *figure*, in a stricter sense, when it stands up as an image and distances itself from the backdrop of mundane reality. Leonardo's *Vitruvian Man* is a figure in this more particular sense. Figures are more specifically bodies, compositions, or pictures that Haraway wants to showcase in her works *as* elements of "[her] game."[152] The cyborg, oncomouse™, several cartoons from *The New Yorker*, science fiction characters, Jim's dog, or the hybrid formed by Cayenne Pepper and Donna Haraway are such figures. Haraway thus incorporates pictures, photographs, paintings, cartoons and exemplary stories in her works and takes them as normal, no less seriously than she does alleged realities. Take the example of the photograph of a redwood stump that looks like a dog. This picture was taken and sent to Haraway by her friend Jim, who knows that she loves dogs. Jim's dog is a frail "reality" for common sense, a mere effect of perspective. But this picture becomes revealing and metaphysically relevant in Haraway's approach. Figures in a stricter sense—I will call them *fantastic figures*—are the *conspicuous* intertwinement of reality and the imaginary. They consequently appear and are irresistibly apprehended *as images*. This does not mean that such fantastic figures appear as copies of an original, but rather that they present themselves with a characteristic "delayed effect" and "distance effect," *a little aside from* mundane reality. They do not blend in established reality. Fantastic figures are floating phenomena, and they *manifest* the imaginary dimension of every reality. Every being is a figure in a broad sense: It is always the result of tropes, displacements, and symbolic alliances. But most of these figures tend to ossify and present themselves as mundane realities. Haraway's fantastic figures can bring to light the uncanny nature of sedimented reality and thus revive imagination. "I emphasize figuration to make explicit and inescapable the tropic quality of all material-semiotic processes."[153]

Figures are essentially pictorial, namely: Analog rather than digital. They are concrete and give us the whole before the parts, whereas language first and foremost, albeit not exclusively, is discursive and operates analytically, step by step. Language divides, circumscribes, and deduces. Figures are "knotted being."[154] They bear the world—although a fractured world—and bring us, all at once, complex dynamic and relational networks. As such, they give us riddles not exactly to be untangled, but to be taken up. "Figures collect the people through their invitation to inhabit the corporeal story told in their lineaments."[155] Figures *invite* us to join them and actively take part in open stories where "inert" bodies, matter, tools, technologies, commodities, and living beings shape each other and invent a new world, a process that Haraway calls *"autre mondialisation* [other-worlding]."[156] Haraway's figures thus display a delayed effect and are hovering entities *also because* they involve dancing partners. Some of their inhabitants can play, namely they can invent, pretend, produce images and symbols, accept to dance, hesitate, or balk. "For me, figures have always been where the biological and literary or artistic come together with all of the force of lived reality."[157]

By referring to the creation of a new world ("autre-mondialisation"), Haraway once again distances herself from Deleuze and Guattari. The latter underscore that the wasp and the orchid had *"nothing whatsoever to do with one another."*[158] Deleuze and Guattari refuse to value or focus on a world *in which* individuals may co-exist. The reference to such a companionship does not make sense, indeed, without a revival of the phenomenological conceptuality of subjects and a certain—at least nascent or haunting—transcendence. Haraway's analyses precisely show that agents cannot be seriously recognized without being associated with, at the minimum, some phantom-subjectivity. Anthropomorphism and the very solicitation of non-human animals as "as if" subjects are the only way of conjuring up their subjectivity. Agents are inseparable from the reference to their *virtual* perspectives and feelings. It is crucial to emphasize that Haraway's ontology cannot be accurately reduced to an object-oriented ontology, despite her interest in networks and her overcoming of the sovereign human subject. As she puts it in "Situated Knowledges," "The imagery of force fields, of moves in a fully textualized and coded world (...) is, just for starters, an imagery of high-tech military fields, of automated academic battlefields."[159] Figures of machines and even, to a certain extent, cyborgs bring us close to an alienating imagery that incapacitates possible agents and leaves the current authority figures in place. Haraway rather calls for a theory of "nonisomorphic subjects, agents, and territories of stories unimaginable from the vantage point of the cyclopean, self-satiated eye of the master subject."[160] Hence, the phenomenological perspective must be preserved in a philosophy of becoming-with.

The choice of a method centered around figures is also directly connected to Haraway's love for science fiction. Science has the unfortunate tendency to veil and overlook its ideological biases and historical extraction. Haraway thus contends that science cannot fulfil its cognitive task outside of a cooperation with science fiction. "SF conventions invite—or at least permit more readily than do the academically propagated, respectful consumption protocols for literature—rewriting as one reads. The books are cheap; they don't stay in print long; why not rewrite them as one goes? Most of the SF I like motivates me to engage actively with images, plots, figures, devices, linguistic moves."[161] Science fiction fluidifies cooperation processes. It reintroduces openness and irony, where authoritarianism and reverence threaten to take over. Like Deleuze and Guattari, Haraway wants to remind the reader that no one will ever speak from an overhanging position: We are struggling and experimenting within a plane of virtuality. In this regard "plateaus" are comparable to figures and, in *Mille Plateaux*, "Introduction: Rhizome" is a vigorous invitation to dance. But in "1730: Becoming-Intense, Becoming-Animal, Becoming-Imperceptible…," a counter mood is at work.

§36 Animal subjectivity and animal imagination in Deleuze and Guattari

A phantasmatic attempt to discard subjects and individual imagination

In "1730: Becoming-Intense, Becoming-Animal, Becoming-Imperceptible…," animals are secondary. Deleuze and Guattari themselves make clear that becomings-animal matter insofar as they enable us to break with the illusion of living in an orderly and obdurate world. Becoming-animal is thus inseparable from becoming-child, becoming-woman, and becoming-molecular. Non-human animals play the role of catalysts. Indeed, since the great divide, animals have been a thorn in the human flesh. Because humans are and are not animals, animality creates a serious imbalance at the heart of the human identity equation. But, Deleuze and Guattari argues, we should avoid focusing too much on animals, on their individuality, their particular body, and their world, for this puts a stop to the dynamism of becoming-molecular. Deleuze and Guattari find a good illustration of this idea in Kafka's *The Metamorphosis*. Precisely because the insect is overwhelmingly present, becoming-animal turns into a deadlock, and Gregor ends his existence crushed under an apple. "Gregor's deterritorialization through the becoming-animal fails; he re-Oedipalizes himself through the apple that is thrown at him and has nothing to do but die (...) Aren't the animals still too formed, too significative, too territorialized? Doesn't the whole of the becoming-animal oscillate between a schizo escape and an Oedipal impasse?"[162] In Deleuze and Guattari's becoming-animal, the animal should just be a passage.

Correlatively, Deleuze and Guattari only highlight and value certain reductive aspects of animality: The pack, proliferation, attack, flight, and the anomalous individual who stands at the margin of the pack and makes it porous and unrecognizable. "What we are saying is that every animal is fundamentally a band, a pack,"[163] namely a political precarious multiplicity, structured and troubled by power relations. Why would animals incarnate multiplicity and proliferation more than human beings? Here Deleuze and Guattari deal with *figures*, but only Haraway makes clear that the concept of becoming in *A Thousand Plateau* relies upon *"fantasy* wolf-packs."[164] Animals are paradigmatic multiplicities in the framework of the myths that have been forged along with the great divide. While humans are essentialized, all the avatars of evolution, and all the phenomena of symbiosis, hybrids, and monsters are thrown on the other side of the gap and ascribed exclusively to animals. From this perspective "the animal" becomes the privileged symbol of the precariousness and the proliferation of vital forms. And, as "the animal" is actually always under the man's skin, animals trouble and undo every common world. But this is only a facet of our human imagery, one possible relation to animals that Deleuze and Guattari have selected *among many others*.[165] The baseline figure for Deleuze and Guattari is thus Moby Dick *and not* my dog or my cat.

> That is what Captain Ahab says to his first mate: I have no personal history with Moby-Dick, no revenge to take, any more than I have a myth to play out; but I do have a becoming! Moby-Dick is neither an individual nor a genus; he is the borderline, and I have to strike him to get at the pack as a whole, to reach the pack as a whole and pass beyond it. The elements of the pack are only imaginary "dummies," the characteristics of the pack are only symbolic entities; all that counts is the borderline—the anomalous.[166]

> Ahab's Moby-Dick is not like the little cat or dog owned by an elderly woman who honors and cherishes it.[167]

Here a shift occurs: The misogynistic remark is philosophically gratuitous. Contempt and even rude, disparaging words ("anyone who likes cats or dogs is a cunt [*tous ceux qui aiment les chats, les chiens, sont des cons*]"[168]) seem to be added as it were as a bonus. But the text makes more sense if we read it as the completion of a procedure of exclusion, the discarding of individuals as "imaginary dummies [*mannequins imaginaires*],"[169] namely as empty fantasies.

This passage is indeed symptomatic: Several of the elements incorporated in this chapter by Deleuze and Guattari betray that some implicit choices and excommunications are at work. The project of unhinging oedipalized animals is perfectly understandable, but, as Haraway shows, one can cherish and honor one's dog without regarding it as a substantial individual or treating it as a slave, a baby, or a fellow human being. Animal existence is much richer than the three options described by Deleuze and Guattari (pets, State animals, or packs). Deleuze and Guattari thus favor a deeply phantasmatic vision of animals in a way that masks the role played by fantasies. *Choosing* this particular path of experimentation in our relation to non-human animals is certainly a legitimate option. But, in this specific case, the strong-willed individual *agency* of the philosopher-experimenter operates in the shade. Deleuze and Guattari's bold and personal imagination cloaks itself in brutal invectives and takes the guise of the return to the traditional distinction between what is real (the becomings, packs, and proliferation) and what is imaginary (the terms of the becomings, "the elements of the pack" defined as "imaginary 'dummies"). In this way, Deleuze and Guattari bring back on board the classic contrast between, on the one hand, the obdurate reality that imposes itself and will remain ontologically authoritative whether I like it or not, and, on the other hand, the figments of the subject's imagination. "Anyone who likes cats or dogs is a cunt" contains no invitation to dance. The other options are ontologically wiped out, declared null and void: What is there to look for on that end since individuals are "imaginary 'dummies'"? The strategy that underpins "1730: Becoming-Intense, Becoming-Animal, Becoming-Imperceptible..." is thus disingenuous.

The dimension of *mise en abyme* and play is much more emphasized in Lautréamont's *The Song of Maldoror* and also justifies maintaining the explicit reference to imagination, metaphors, and the imaginary. Becomings-animal in Lautréamont's work and Bachelard's essay do imply real transformations, while simultaneously bringing to light how the two authors openly choose and orientate these becomings. *The Songs of Maldoror* indeed contains, besides hallucinatory passages (moments when words become pure energy, pure activity, and carry us away), playful pieces where the narrator invites the reader to imagine with him. The narrator promises to intoxicate the reader with haunting and deeply unsettling images, but he also stages himself as an image-maker, an illusionist. He regularly addresses the reader directly and invites her to join the field of self-distanciation and active imagination, where new worlds can be invented or negotiated. "Today I am about to invent a little novel of thirty pages. (...) I believe I have found at last, after several feelers, my final formula (...) This hybrid preface has been exhibited in a manner that will perhaps appear not natural enough, in the sense that it surprises the reader, so to speak, who does not clearly see where he is being led at first."[170] Imagination has the remarkable capacity to move back and forth between hallucinatory daydreaming or even possession (quasi-*presence*) and the distant evocation of a virtual dimension acknowledged as such (*quasi-presence*). Becomings-animal develop between these extremes: They are both represented *and* experienced, simultaneously invented *and* achieved, symbolic *and* real, ideas *and* bodies.

The definition of individuals as imaginary dummies is also to be accounted for in the light of Deleuze's complex relation to phenomenology. In *Cinéma 1*, Deleuze reproaches phenomenology for remaining focused on subjectivity. Phenomenology indeed, Deleuze contends, even attempted to define the transcendental subject as the ultimate principle of absolute science, and it too often overlooked anonymous processes through which subjects emerge.[171] Yet, this critique targets a caricature of phenomenology. I have highlighted the growing role ascribed to anonymous processes in Merleau-Ponty's works as well as in Husserl's genetic phenomenology.[172] On the one hand, Deleuze may be right to denounce a recurring tendency to focus on consciousness and freedom in phenomenology, but it is fair to add that phenomenology came to learn that subjectivity is an emerging, fragile, and intermittent agency. On the other hand, Deleuze's propensity to wipe subjectivity out of the picture proves problematic in the "becoming-animal" plateau. It is not true that becomings concern "an affectability that is no longer that of subjects,"[173] or that the wasp and the orchid "have nothing whatsoever to do with one another."[174] Here, a dangerous fragmentation predominates, whereas, in fact, subjective agencies involved in these becomings play a crucial role and may conquer a greater role depending on the generosity of their partners. Moby-Dick *and not* my little cat, molecular becomings *and not* imagination, Bergson *and not* phenomenology:[175] Deleuze's excommunications are without foundation in his theory.[176] These exclusions constitute the symptoms of extreme voluntarism, in which the individual agency of Deleuze and Guattari operates undercover, while the subjectivity of non-human animals in becomings is overlooked and explicitly discarded. There is a God behind the machine: The negation of subjectivity to the benefit of rhizomes, networks, or machines must be the deed of subjects whose sovereignty becomes incontestable by being made invisible. A theory of many "non-isomorphic"[177] subjects-agents—to borrow from Haraway—is more generous and not less rigorous.

With some necessary adjustments, a philosophy of becomings-animal can and should become interested in experimenting in the direction of becoming-with, which gives a much more significant role to human and non-human personal imagination. In this regard Haraway beautifully develops an original molecular-phenomenological approach.[178] She does not intellectualize or substantialize the self, but she rethinks its necessity as an effective fiction that emerges through molecular processes *as well as* invitations to dance. This subject, as fact-fiction, institutes itself just when an agency develops as a response to such invitations.

For the sake of fairness, I will eventually turn to two concepts in Deleuze's work, that provide bridges between becoming-animal and becoming with animals: *Aberrant communication* and the *refrain*.

Animal intersubjectivity: A marginal route in Deleuze's work

Beyond the rather unsubtle assertion that the wasp and the orchid "have nothing whatsoever to do with one another," Deleuze also develops the idea that a form of communication unfolds in such a double becoming. In this "aberrant communication,"[179] the two parts remain mysterious to each other. A similar form of paradoxical communication is a central topic in Proust's *In Search for Lost Time*. The characters persist in misunderstanding each other, while confusedly sensing the bizarre musical themes instituted by each of them as their personal style. They do feel each other's opaque presence and alien worlds, but they remain separate and never cease seeking ways of fathoming the atmosphere of these distant planets, in vain. Such an aberrant communication, following the model of the wasp and the orchid, is, according to Deleuze, the paradigm of a full-fledged encounter. In effect, Deleuze highlights in *Proust and Signs*, a sign worthy of the name should not be processed, deciphered, understood, reduced. A reductive communication negates the other and fails to properly receive the inexhaustible inspiration instituted by a genuine sign. At stake here is what we have defined as institution, following Merleau-Ponty, in Chapter 4, and "invitations to dance" following Haraway and Savage-Rumbaugh, in other words a non-fusional, non-mastered form of intersubjectivity.

To be sure, in *Difference and Repetition*, Deleuze denies animals imaginative subjectivity by claiming that "the animal is protected [*garanti*: guaranteed] by specific forms which prevent it from being 'stupid' [*bête*]."[180] In other words, according to Deleuze in this text, only humans become subjects in such a way that they develop into egos. And only egos are essentially nothing and can consider virtually anything,[181] whereas non-human animals remain stuck in rigid patterns defining their species.

Even so, in "1837: Of the Refrain," a much more central place is assigned to non-human animal agency. In this chapter, Deleuze and Guattari somehow describe—although not exactly in these terms—subjects *with* whom it would be possible to become. They point out that animals acquire a territory by producing signs that serve as "posters" and "placards."[182] When used to circumscribe a territory, birds' songs, chemical signs, visual signs including outward appearance[183] can function as what Deleuze and Guattari call "refrains [ritournelles]." Likewise, "a child in the dark, gripped with fear, comforts himself by singing under his breath."[184] A territory is actively created by markers whereby a subject signals to itself and to every perceiving animal around that new limits have arisen here. Such boundaries are fictions built through versatile signs, and they allow for dialogue and negotiation. As a result, these territories are inseparable from the opening onto the "outside" world. This new "interiority" is in constant interactions with other territories and subjects. With the concept of territorialization, Deleuze and Guattari describe the process through which the flesh folds on itself and becomes a self-instituting agent. Subjectivation starts through anonymous repetition, but, in what could be described as a leap of auto-poiesis, the mutable matter hatches an individual that exists as that which wants itself, circumscribes itself, and *shelters* itself into a "place of its own [*chez-soi*]."[185] This subject, as a form of self-reflection and self-distance, experiences fear, a diffuse fear that existentialism rather called *anxiety*. The anxious subject needs to reassure itself and thus strangely communicates with itself, by singing to comfort itself.

Assuredly, Deleuze and Guattari insist that what comes first is the mate-rial mark or signature, such as vivid colors, urine, pheromones, or cries. "The territory is not primary in relation to the qualitative mark; it is the mark that makes the territory. (...) The signature, the proper name, is not the constituted mark of a subject, but the constituting mark of a domain, an abode. The signature is not the indication of a person; it is the chancy formation of a domain."[186] Here, however, the idea that a subject is the *result* of a subjectivation process comes to its limit. A subject also implies a self-creative *agency* and a growing self-poiesis, although the latter is inseparable from material signs and negotiation processes.

Notes

1. The English word "phantasy," rather than fantasy, is a good translation for the French *fantasme*, which is, in turn, a classic translation of Freud's *Phantasie*. A phantasy is more powerful than a fantasy. A phantasy provides the obsessive and haunting imaginative fulfilment of deep desires and profound urges. If becomings are not phantasies, they are *a fortiori* not fantasies.
2. Deleuze, Gilles and Guattari, Félix. *Mille plateaux*. Paris: Minuit, 1980, p.291. Trans. B. Massumi. *A Thousand Plateaus*. London: Athlone, 1988, p.238.
3. Deleuze, Gilles and Guattari, Félix. *Kafka pour une littérature mineure*, Paris: Minuit, 1975, p.127. Trans. Dana Polan. Minneapolis: University of Minnesota Press, 1986, p.70.
4. Deleuze, Gilles and Guattari, Félix. *Mille plateaux*, p.66 (Trans. p.49).
5. Deleuze, Gilles and Guattari, Félix. *Qu'est-ce que la philosophie*. Paris: Éditions de Minuit, Paris, 1991, p.175. Trans. H. Tomlinson and G. Burchell. New York: Columbia University Press, 1994, p.184.
6. See Deleuze, *Cinéma 1, L'image-mouvement*, Paris: Minuit, 1983, and *Cinéma 2, L'image-temps*, Paris: Minuit, 1985.
7. de Lautréamont, Comte (Isidore Ducasse). *Les Chants de Maldoror*, Paris: L. Genonceaux, 1890, p.47. Trans. Guy Wernham, New York: New Directions, 1965, p.36–7.
8. *Ibid.*, p.63, Trans. p.49 (I have modified the translation).
9. *Ibid.*, p.284, Trans. p.223 (I have modified the translation).
10. *Ibid.*, p.153, Trans. p.119.
11. *Ibid.*, p.153, Trans. p.118.
12. *Ibid.*, p.3, Trans. p.2–3.

13. Bachelard, Gaston. *Lautréamont*, Paris, José Corti, 1939, p.9. Trans. by James Hillman and Robert S. Dupree. Dallas: Dallas Institute, 1986, p.2. See also p.12 (p.4) "Some pages are unbelievably dense in animals. Furthermore, this density consists of a group of impulses rather than images."
14. *Ibid.*, p.10 (Trans. p.2).
15. *Ibid.*, p.12 (Trans. p.4).
16. *Ibid.*, p.13 (Trans. p.4).
17. *Ibid.*, p.17 (Trans. p.7).
18. *Ibid.*, p.17-19 (Trans. p.7–8).
19. *Ibid.*, p.14. (Trans. p.5) (Trans. slightly modified).
20. Voir notamment Bachelard, *L'eau et les rêves*, Paris, José Corti, 1942 et *L'air et les songes*, Paris, José Corti, 1943.
21. Bachelard, Gaston. *Water and dreams*. Trans. Edith R. Farrell, Dallas: Dallas Institute of Humanities and Culture, 1999, p.1
22. *Ibid.* p.1–2. See also in *Air and dreams*, Chapter 5: Bachelard shows that the superficial image of the wing remains ineffective unless it is accompanied by a process of elevation and *becoming-lighter* expressed through an ethereal and lively tonality in the poem, swift movements, increasing brightness, in short: a dynamism that must pervade the poem and contaminate the reader.
23. Bachelard, Gaston. *L'air et les songes*, p.7 (Trans. Edith R. Farrell. Dallas: Dallas Institute of Humanities and Culture, 1988, p.1).
24. There are many connections between Bachelard's theory of material imagination and Merleau-Ponty's ontology, as well as, on this specific point, with Deleuze's ontology. The model of eternal ideas applied to an inert matter cannot account for our world. Clear-cut forms are barren. Elusive and ambiguous forms emerge only through contingent, sensible connections and their repetitions: Deleuze speaks of *habits* and Merleau-Ponty of *musical themes, musical ideas*, or *wild essences*. Material imagination, because it draws on such themes, manages to give a genuine quasi-presence to imagined objects. Formal imagination, on the contrary, gives birth to empty shells that can only be entertainingly contemplated from afar, but that never transform us or transport us to the birthplace of beings.
25. *L'air et les songes*, p.7.
26. Bachelard, *Lautréamont*, p.50 (Trans. p.27).
27. *Ibid.*, p.103 (Trans. p.59).
28. See for instance *L'Eau et les rêves*, p.208: through art, we practice a certain "imaginary gymnastics."
29. What is at stake, at least in a first stage, is to "galvanize" or "dynamize" [dynamiser] the complex (Bachelard, *Lautréamont*, p.156 (Trans. p.91)).
30. Merleau-Ponty (about Bachelard) in *Signes*, Paris: Gallimard, 1960, p.72 (Trans. 1964, p.57).
31. Bachelard, *Lautréamont*, p.109 (Trans. p.62).
32. *Ibid.*, p.51 (Trans. p.27).
33. See also Mattias Preuss' beautiful analysis of arachnopoetics and the way webs, threads, spinning, weaving, and excretion all provide underlying tenacious models for expression, from Ovid to postmodernism. "Spinning Theory: Three Figures of Arachnopoetics". In *What is Zoopoetics? Texts, Bodies, Entanglement* (Kári Driscoll, Eva Hoffmann Eds.)
34. This idea was already sketched by Aristotle (*On the Soul*, III 7: thinking without images is impossible), and by Kant in the theory of schematism in *Critique of Pure Reason*. Kant takes a crucial step further by making it clear that imagination essentially thinks through dynamic schemas rather than through static mental pictures. Merleau-Ponty has also argued that no idea can make

sense without dynamic images and he connects the latter to the potentialities of the body schema. See for instance *Phenomenology of Perception*, 2012, p.406.

35. See for instance Merleau-Ponty's analysis of habits in *Phenomenology of Perception*, Part I, 3 "The Spatiality of One's Own Body and Motricity."
36. Bergson, Henri. *Matière et Mémoire*. Paris: Alcan, 1896. Reprint Chicoutimi: Les classiques des sciences sociales, 2003, p.66, 68, 70, 73.
37. Bachelard, *Lautréamont*, p.106 (Trans. p.60–1).
38. Lautréamont, *The Songs of Maldoror*, p.16.
39. Bachelard, *Lautréamont*, p.111 (Trans. p.63).
40. *Ibid.*, p.109 (Trans. p.62).
41. *Ibid.*, p.24 (Trans. p.11). Moe, Aaron, in *Zoopoetics: Animals and the Making of Poetry* (2014), also refers in an illuminating manner to many poets who emphasized the close relation between poetry and body movements (p.8 sqq) as well as to Kennedy's "A Hoot in the Dark: The Evolution of General Rhetoric." (in *Philosophy and Rhetoric* 25, 1992, 1-21). Moe thus institutes the field of zoopoetics, defined as a discipline that "exposes the places [in poetry] where the gestures of poetic form depend on, mime, or play with the gestures of animals" (p.24). He concludes that poetry cannot be regarded as "a mono-species event", for it is deeply and fundamentally defined by gestures and gesticulations. "Innumerable other animals have crept into the forms of many poems, much like the gestures of hands migrating across the body into the gestures of human speech." (*Ibid.*). Even more radically, muscular lyricism can be defined as an essential root of human thought overall. Bachelard provides the general bases for a theory of animal dynamogenism as the source of human thought, but focuses almost exclusively on Lautréamont's poetry. Zoopoets on the other hand, develop a wealth of thought-provoking analyses of many texts where animal movements, trajectories, styles, and moods animate the flow of words.
42. Aaron Moe also ventures the provocative idea that "for zoopoetics the animal precedes the plant", (*Zoopoetics*, p.23). This phrasing is subtly chosen since, obviously, in the imaginary realm, rigid boundaries, whether between humans and animals or between animals and plants, are never in order.
43. Bachelard, *Lautréamont*, p.106 (Trans. p.61).
44. "Imagination motrice" (Bachelard, *Lautréamont*, p.41), which was translated as "driving imagination", in the published translation of Bachelard's book (p.21).
45. Unicorns have long been regarded as male. They became feminine or queer more recently.
46. Freeman, Margaret B. *The Unicorn Tapestries*, New York: The Metropolitan Museum of art, 1976, p.15.
47. Saint Basil. *Homily on Psalm, 28*. In *Exegetic Homilies*. p.204–5. Trans. Sister Agnes Clare Way. Washington: Catholic University of America Press, 1963. Quoted by Freeman, *The Unicorn Tapestries*, p.17.
48. Eason, Cassandra. *Fabulous Creature, Mythical Monsters, and Animal Power Symbol: A Handbook*. Wesport CT: Greenwood Press, 2008, p.57.
49. Freeman, *The Unicorn Tapestries*, p.34.
50. Fischer, Alice. "Why the Unicorn has Become the Emblem for Our Times," In *The Guardian*, Sun 15 Oct 2017.
51. See Hallowell, A. Irving. "The Role of Dreams in Ojibwa Culture." In *Contributions to Anthropology: Selected Papers of A. Irving Hallowell*, Fogelson, R.D. et al. (Eds.). Chicago: University of Chicago Press, 1976.
52. Ingold, *The Perception of the Environment*, 2000, p.94.

53. *Ibid.*, p.94. See also Bourgeois, A.P. (Ed.). *Ojibwa Narratives of Charles and Charlotte Kawbawgam and Jacques LePique, 1893–1895.* Detroit: Wayne State University Press. 1994, p.69

54. See Ingold, *The Perception of the Environment*, p.100: The self, according to the Ojibwa people, emerges through its "ongoing engagement with the environment." It is not "enclosed within the confines of a body."

55. *Ibid.*

56. Deleuze and Guattari, *Mille plateaux*, p.17 (Trans. p.10).

57. Bachelard, *Lautréamont*, p.24, Trans. p.11. In the same vein, Bachelard (p.144, Trans p.84) refers to Caillois' theory in *Le mythe et l'homme*. Caillois asks: how can the macabre eroticism of the praying mantis perfectly mirror the coalescence of love and death in so many human myths? Following a Bergsonian inspiration, Caillois contends that what is expressed through myths in the human realm is expressed in animals through instinct, under the guise of behaviors and morphologies (Roger Caillois, *Le mythe et l'homme*, Paris: Gallimard, 1938, p.81). The flaw of this approach lies in the concept of blind mechanical instincts on the animal side. The capacity of such a purely instinctive animal life to mirror ek-static and open fantasies on the human side cannot but remain a mystery.

58. Bachelard, *Lautréamont*, p.148 (Trans. p.154–5).

59. *Ibid.*, p.16, Trans. p.6: "...Casanova is no better than Sade at moving beyond human limits. (...). All his fieriness is human; expressed solely in metaphors, it never achieves any metamorphosis."

60. *Ibid.*, for instance, p.23–4 (Trans. p.10–1).

61. Deleuze and Guattari, *A Thousand Plateaus,* p.294.

62. *Ibid.*, p.237.

63. *Ibid.*

64. *Ibid.* See also note 5, p.538.

65. *Ibid.*, p.538.

66. *Ibid.* See Lévi-Strauss, Claude. *La pensée sauvage*, Paris: Plon, 1962, p.152 sq.

67. Deleuze and Guattari, *A Thousand Plateaus,* p.235.

68. Jung, Carl Gustav. *Archetypes and the Collective Unconscious*, Trans. R.F.C. Hull, *The Collected Works of C.G. Jung.* Volume 9, London: Routledge and Kegan Paul, 1931, p.76.

69. *Ibid.*, par 309 note 1 p.133.

70. Bachelard's theory cannot in fact be reduced to the analogy of proportion. To be sure, this path is present in *Lautréamont*. Quite like Jung, Bachelard was looking for psychological lines of evolution. Moreover, he clearly conceives of the imaginary in terms of hierarchy and progression/regression (more or less intense or pure manifestations of aggressiveness for instance). But Bachelard pays particular attention to the transformation process at work where the terms of the "analogy" encroach upon each other. Muscular lyricism lies below the realm of distinct individuals. Bachelard's analyses also allow for an interpretation that brings us towards a new ontology. It is only through caricature that one finds in *Lautréamont only* a philosophy of Platonic pure essences and participation. Bachelard's works about the imaginary contain at least to the same extent the seeds of a philosophy of alogical essences that Deleuze developed in *Proust et les Signes,* Paris: PUF Quadrige, 1998, p.50.

71. Deleuze, *Différence et répétition*, Paris: PUF 1968, p.82. Trans. Paul Patton. New York: Columbia University Press, 1994, p.59

72. "Tode ti" is a key concept in Aristotle's *Metaphysics*. "Tode ti" designates a "this" that is, on the one hand, recognizable, determinate—thus connected to an essence, a general species—and, on the other hand, a singular instance through which this essence takes on a unique form.

73. Schelling, Friedrich Wilhelm Joseph. *Philosophical Investigations into the Essence of Human Freedom.* Trans. Jeff Love and Johannes Schmidt. New York: State University of New York Press, 2006, p.17–8.
74. Deleuze, *Différence et répétition*, p.36 (Trans. p.23).
75. What Deleuze also calls "plane of consistency," and "plane of immanence" contains multiplicities formed through repetitions.
76. Deleuze, *Proust et les Signes*, p.50.
77. See Deleuze. *Empirisme et subjectivité.* Paris: PUF, 1953. See also *Différence et répétition*, p.98 (Trans. p.71).
78. Deleuze, *Cinéma 1*, p.88.
79. Deleuze, *Différence et répétition*, p.102 (Trans. p.75).
80. *Ibid.*, "We speak of our 'self' only in virtue of these thousands of little witnesses which contemplate within us: it is always a third party who says 'me.' These contemplative souls must be assigned even to the rat in the labyrinth and to each muscle of the rat."
81. Ibid., p.128 (Trans. 96).
82. Ibid., p.10 (Trans. 19).
83. Ibid., p.329 (Trans. 256). The concept of molecular processes becomes central in *Mille plateaux.*
84. With his theory of molecular becomings and machines, Deleuze endeavors to overcome the dichotomy between matter and thought. In a phenomenological approach, the concepts of *hylé*, *Abschattungen*, and passive syntheses struggle with similar difficulties and have a joint ontological and transcendental nature as well.
85. Simondon, George. *L'individu et sa genèse physico-biologique*, Paris: Presses Universitaires de Frances, 1964.
86. Deleuze, *Différence et répétition*, p.305, 321–2, and 102 (Trans. p.237, 249–50).
87. *Ibid.*, p.155, Trans. p.118.
88. Deleuze and Guattari, *A Thousand Plateaus*, p.40.
89. The schematism of our understanding is "an art concealed in the depths of the human soul, whose real modes of activity nature is hardly likely ever to allow us to discover." I. Kant, *Kritik der reinen Vernunft*, in *Gesammelte Schriften*, Vols. 3 and 4 (Berlin, 1903/04), A141, B180; translated by P. Guyer and W. W. Wood, vol. 3 (Cambridge: Cambridge University Press, 1998), p.136.
90. Deleuze *Différence et répétition*, p.97 (Trans. p.71).
91. Deleuze, *Différence et répétition*, p.269 (Trans. p.208).
92. Deleuze and Guattari, *A Thousand Plateaus*, p.257.
93. *Ibid.*, p.260.
94. *Ibid.*, p.336.
95. *Ibid.*, p.11.
96. *Ibid.*, p.240.
97. *Ibid.*, p.274.
98. *Ibid.*, p.258.
99. *Ibid.*, p.257.
100. The individual arises "in a matter that is no longer that of forms, in an affectability that is no longer that of subjects", *Ibid.*, p.258.
101. *Ibid.*, p.257, my emphasis.
102. The notion of theme is, as I have indicated, absolutely central in Merleau-Ponty's philosophy and pertains to an ontology of the imaginareal. Merleau-Ponty borrowed it from Uexküll and Buytendijk on the one hand and Proust on the other hand. Deleuze avoids mentioning Merleau-Ponty, but draws on the same sources. When, in *Proust et les Signes*, Deleuze puts forward the concept of theme (G. Deleuze, *Proust et les signes*, p.42) and of "alogical essence" (p.50), the opportunity arises to connect the dots and unmask the

218   *Metamorphoses and corporeal imagination*

influence of Merleau-Ponty (see my article "Sous les masques il n'y a pas de visages: l'éthique merleau-pontyenne entre problème de l'altérité radicale, foi et institution" in *Chiasmi International*, No. 17, 2015).

103. Deleuze and Guattari, *A Thousand Plateaus*, p.258.
104. *Ibid.*, p.10.
105. Deleuze, Gilles and Parnet, Claire. *Dialogues*. Paris: Flammarion, 1977, p.8–9. Trans. Hugh Tomlinson and Barbara Habberjam. New York: Columbia University Press, 1987, p.2–3, my emphasis.
106. Deleuze and Guattari, *A Thousand Plateaus*, p.386 (they here refer to Uexküll).
107. See infra, §36.
108. *Ibid.*, p.238.
109. *Ibid.*
110. Haraway, Donna. W*hen Species meet*. Minneapolis, London: University of Minnesota Press, 2008, p.314
111. *Ibid.*, p.27.
112. *Ibid.*, p.9.
113. *Ibid.*, p.29, p.314.
114. Haraway, Donna. *The Companion Species Manifesto: Dogs, People, and Significant Otherness*, Chicago: Prickly Paradigm Press, 2003, p.19–20.
115. Haraway, *When Species Meet*, p.4–5.
116. *Ibid.*, p.20. "Derrida failed a simple obligation of companion species; he did not become curious about what the cat might actually be doing, feeling, thinking, or perhaps making available to him in looking back at him that morning."
117. *Ibid.*, p.3–4.
118. Margulis, Lynn and Sagan, Dorion. *Acquiring Genomes: A Theory of the Origins of Species*, New York, Basic Books, 2002.
119. Haraway, *The Companion Species Manifesto*, p.9.
120. Margulis and Sagan, *Acquiring Genomes*, p.72.
121. *Ibid.*
122. Haraway, *When Species Meet*, p.15.
123. *Ibid.*, p.229. Haraway refers to Vinciane Despret, "The Body We Care For: Figures of Anthropo-zoo-genesis," In Akrich, M. et Berg, M. (Eds.) *Body and Society*. Special issue on "Bodies on Trial" 10 (2–3), 2004, p.115.
124. "A Cyborg Manifesto: Science, Technology, and Socialist-Feminism in the Late Twentieth Century." In *Simians, Cyborgs and Women: The Reinvention of Nature*. New York: Routledge, 1991, p.150.
125. *Ibid.*, p.154.
126. Haraway, *When Species Meet*, p.226. See also "'We' did not originally choose to be cyborgs, but choice grounds a liberal politics and epistemology that imagine the reproduction of individuals before the wider replications of 'texts.'" (*Cyborg manifesto* p.176). In Haraway's philosophy, "we" live and philosophize within a plane, as fragile concrescences or individuation processes where ambiguous meanings and bodies arise.
127. Haraway speaks of "subject-forming entanglements" (*When Species Meet*, p.313) and "subject-making connection" (p.227) for instance.
128. Haraway, *The Companion Species Manifesto*, p.41–2.
129. *Ibid.*
130. *Ibid.*, p.242.
131. *Ibid.*
132. *Ibid.*, p.220.
133. *Ibid.*, p.212–3.
134. *Ibid.*, p.212.
135. See Pryor, Karen, and Ramirez, K.R. "Modern Animal Training: A Transformative Technology." In McSweeney, F. and Murphy, E. (Eds.), *A Handbook of Operant and Classical Conditioning*. New York: Wiley and Blackwell,

2014. Karen Pryor found a way of raising the call for creativity borne by the method of positive reinforcement to the second power. During each training session, she marked only behaviors that had not been reinforced before. After exhausting her repertoire, Malia, female rough-toothed dolphin, went through a phase of deep frustration, showed signs of strong irritation, but eventually seemed to "get the idea" and started displaying more and more new behaviors (see also Pryor, K. *Reaching the Animal Mind: Clicker Training and What It Teaches Us About All Animals.* New York: Scribner, 2009; and Pryor, K.W., Haag, R., & O'Reilly, J. "The Creative Porpoise: Training for Novel Behavior." In *Journal of the Experimental Analysis of Behavior,* 12, 1969).

136. Haraway, *When Species Meet,* p.211.
137. *Ibid.,* p.224.
138. Smuts, Barbara. "Gestural Communication in Olive Baboons and Domestic Dogs." In Bekoff, Marc, Allen, Colin, and Burghardt, Gordon (Eds.) *The Cognitive Animal Empirical and Theoretical Perspectives on Animal Cognition,* 2002, p.306.
139. Savage-Rumbaugh, E. Sue et al. "Language Comprehension in Ape and Child". *Monographs of the Society for Research in Child Development,* Vol. 58, 1993.
140. *Ibid.,* p.27.
141. *Ibid.*
142. Boesch, Christophe. "Teaching Among Wild Chimpanzees," *Animal Behaviour,* 41, 1991.
143. Savage-Rumbaugh et al., "Language Comprehension in Ape and Child," p.28.
144. *Ibid.*
145. *Ibid.,* p.28.
146. See also my analysis of the concept of flesh in Sartre's *The Family Idiot* in "De la chair à la révolte: l'activité passive dans *L'idiot de la famille*", in *Horizon. Studies in Phenomenology,* 3 (2), 2014.
147. For instance, Haraway, *The Companion Species Manifesto,* p.41–2.
148. See supra, p.100 on autopoiesis.
149. Bekoff, Mark and Allen, Colin. "The Evolution of Social Play: Interdisciplinary Analyses of Cognitive Processes." In *The Cognitive Animal,* Bekoff, M., Allen, C., and Burghardt, Gordon (Eds.), 2002, p.432.
150. Félix Gaffiot, *Dictionnaire Latin-Français,* Paris: Hachette, 1934.
151. Haraway, *When Species Meet,* p.4.
152. *Ibid.,* p.32.
153. Haraway, Donna. *Modest_Witness@Second_Millennium. FemaleMan_Meets_OncoMouse: Feminism and Techno-Science.* New York: Routledge, 1997, p.11.
154. Haraway, *When Species Meet,* p.5.
155. *Ibid.,* p.4.
156. For instance, *Ibid.,* p.3.
157. *Ibid.,* p.4.
158. Deleuze and Parnet. *Dialogues,* p.8–9 (Trans. p.2–3), my emphasis.
159. Haraway, Donna. "Situated Knowledges: The Science Question in Feminism and the Privilege of Partial Perspective." In *Feminist Studies,* Vol. 14, No. 3, 1988, p.577
160. *Ibid.,* p.586.
161. Haraway, Donna. "The Promises of Monsters: A Regenerative Politics for Inappropriate/d Others." In *The Haraway Reader,* London New York: Routledge, 2004, p.107–8
162. Deleuze and Guattari, *Kafka. Toward a Minor Literature,* p.15.
163. Deleuze and Guattari, *A Thousand Plateaus,* p.239.
164. Haraway, *When Species Meet,* p.27, my emphasis.

165. Deleuze and Guattari do not make great efforts to clarify this phantasmatic nature of their favorite *figures*. However, they sometimes outline a more nuanced approach. They make a distinction between three kinds of animals (1) "individuated animals, family pets, sentimental, Oedipal animals each with its own petty history, "my" cat, "my" dog; (2) great myths, archetypes and models, State animals; (3) "pack or affect animals that form a multiplicity, a becoming, a population", but they also immediately note that, perhaps, these should rather be regarded as *"three ways of treating"* every animal (*A Thousand Plateaus*, p.240–1).
166. Deleuze and Guattari, *A Thousand Plateaus*, p.245.
167. *Ibid.*, p.244.
168. *Ibid.*, p.240 (*Mille Plateaux*, p.294), I have modified the translation. "A fool" partly emphasizes the main meaning (moron, stupid), but this term remains quite soft and polite in comparison with the French "con."
169. Deleuze and Guattari, *Mille Plateaux*, p.299 (Trans. 245). In fact, Deleuze and Guattari borrow the word "mannequin" from Melville, more specifically from the French translation by J. Giono, L.J.-Jacques, and J. Smith, Gallimard, 1941: "tous les objets visibles ne sont que des *mannequins* de carton." The original sentence in English is: "All visible objects, man, are but as pasteboard masks". The adjective "imaginaire [imaginary]" is nevertheless, interestingly enough, added by Deleuze and Guattari.
170. Lautréamont, *Songs of Maldoror,* p.257–8.
171. Deleuze, *Cinema 1*, p.84–5, 89–90.
172. See chapters 2 to 4.
173. Deleuze and Guattari, *A Thousand Plateaus*, p.258.
174. Deleuze and Parnet. *Dialogues*, p.8–9 (Trans. p.2–3).
175. Deleuze, *Cinéma 1*, p.89–90, Trans. by Hugh Tomlinson and Barbara Habberjamp, University of Minnesota Press, 1986, p.60–1. Deleuze exaggerates the contrast between Bergson's philosophy (reduced to "all consciousness is something" *Cinema* trans. p.56 and "matter is light", p.60) and Husserl's (allegedly reducible to "all consciousness is consciousness of something," p.56, or "consciousness is a light that reveals things," p.60). He does so by overemphasizing the role played by subjectivity in phenomenology and overlooking the fact that Bergson's use of the word "images" to define alleged pure asubjective matter is problematic. Bergson implicitly refers to our perception, and even to a perception that consists in "images," namely a set of facets and appearances. Thus, claiming that Bergson's "pure matter" has no link with personal subjectivity is wrong. Thought does not exist without centers, interpretations, and imaginative representations. Phenomenology, for its part, had to learn that subjectivity emerges from folds in the matter, and that a subject thinks only through partial images. But even Husserl, in genetic phenomenology, makes significant progress in this respect.
176. And, in fact, the borrowings from phenomenological conceptuality, terminology, and approach are legion in Deleuze's works, however (for a scrupulous analysis see for instance Tejada, Ricardo. "Deleuze face à la phénoménologie," *Papiers du Collège International de Philosophie*, No. 41, 1998).
177. Haraway, "Situated Knowledges," p.586.
178. Phenomenology as a school of philosophy does not seem to belong to Haraway's field of interest. And the idea that the subject is to be invoked through fiction is not a classic phenomenological approach either. Nonetheless, in an age where molecules, cyborgs, and machines, more fundamentally new materialisms, have taken center stage, I find it crucial to emphasize the irreducible role that the first-person and second-person perspectives still have to play. Although Haraway, a key figure in new materialisms, and phenomenologists

have not traveled from the exact same conceptual homeland and through the same paths, they still have this special idea in common: subjectivity and inter-subjectivity are not vain words and even a philosophy of a plane of immanence still has to carve space for them. Similarly, Merleau-Ponty is interested in emergence and the spirituality of the body, but defining him as a materialist, old or new, would be a manner of sadly flattening his ontology.

179. Deleuze, *Deux régimes de fou. Textes et entretiens, 1975–1995.* Paris: Minuit, 2003, p.38. Trans. Ames Hodges and Mike Taormina, New York: Semiotext(e), 2007, p.39.
180. *Différence et Répétition*, p.196, Trans. p.150.
181. *Ibid.*, p.197 Trans. p.151. Deleuze claims that the norms and instincts defining a species would remain at an explicit state in animals and would constitute a shield strong enough to *protect* them and even *guarantee* them against the opening onto the formless ground. See Derrida's critique of this passage in *The Beast and the Sovereign* and my article "Who/What is Bête? From an Uncanny Word to an Interanimal Ethics", In *Environmental Philosophy*, Vol. 16, Issue 1, 2019.
182. Deleuze, *Mille Plateaux*, p.389, Trans. p.316.
183. "What defines the territory is the emergence of matters of expression (qualities). Take the example of color in birds or fish: color is a membrane state associated with interior hormonal states, but it remains functional and transitory as long as it is tied to a type of action (sexuality, aggressiveness, flight). It becomes expressive, on the other hand, when it acquires a temporal constancy and a spatial range that make it a territorial, or rather territorializing, mark: a signature." (*A Thousand Plateaus*, p.315). In fact, ascribing certain animal appearances to a pure mechanical state or a functional role is impossible. Hence, as I have argued, in line with Portmann, expressivity and territorialization begin with life. Subjects are more precocious than argued by Deleuze.
184. Deleuze and Guattari, *A Thousand Plateaus*, p.311.
185. Deleuze and Guattari, *Mille Plateaux*, p.382, Trans. p.311. I have modified the English translation ("A place of one's own" instead of "at home") in order to emphasize the reference to the self (soi) that the French "chez-soi" explicitly involves.
186. Deleuze and Guattari, *Mille Plateaux*, p.388–9, Trans. p.315–6

Conclusion
Why imagine (with) animals?

§37 The misadventures of the imaginary of animals: Reified images and simulacra

An archaic affinity ties together animals and imagination, as the figures of the chimera (a young goat/a fantastic creature composed of parts from different animals/an organism with genetic material from two or more sources/a fantasy) and zoographia (painting/animal-drawing) suggested in a sibylline mode. The human imaginary thrives on the animal imaginary as animals proves to be imaginative beings instituting human imagination.

While the connection between the questionable human-animal dichotomy, the exploitation and destruction of animals, and the racist and sexist use of animalizing metaphors has been thoroughly analyzed in the groundbreaking works of Marjorie Spiegel,[1] Carol Adams,[2] and Jacques Derrida[3] in particular, I have argued that the imaginary-real dichotomy constitutes another key dimension in the anthropocentric system of exclusion. Overcoming this dichotomy thus becomes a most promising route toward the dissolution of this oppressive system. And no strategy to build a fair relationship with non-human animals can be successful if it glosses over the problem of the imaginary of animals. The complicity between the human-animal and the real-imaginary dualities indicates the path to follow: We can heighten the understanding of animals through our imagination, and, conversely, our imagination develops as a fundamental animal ability to create worlds through a dialogue with animals as imaginative beings. The phenomenological approach that I have defended provides a general ontological framework that legitimates and encourages the use of phantastic empathy, subjectification processes, and cooperation between art, literature, philosophy, and sciences in developing a fruitful human-non-human imagination

We have thus every reason to rejoice about the fact that images of animals thrive in the media and popular imagery ... But wait, really? I do indeed advocate this claim, eventually to my own surprise since the overabundance of representations of animals and animal metaphors in our societies has been famously and convincingly diagnosed by John Berger and

Carol Adams in particular as the sure sign of an oppressive attitude toward non-human animals. Admittedly, the imaginary of animals can take on different forms, including dangerous ones. A final clarification is thus needed to prevent misinterpretations of the invitation to embrace the human-animal metaphor.

Ethical issue: Imagine ... in any manner?

In this book, my intention was first to show the decisive role played by imagination and the imaginary in our human relation to non-human animals and moreover to contend that, *under certain conditions*, we should embrace and even reinforce the imaginative dimension of our engagement with the animal realm.

By "under certain conditions," I do not mean that imagination should be kept under strict control. There are no *de jure* eternal order, concepts, or laws. The reality principle and "reason" emerge and morph through phantom-like processes at the imaginareal level[4] and via negotiations and adventurous attempts undertaken by the imagination of individuals. Invitations to dance (see §35) are among such attempts ventured by some human and non-human animals to create a common world. But the great divide is another fantasy stemming from the imagination of some humans, and so are sugary and silly pictures of animals in Disney movies, cuddle toys, clothes, trinkets, and ornaments. How are we to define a distinction between fruitful and detrimental imagination in our relation to animality?

My contention is that imagination cannot *fail*, although it may *falter*. The analyses presented in this book show that the center of gravity of ethics should shift *from* the ideal of absolute rules *to* risk, instability, and manifoldness embraced as such. Let us rephrase the fundamental ethical and epistemological question: To what can one be more or less faithful? Certainly not to human nature or reason, but rather to the complexity and the depth of the imaginareal, which includes countless other nascent subjectivities. Animal subjectivities, in the framework of this ethical task, represent a most crucial challenge, with significant consequences for the shape the world(s) will take. Indeed, animal subjectivities refer us to processes of meaning-making and world-making beyond the false sense of security granted within the realm of *logos*. Imagination is always particularly solicited whenever understanding or representing animals is at stake, but when does fantasy "miss" them and when is it faithful to them?

When imagination falters: Dead images, alleged stubborn reality, alt-reality (simulacra)

The imaginareal essentially implies a tension between the hovering in the virtual, on the one hand, and recurring themes, on the other. Western rationalism has radicalized this distinction by instituting an ontological divide between the real (allegedly obdurate) and the imaginary (allegedly insubstantial and recreational). In other words, *some* fantasies, myths, and constructions have presented themselves as *the real*. The meandering and multithreading emergence of such fantasies is concealed through the creation of consistent narratives combined with consistent power takeovers and subjugation processes. In *The Animal That Therefore I Am,* Derrida thus describes how the baroque myth of the great divide between humans and animals could, through history, texts, acts, and institutions, take on the appearance of the plain description of what simply is. This reference to an authoritative reality is the first dangerous mistake made by human imagination. It disastrously kills two birds with one stone: Imagination is largely devitalized and non-human animals are confined to the figure of mute beasts.

In his famous *Why Look at Animals?*, John Berger depicts what may be understood as a complexification of this process and a new avatar of the tricky relation between the real and the imaginary. In Western societies, the "culture of capitalism"[5] turns non-human animals into images. Berger describes a conjunction between two processes that started after the Industrial Revolution, and can reasonably be regarded as correlative.

1 On the one hand, non-human animals are more and more confined to very specific spaces (like game reserve and national parks), poisoned, hunted, turned into productive units and commodities, and slaughtered on an industrial scale.

2 On the other hand, images of animals flow and flourish in home decoration and, for instance, print patterns in clothing products and accessories. Berger also refers to non-human animals in cartoons: Overlaid with anthropomorphic traits, they speak and behave like human beings. Their preoccupations and adventures are directly borrowed from human social situations.[6] In this context, non-human animals become free-floating signifiers:[7] They can be used for any advertising purpose and to represent anything. The broad and systematic phenomenon Berger describes also includes the abundance of zoological charts and the growth of animal studies, in sum, the development of re-presentations of non-human animals. The overabundance of animal videos on YouTube is part of the same phenomenon. Even real animals are somehow turned into images. Zoos transform animals into spectacles: Isolated, encaged, cut off from a great number of exploratory, hunting or gathering activities, and interspecific interactions, they become diminished, to a great extent, devitalized versions of themselves.[8]

In the light of my analysis of the imaginary of animals, it appears that the images Berger refers to are not *any* images and, for instance, not living-images (see supra §15). They belong to the framework of the dichotomy between the real and the imaginary and are the descendants of copy-images. They are copies claiming to supersede the original—which is a way of maintaining this original-copy duality. Such images function as *mere figments* of human imagination, instead of being the result of the interaction between our imagination and a deep anonymous imaginary. Consider animals in Disney movies, for instance. They talk, show human expressions, and make exaggerated faces: They smoothly enter our world. As familiar and rather schematic or simplistic patterns, the images described by Berger offer restrained meaning and provide easy pleasure. They remain *under the control* of a strong-willed imagination. Animals turned into such images become more "fun," more manageable, and more obviously expressive.

I have recently encountered another striking example of these bizarre images. Colorful or, even better, neon-colored fiberglass resin statues of animals (often rhinoceroses, crocodiles, giraffes, or lions, among others) have become quite fashionable lately. Neon colors hit us in full force, without much room for fine modulations and nuances: They stake everything on over-saturation and its power to overwhelm us and make all the rest temporarily disappear. Neon colors are simultaneously joyful and up in arms, fighting some weird battle for attention. A sizable neon pink crocodile stands on the main square in front of our faculty building (Figure 6.1). It is chained up to a bin and turns out to be the mascot of the campus supermarket.

Figure 6.1 Pink fiberglass resin crocodile on Erasmus square, Nijmegen (April 2018).

This image does not say much about crocodiles, who are among the most endangered animals in the world. The pink crocodile vaguely refers to real crocodiles, but its jubilant hot pink color supersedes everything. In this case, at least, the image succeeds in floating freely, while it would most likely fail to do so if a fiberglass hot pink statue of a human being in actual size were chained to a bin on the campus main square. The eeriness of the image would strike us and make us cringe, whereas people just walk past our pink crocodile several times a day and hardly notice it. For months I have not heard anybody commenting upon the pink crocodile, wondering what it represents or if it represents its "model" in a faithful, inspiring, or provocative manner. The image seems to have replaced the original, silently.

Such oppressive images operate within a logic of competition (image/original) and paradoxical reduction (of the original to its image). As a result, these fantasies can work hand-in-hand with a process of negation and actual extermination. Colorful and striking, but also cloying and super-ficial, such images are doomed to possess a mixed effectiveness: They need the backup of the destruction of their opponents in order to simply win by default. Conversely, the negation of non-human animals is so violent that it appears to require the introduction of surrogates, namely a compensation process. Thus, such innocent-looking images actually take a crucial part in the destruction of animals, by providing us with substitutes without which the enterprise of mass destruction would appear as unbearable.

In these new avatars of copy-images, one can recognize what is tradition-ally called *simulacra*.[9] Engaged in a fight for power, operating within a logic of control and concealment, simulation does not reveal the potentialities and imaginary dimension of "the real" (at heart: The imaginareal). Instead, simulation simply erects a lame alt-reality. Yet, simulacra essentially fall to pieces.

Why simulacra cannot stop imagination

Berger's diatribe against the modern "imagery"[10] of animals is merciless. Berger thus successfully highlights the real danger posed by these simula-cra, but he does not do justice to their complexity and potentialities.

The relation between animals and our imagination cannot simply come down to the modern substitution of images for real animals. Indeed, such a substitution is actually a failure. Simulacra bear the trace of the crime they attempt to commit. Many people feel at least ill at ease in zoos. Contemporary zoos—they still exist, which is, to begin with, rather incred-ible and demonstrates our ability to withstand cognitive dissonance to a high degree—are more and more keen to maintain high standards of animal welfare and to respect the relation of animals to a meaningful environment, but the malaise subsists in front of a setting that is primarily the result of the scopic drive. It is important not to refrain such a malaise. Similarly, the pink crocodile has a lot to say to those who will take even a minute to consider it. Its eeriness as well as its colonial aftertaste are unmistakable. I have doubts concerning the pink crocodile's capacity to look unambiguously cool, fun, and somehow wild. The fact that it is chained up to a bin certainly does not increase its simulation power, but there are many other aspects that give its violence away. In fact, as I eventually found out, one article was published in a satiric webzine about the pink crocodile on my campus:[11] It is entitled "the pink crocodile of Spar University demands better working conditions." The short piece, presented as an interview of the crocodile, makes her talk in the style of a labor activist. It astutely combines references to animal rights, to the crocodile's wild life ("I am not even allowed to say hi to the customers, let alone to bite them into pieces"), and to the general (usually human) issue of working conditions.[12] Discovering this article cheered me up: The pink crocodile, that I failed to even notice for weeks, was still capable of arousing high-spirited fantasies. This simulacrum does not consist in a completely dead meaning, it does not really achieve the "world without animals" that Derrida regarded as the secret goal of civilization.[13]

This relative failure of simulation was actually to be expected: Because of their very structure, simulacra cannot fully succeed in their endeavor to simply replace the original and fully destroy real animals. In effect, firstly, simulacra obviously belong to the realm of images. In other words, they are innervated by a tension between different modes of presence (in person *vs* by proxy, in fantasy *vs* perceptive, this image *vs* many other possible images). And indeed, simulacra also *manifest* an enterprise to manipulate these modes of presence and to impose images as the new reality (think of the combative neon pink). As a result, the structure of simulacra is unhinged and, at least, slightly unsettling: Images cannot function as plain, well-established realities.

Secondly, simulacra exaggerate the gap between the original and the copy, which weakens the copy (as a mere substitute). The attempt to *replace* the original by the image-*of-the-original* cannot but fail. As Akira Lippit points out, simulation is a process, never a completed enterprise: "Animals never *entirely* vanish. Rather they exist in a state of *perpetual vanishing.*"[14]

Third, the project to replace animals with fully-controlled images is bound to fail also because images essentially draw on the ubiquity of the original, which can appear either as fully present or as mysteriously quasi-present *through* countless *analoga*. This fundamental floating nature of beings makes images (fantasies and pictures) possible and sustains their ability to endlessly inspire new variations. The imaginareal can never be tamed and cast into a limited set of clichés. To be sure, images possess their own strength and simulacra indeed tend to effectively blind us. Yet, such images harbor a depth of wild associations, references, and possible metamorphoses; they tickle our imagination and keep it bound to the inexhaustible imaginary. As images are never mere blind forces, but depend also on the way the subject apprehends them, simulacra always allow us to fantasize their meaning in a more creative and genuine manner.

Berger does not emphasize the irreducible profundity of simulacra, but his analyses do not clearly call for a return to real animals, beyond images and imagination. And for good reasons, since images of animals started to thrive long before the Industrial Revolution. Myths about animal ancestors of human, narratives about marriages between human and non-human animals, or about metamorphoses and hybrids did not appear with the Industrial Revolution. Moreover, as Berger highlights, "the essential relation between man and animal was [already in pre-industrial societies] "*metaphoric*."[15] Therefore, the key to the imaginary of animals lies neither in simulacra nor in a concept of obdurate reality. It is pointless to look for the alleged ultimate true nature of animals, defined independently of the way human animals imagine them. And the imaginary cannot be reduced to the mere figments of human imagination. The key to the imaginary of animals can thus be found beyond the imagination-reality dichotomy that simulacra try, in vain, to radicalize.

By the same token, Western rationalist philosophies, objective sciences, or, more broadly, all the narratives and figures of the human-animal war, remain haunted by the imaginary of animals. The latter vastly exceeds them but, also, always manifests in them. The imagineareal overflows the bounds of particular, even reified figures, and always reappears through experience. Hence, animal sciences struggle with the tension between animals as objects and animals as subjects, fact-checking and invention, or between the ambition to fully and clearly know animals and the will to respect them. But the possibility to make world with non-human animals actually always looms at the ragged edge of the general system of exclusion that still prevails nowadays or may even appear as more brutal and divided than ever: On the margins, where fictions and fantasies are allowed to multiply, new alternatives form. It is through the imaginary aura of the concept of "pure imitation" (Chapter 1), or through the phantom dimension of Gestalten, or in the imaginary of Darwin (Chapter 3), for instance, that we can find troubling openings toward animal worlds, and, simultaneously toward other cultures and alternative paradigms, as Ingold's work shows.

§38 The stubborn metaphoricity of animals: Animal imagination and animal agency

Dynamic transpositions (metaphors) between human and non-human animals

Berger, like many other thinkers of issues related to animals and animality, stumbles over the strange metaphoricity of human and non-human animals. "Animals are born, sentient and are mortal. In these things they resemble man. In their superficial anatomy—less in their deep anatomy—in their habits, in their time, in their physical capacities, they differ from man. They are both like and unlike."[16] I would not go as far as claiming that "[The animal's] lack of common language, its silence guarantee its distance, its distinctness, its exclusion from and of man,"[17] but I have also argued that a theory of pure continuity or identity between humans and animals would be inaccurate (Chapter 1). Berger thus contends that "it is not unreasonable to suppose that the first metaphor was animal."[18] He does not refer to animals' poetry, as researchers in zoopoetics do, but, at least, he emphasizes that humans have been forced by their similarity/difference with other animals to think themselves through the issue of their relation to animals, and, conversely, to think animals through fantasies and wild transpositions. "The parallelism of their similar/dissimilar lives allowed animals to provoke some of the first questions and offer answers."[19] Indeed, unfamiliarity with non-human animals helps arouse puzzlement, while our kinship with animals fosters the articulation of questions along with tentative illuminating comparisons/transpositions.

In this book, I have contended that it is crucial to embrace and enact the human-non-human animal metaphor instead of trying to freeze it and reduce it. This means that (human and non-human) animals should be regarded as being intrinsically beyond themselves, tantalized by each other. They live, think, and become for their own sake, as autopoietic, intentional individuals, but they also exist way ahead of themselves, at a distance from themselves, in their privileged other.

The key to the human-animal metaphoricity is, as I have argued, a phenomenological perspective on animal subjectivity. The fundamentally descriptive nature of phenomenology opens up the possibility of a pluralist approach to the complexity and manifoldness of our relation with animals. In particular, a phenomenological approach has the ability to foster the most straightforward and the most generous recognition of non-human animal subjectivity and agency. As I have argued, a phenomenological approach to imagination permits us to overcome the understanding of imagination as a purely psychological phenomenon and a human mental faculty. Subjectivity begins in the world; imagination stems from the imaginary, which stems from the imaginareal—the intrinsically ubiquitous and floating nature of real beings. The spectrum of subjectivity has to be expanded. Lucid consciousness is just one facet. Ambiguous meaning arising from the world or body intentionality, for instance, are significant dimensions of subjectivity. Furthermore, genetic phenomenology decenters human subjects and forces them to look for companion-subjectivities without which their thoughts ossify and deteriorate into obdurate clichés. *Einfühlung* [empathy]—defined as a form of phantasy that is nonetheless the accurate way towards real animals—is an integral part of any thought, a fundamental aspect of transcendental subjectivity, or, more accurately, transcendental interanimality (Chapter 2), an interspecific community without which we could not exist or makes sense of anything. Non-human animals

think and dream through our thoughts, even in the most rational meditations: We *constantly* dream *with* them and we can learn a lot from attending
to expressiveness in non-human animals (Chapter 4). Although the role
played by actual metamorphoses and molecular becomings proves crucial
in overcoming the human-animal dichotomy, an ontology of machines or
an object-oriented ontology cannot but miss a decisive dimension of animality: The reference to and the solicitation of other animals' agency and
perspectives, namely what I have described as *invitations to dance,* essentially involve fantasies and images (Chapter 5). It is through such invitations that we can make the most of the dimension of becoming-animal,
without stifling individual creativity and alternative subjectivities.

On this basis, the scope of "the imaginary of animals" expands from the
human imagination of animals to the anonymous imaginary of animals
and, even further, to animals' imaginations.

Animal imagination

As I have mentioned in the introduction of this book, humans are classically
defined in the Western rationalist philosophical tradition as imaginative
beings, while imagination was first and foremost understood as the power
to connect to the indefinite. Sartre thus identifies imagination as freedom,
the being-beyond-oneself that opens up a subject to the world.[20]

I have contended that the same correlation between existence and imagination can in fact be found in animal life. As a first step, I have studied
the imaginary dimension of the *being* of animals (Chapter 3). Living beings
as Gestalten are essentially open to an abysmal, unknown future. Living
beings intrinsically consist in a relation to the virtual. They are the emergence of a general plan, a set of general functions, and original goals. As
such living beings are at once autopoietic and fanciful. A force of orientation, selection, and organization bursts forth while nothing in the world
guarantees the persistence of this emerging being. Each living being stems
from a phantom-like theme that is immanent in a material structure and
becomes, so to say, its own active support, shaping organs, behaviors, and
even the world to ensure its own perpetuation. Nothing defines in advance
what such a phantom will create. Gestalten do not quite *know* where they
are heading. They cannot be ascribed to a place or a concept. This phantom-
like being is the key to the *imaginary* of animals: An *ontological ubiquity,* but
not yet an imagination.

In animal expression, more active and subject-centered forms of imagination develop. Autopoietic processes in animals further unfold in the
form of expressive and even symbolic phenomena (Chapter 4). In a more
active way, animals get hold of the meaning that emerged with vital
Gestalten and turn this meaning into the theme of some of their operations, at the morphogenetic or behavioral level. Meaning thus becomes
that which can be staged, dramatized, fine-tuned, interpreted, or faked.

This new development of autopoiesis also follows a process of chimerical creation immanent in a series of accidents and derailments; it includes enacted ubiquity and even symbolic ubiquity, when some images become relatively stable signs and even cross-specific signs (like eyespots, for instance, or the hiss). The very formation of animal appearance, conceptualized as *Selbstdarstellung* by Portmann, is a process of image creation. This opens up a space for hesitation, play, negotiation, as well as for a more active and individualized creativity (Chapter 4), and subjectification processes (Chapter 5).

Some non-human animals relate to absent, as it were fictitious, entities. Uexküll gives the example of the starling chasing absent flies.[21] Sue Savage-Rumbaugh reports that Kanzi plays with imaginary objects. Indeed, Kanzi—a bonobo who was immersed, as an infant, in the human-ape communication protocol that Savage-Rumbaugh created for his mother— pretends to hide invisible objects in a pile of blankets and even engage human or non-human partners in the game, "by giving them the pretend object and watching to see what they do with it."[22] Sherman and Austin, two pan Troglodytes Chimpanzees, "pretended that a fearsome animal was housed in an empty cage" after seeing the film King-Kong. More broadly, non-human animals who play perform a reference to *that which is not but could be*, as demonstrated by Bateson. These examples are important of course, but are the tip of the iceberg and they should not divert our attention from a more fundamental relation to the imaginary realm that unfolds in human and non-human animals in general. Primordial forms of imagination develop widely in the animal realm, as a result of the non-human animals being haunted by elusive and uncertain functions and meanings. Forms of a more focused and subject-centered imagination are a response to a more diffuse animal imaginary that is already an active creativity and upon which this book primarily focuses. Seeking to draw a clear-cut boundary or, worse, to establish a hierarchy among different forms of imagination—including human imagination—would amount to negating the inspiring force that sustains imagination overall, in human and non-human animals.

Animal agency: The myth of sovereign agency vs oneiric viscous agency

During a discussion session after a lecture, in which I explained that the famous existentialist slogan "existence precedes essence" applies to non-human animals, I was asked the following question: "But can the spiders who mimic ants *choose not to* mimic ants?" Where does individual animal imagination begin? Is animal imagination as free as human imagination? I will highlight several key points to reframe this issue.

First, as aptly stressed by Aaron Moe, the behavior of many non-human animals grapples with time and the possibility of seizing *the right moment*.

Many animals exhibit the individual ability, also depending on stress and fatigue, of grabbing the chance, feeling the *Kairos*,[23] and launching a gesture—for instance, a mimetic behavior—with good or bad timing. Karen Pryor, in "Creating Creative Animals,"[24] and Haraway in *When Species Meet*, show how the framework of positive reinforcement gives animals the opportunity to propose original and surprising new behaviors. I have also referred to examples put forward and analyzed by Hearne, Haraway, and Pereira to demonstrate that non-human animals sometimes seize the opportunity to, so to say, hijack a training process and divert a set of signs learned through regular practice. Indeed, they sometimes perform movements as they learned them, but they may also exploit the ambiguity and introduce actually meaningful new variations.[25] In this case, a more active and individualized imagination is looming.

With the concept of the imaginary of animals, I wanted to foreground an oneiric thought that forms below the conscious-unconscious duality and constitutes the living heart even of the highly lucid and reflexive forms of human thinking. In the oneiric, iconic, holistic thought, consciousness is unavoidably muddled and multicentered/ubiquitous. But, already, all the aspects that will expand and thrive in high levels of consciousness and reflective consciousness, namely differences, self-distance, various perspectives, selective focus, hesitation, and the relation to specific and open possibilities, are already at work. The question of the level of consciousness that can be—and is in fact—developed by specific animals is not dismissed but should be appended to, first, the non-binary exploration of the oneiric thought and, second, to an existential approach to consciousness and subjectivity that refuses to regard them as substantial beings. As I have argued in Chapter 5, the imagination of animals cannot be treated as a pure matter of fact, namely as a reality whose presence or absence should simply be noticed and acknowledged: It is only when one invites the other animal to dance, makes room for her perspective, and creates means through which this perspective and this personal imagination can blossom, that one gives the other the opportunity to develop and take over her perspective and the bold initiatives of her active imagination.

At stake here are also the definition of agency and the myth of human sharp and glorious decisions. Such decisions may occur, but only under certain favorable circumstances, so much that it is quite unsatisfactory to depict these decisions as the exclusive manifestation of a personal will. Staging absolute resolutions and fantasizing a sovereign will are indeed activities that many humans fancy and that were turned into high art by some philosophical and political traditions, Western Rationalism having a place of honor at the head of the procession. Nevertheless, the ability to stage power and forge the luring image of monstrous and impressive capacities already constitutes a major animal skill, as Chapter 4 explained.[26] Hence, even the god-like pure will, or its image—but it is inseparable from its image—has some deep kinship with animal strategies.

"We considered ourselves consciousnesses free and naked before the world," Merleau-Ponty writes in "The war has taken place," describing a group a young students and respectable professors who advocated pacifism on the eve of the Second World War. But, far from being a purely intellectual vision, this idealism was instead the artificial and very particular product of exceptional circumstances. In a peaceful and rich country and in some social privileged situation, it becomes possible to have an effective will and fancy oneself as a free consciousness. Oppressed people do not enjoy the same latitude. They "know"—actually, they confusedly sense—that agency implies first and foremost to go with the flow. They parasitize existing predominant currents in order to introduce some slight inflexions through manipulation or negotiation with those in power. Their tools, their abilities, and their intellectual means of understanding are defined by mysterious and diffuse others: This estrangement constitutes their very flesh as well as their mind.[27] When authority figures do not pay attention to the subordinates, when they stare into space and change the subject as soon as the latter speak, when one is consistently defined as stupid or when educating tools are lacking, agency must unfold through a viscous flesh. Those who were hardly instituted as subjects (see §36) experience the dark depths of agency and consciousness.

Understanding agency primarily requires paying attention to slight inflexions in usual processes, surprising outcomes, resistance, negotiation, and manipulation in power-relations, as well as to timid invitations to dance. In this respect, the study of animal agency is illuminating and obliges us to aim at a complexified concept of agency. The issue of conscious and sharp decisions is not devoid of interest, but why should it be our paramount concern? Many works of art thus play on this affinity between the agency of muzzled people and the agency of non-human animals.[28] After presenting a legal, polished, and apparently legitimate world on the surface, they suggest that, under the masks, animals are operating. Kafka's *Metamorphosis* is an obvious example. We have also studied the role played by oneiric meaning in *The Night of the Hunter* (see §29). More recently, Jane Campion's *Top of the Lake* has beautifully brought together these themes of stifled agency and animal expressiveness. Campion depicts a world ruled by men, where legality is nothing more than the institutional face of this domination (the local police station harbors a boys' club atmosphere, and the police chief uses a charity program of rehabilitation as a cover to run a child prostitution ring). In Campion's world, the consistent move toward forces, sensations, malaise, and animality—namely the level of what I have called oneiric expressiveness—functions as a means of subversion and a means of resistance. In the series, female characters subvert the game. Tui, a 12-year-old girl who was the victim of a collective rape and is treated as a commodity by her father and his fellows, does not behave like a sobbing victim and a reserved little girl. She escapes into the forest, turns feral, and hisses furiously when shooting her father in the back. Tui demonstrates a strong agency, but she does so by moving to an animal level of existence that provides her with powerful support. Robin, the main character, gives a good example of viscous agency: As a police detective, she engages in a subtle power-negotiation process with her male colleagues and superiors, and struggles with her confused understanding of this many-layered world. Following in her footsteps, we are led into a deepening sensitivity to animal agency and the hope for new worlds.

Why look at animals? Why imagine with them?

We may look at animals to reduce them to simulacra. But even the latter bear the mark of a more fundamental imaginary of animals. We first and foremost look at animals because they are, produce, and enact images. In other words, they are emerging subjectivities (Chapter 3) and develop expressive appearances (Chapter 4). Phantom-like forms manifest through their morphologies and behaviors. The discrepancy between essentially elusive themes (animal subjectivity, intentions, moods, and emotions) and their sensible manifestations through spatial and temporal structures arouse inexhaustible curiosity for animal forms and gestures. What does this dog want? Where are these ants heading? What does this fish feel? These are normal questions raised by the observation of animals. And it is sufficient for such a questioning to begin that an animal simply *shows up*.

The observation of non-human animals certainly may be driven by the scopic drive and the project of explaining or even holding non-human animals in thrall. But, in any case, this observation possesses a fundamental and irreducible imaginary dimension that opens it to less oppressive relations to other animals. The imaginary dimension of "looking at animals" is entailed by the very *imaginary of animals:* Their *own* imaginary. Phantoms are unfinished beings; we can never fully *know* what their upcoming developments are. The most adequate apprehension of these living images is to imagine *with* them, under their guidance and in dialogue with them. Empathy, transpositions, myths, and fantasies, but also interaction, cooperation, and becoming-with are different forms of what Vicki Hearne has described as *risky sorcery.*[29] These relations to animals all imply muscular lyricism and the transformation of our existence (Chapter 5). We look at animals because they solicit and enhance our imagination. And they do so because they imagine in many ways. We cannot but imagine with them, namely, inseparably, try to create with them new ways of communicating and living together.

When human imagination loses touch with these imaginative worlds, or fails to engage them, it becomes the instrument of an anthropocentric thought and a project of domination. It also moves away from the life and the multiplicity of perspectives without which images turn into polished, predictable, and fleshless forms.

But the longing for a lost fusion with nature, such as displayed in many Romantic works[30] or in Hugo von Hofmannsthal's *Letter of Lord Chandos*[31] corresponds to a fantasy of separation, rather than to fantasy as a result of real, radical separation from nature. In fact, *nothing* could be phrased, thought, or could even make sense, if our thoughts were not rooted in animal imagination, which implies that imagination did not begin with humans. One of the crucial consequences of the chiasm between human and non-human imaginations is thus a change of approach in the history of animal symbols in human cultures and the history of human-non-human animal relations. The oriented and linear temporality of origins or evolution does not any longer apply to the relation between non-human and human animals. I will elucidate this concept of alternative temporality in the last paragraph of this book to further clarify how imagining with animals allows us to let go of our old habits of mind.

§39 What happened at the beginning? Original symbols, sacred bulls, and an alternative temporality

Why have human beings represented animals so abundantly at least since Middle Stone Age and Paleolithic times? What do ancient mythic creatures such as sirens, the minotaur, or dragons symbolize exactly? What were their original function and meaning? These were questions that tormented me when I began research on the imaginary of animals. Archeology, art history, and anthropology do provide some answers. The analyses developed in the present book establish a general framework in which these answers can be put into perspective. My contention, in light of Merleau-Ponty's concept of institution and Louise Westling's work on sedimentation, is that the quest for univocal answers and *the* original meaning is inadequate.

Original symbols: The key role played by the imaginary

Let us play a little with symbols. It is widely agreed that, in the Neolithic, the figures of the bull and the goddess became pivotal. They kept flourishing in various Bronze Age civilizations in the Mediterranean and Middle East areas. In *The Birth of the Gods and the Origin of Agriculture*,[32] Jacques Cauvin puts forward a daring and still influential theory.[33] Against a then-dominant materialist model, Cauvin contends that the Neolithic revolution cannot simply be explained in terms of economic and social changes, or as an adaptive response to environmental constraints (climate or demographic changes for instance).[34] Invoking Leroi-Gourhan's concept of "interior environment,"[35] Cauvin highlights the decisive role played by the imaginary and symbols in Neolithization.

As Cauvin argues, it is necessary to acknowledge the extra dimension that constitutes the imaginary field in pre-Neolithic and Neolithic societies, since a striking discrepancy can be noticed between the consumed animals (sheep, goats, pigs) and the animals represented in statuettes, high reliefs, and frescoes (wild animals, with a remarkable abundance of bulls, whereas the latter were at that time economically insignificant and hardly hunted yet[36]). Cauvin also contends that the sudden "love affair"[37] of humans with weapons in Pre-Pottery Neolithic B, the emergence of rectangular houses,[38] and the new "*zeal* for domestication"[39] cannot be explained by mere utilitarian motives.[40] Cauvin points out the irruption, at that time, of an unprecedented verticality that accentuates the gap between the imaginary and the actual reality: Figures of suppliants appear, while the feminine character seems to be "seen on a transcendent plan where fears and conflicts are resolved."[41] Dangerous animals become the retinue of the "goddess," her allies, the figures of her powers. She is represented with a "surreal" combination of attributes referring to 'royalty' (an anachronical or precursory concept), fertility, and death.[42]

According to Cauvin, a revolution takes place precisely through the combination of the following spectacular phenomena: (a) The choice of a couple of prevailing and complementary symbolic values: The bull and the goddess, (b) a boom in symbolism, weaponry, and domestication, (c) an unprecedented self-assertion of a full-fledged and power-hungry human realm. Cauvin claims that symbols acted as driving forces in this process and opened up the possibility of a radical transition from a shamanic horizontal relation to nature to a humanity defined alongside the model of virile power. The bull is indeed, in the first instance, in Çatalhöyük for example, subordinated to the goddess,[43] but he also simultaneously receives a leading role and becomes the symbol of brute force and masculinity.[44] This violent force can be fought and defeated through actual bullfighting, as well as through symbolic sacrifices. The next step, as shown by many myths in Bronze Age, replaces the goddess with male heroes and gods who overcome the bull: "Through the image of Baal of Ugarit who was the master of the Bull or could himself become the Bull, and was represented as warrior-hero and civilizing force, by braving the wild bull-animal man could learn to seek the opportunity to prove his mastery over it, his courage, and his effectiveness in combat. It is thus not surprising that, from Çatalhöyük, through Minoan Crete to Iberia, the confrontation between man and bull should be as a solemn and ritual act, whose still vivid reverberations are preserved in the brilliant costume of the matador of today."[45]

So many bulls: A protean imaginary

As the above quotation shows, Cauvin's analyses contain several sweeping statements and idealist interpretations, but they also unveil an imaginary field that cannot be reduced to this idealist view. Is Baal the master of the bull or does he become the bull? When Cauvin understands the Neolithic bull in the light of the Spanish corrida, and, even narrower, in the light of the slaughter of the bull by the matador, he condemns himself to a partial comprehension. Cauvin does not fear to put forward bold trenchant claims such as: "The idea that the image of the wild bull signifies a brute force, instinctive and violent, is spontaneous in us and is without doubt universal."[46] It is difficult to read this without flinching. The term "instinctive," for instance, is ideologically-laden and the concept of instinct is at best questionable. The bull may also connote cosmic vital forces. Moreover, the verb "signifies" thus employed solidifies a floating meaning. The positivist model of a correspondence between a sign and a clear-cut signified seems to be taken for granted. "Evokes" would be more appropriate than "signifies." And "universal" is simply brazen.[47] Cauvin tends to essentialize what he calls "the elemental imagination," an imagination defined in terms of quasi-platonic archetypes. To be sure, the "virile" line of meaning in the symbolism of bulls has formed and was resumed through centuries. This bellicose symbol is indubitably still extremely powerful for us nowadays. However, being the bull, being inspired by it, beating it, and killing it cannot constitute one and the same clear signified. If humanity proved and still has to prove, over and over again, its civilizing force by sacrificing bulls, it foolishly cripples itself.

The existing myths and symbols are in fact denser and more complex than that. For instance, Minoan art famously depicts numerous scenes of what appears to be bull-leaping: The human figures are acrobats who do not attack and stab the bull, but become skillful enough to attune to its moves, cling to its horn, and use its impulse. The actual practice of corrida itself is manifold: The torero "heroically" *confronts* the bull but also learns its tricks, adjusts to its style, and becomes a hybrid.[48] While the myth of the *monstrous* to-be-massacred Minotaur is also shaped by Athene's rivalry with Crete, in the myth of Europa, on the other hand, the bull is first a gentle creature, affectionate and calm,[49] which a bull can indeed be in fact. And in Euripides' *Bacchae*, whereas the bellicose Pentheus wants to hobble and subdue the bull, Dionysus *becomes* a bull. Euripides' play does not provide any univocal moral: Both Pentheus and Dionysus are, in their turn, deeply frightening. On the one hand, the reader understands Pentheus' will to protect the city against violence, madness, and sexual license, but she also discovers the horrid dark side of reason in him. On the other hand, Dionysus is depicted as admirable in his ability to maintain a non-exclusive relationship to nature, animals, and, what is more, femininity.[50]

The imaginary of animals: An elusive origin and a field of endless exploration

To a certain extent, Cauvin misconstrues the nature of the imaginary. Beyond the essentialized duality between the Bull and the Goddess, actual and virtual metamorphoses keep shaping, reshaping, and subverting the symbols that have been forged at a particular time and place. Symbolic structures, with their formal systems of values, should not *cover up* this living imaginary. Cauvin courageously undertakes the difficult task of looking for key themes and central axes that emerged in various cultures during the Neolithic. "Throughout the total duration of the Neolithic across the whole of the Near and Middle East, a unique 'ideology' is found, expressed through different modes and artistic styles."[51] Such an idealist phrasing has unsurprisingly earned criticism,[52] yet, myths, figures, imagery, and images in different civilizations indeed strangely echo each other, distort each other, and respond to each other. Still, the paradigm of sedimentation Louise Westling puts forward[53] is more accurate than that of archetypes and universal symbols. Westling contends that, in our body's memory, deep sedimentary layers of asleep and dreamlike interpretations of the world, in other words, asleep *Umwelten*, are secretly at work and are always liable to be more actively resumed. The imaginary is profound and manifold. It is not systematic but cumulative, inclusive, and downright inexhaustible. Its *"logic"* takes the form of metamorphic themes immanent in series of symbols that are never at peace with themselves and, consequently, keep proliferating. Separating and isolating a fragment from this organic whole is impossible. The concepts of institution and sedimentation account for such a structure, without reducing many-layered symbols to essentialized archetypes. These two concepts indeed deflate the idea of a localizable birth of humanity along with the fascination for origins.

If I am looking for the origins of animal symbols, if I find them equivocal and confusing in their present form, I should not be too impatient to find a key in some remote past: As far as it may go, my question will find its reflection in every older avatar of these symbols. Because the latter evolved in such a confusing fashion, they lent *themselves* to countless different variations and inflexions and allowed them to blossom: On no account could they originally be crystal-clear. There cannot be any fully ascribable original meaning. In other words, as Merleau-Ponty claims, "Every origin is myth [*toute origine est mythe*]."[54] The ultimate original institution must be imaginary. Such an origin does not exist as one unique real localizable event, but it lies in the dynamism that unhinges every particular image or symbol and integrates it in a fractal structure of interlocked images and symbols that reflect each other.

And indeed, what can we find at the origin, wherever the latter may be hidden? The metaphorical relation between humans and animals, as well as between individuals and species. Forms have developed: Different species, different individuals, different symbols constitute an actual framework of relatively stable and identifiable incarnations of themes that can thus become more and more recognizable. But they have emerged through accidents, variations, and creative attempts at various levels: In the evolution of species, in the existence of individuals, and in the misadventures of images, symbols, and signs.[55] Accidents become necessity, in the sense that these facts are then irreducible parts of the sedimentation of the imaginareal.

As suggested by Westling, we can find in our cultural institutions layers of meaning that do not simply refer us back to older *cultural* institutions: Sedimentation already started with the history of animal species and makes even our bodies a palimpsest of animal experiences. Each symbol that we will find in our study of contemporary cultures, Ancient civilization, or non-human animal life, is essentially scattered and tormented by questions. Each living body, each image, each work of art consists in an institution, namely a process that will keep taking on new forms. An author's thought is always instituted by old questions and old unfinished poetic ideas that turned into asleep texts, phrases, and words; the latter regain life while sustaining the creation of new works of art, which in their turn sediment again into sleepy ossified ideas in dictionaries, libraries, museums, traditional patterns of speech, forgotten word-roots, old practices and habits. Layers of meaning thus pulsate beneath the surface of each cultural product. The same sedimentation already occurs at the level of living beings and in the images and symbols these living beings develop. As such, animal meaning institutes a culture, teases us, and inspires us. Moreover, many non-human animals do actively try on and play with such forms and metamorphoses, via the creation of new images and symbols, through symbiotic relations, in communication and play. We are not determined by our animal nature and heritage: The moves, symbols, and thoughts of innumerable non-human animals still resonate in human individuals and are taken up and re-instituted by them. The relation between human and non-human institutions and re-institutions is circular: Every event is ahead and behind itself. Thus is revealed the instituting power of non-human animal existences, namely their essential dimension of desire and creation.

Whereas Cauvin was looking for obdurate archetypes, I am interested in imaginary themes. While he intends to pinpoint the revolutionary advent of the human imaginary, I contend that the latter is inseparable from the imaginary of non-human animals: It emerges as a development of and a response to this imaginary.

"Instinct does not know what it wants," Merleau-Ponty has argued. "Instinct" actually takes the form of an open-ended and creative desire. Dolly, the bottlenose dolphin who threw the metaphor of milk-smoking into the blue, as a provocative way of communicating *through the unsettlement of meaning*, should be regarded as a paradigm of animality, rather than as an exception. Animals question the world and, more specifically, all the other living beings who are able to provide models, inspirations, and tentative answers. Our quest for origins leads us to the imaginary of animals, and, from there, to a new serious responsibility and a threefold task: To take fixed symbolic patterns in non-human and human realms with a significant pinch of irony, to embrace the dynamism of human-animal metaphors, and to never stop using our imagination to understand and provoke imagination in other animals.

The challenge today for humans is to find a way out of the divisive, oppressive system that has been prevailing for centuries and in which humans fantasize themselves as separate from the animal realm. Such fantasies have had a devastating effect on human and non-human animals, which demonstrates that, indeed, a fantasy of separation can be instrumental in shaping reality and causing real suffering. However, I have not exactly defined a way *out* of this system, which would be a new avatar of such a divisive mode of thinking. Instead, I have undertaken to turn this system upside down and inside out, to unfold it and fold it again differently. This book has foregrounded chiasms, interfaces, and circular co-institutions between the human imaginary and the animal imaginary. At stake for us is consequently less to re-connect with the imaginary of animals than to welcome its risky manifoldness and foster its cumulative and integrative power.

Notes

1. Spiegel, Marjorie. *The Dreaded Comparison: Human and Animal Slavery.* Philadelphia: New Society Publishers, 1988.
2. Adams, Carol. *The Sexual Politics of Meat,* New York: Bloomsbury Academic, 1990.
3. Derrida, Jacques. *L'animal que donc je suis,* Paris: Galilée, 2006.
4. See supra, §15.
5. Berger, John. "Why Look at Animals?", In *About Looking,* London: Writers & Readers, 1980, p.28.
6. *Ibid.*, p.19.
7. Burt, Jonathan. *Animals in Film*, London: Reaktion Books, 2002, p.27. Burt refers to Lippit. Akira. *Electric Animal: Toward a Rhetoric of Wildlife*. Minneapolis: University of Minnesota Press, 2000, p.21.
8. Berger, "Why Look at Animals?", p.25.
9. See Baudrillard, Jean. *Simulacre et Simulation.* Paris: Galilée, 1981.
10. Berger, "Why Look at Animals?", p.22.
11. Vercammen, Maurits. "Roze krokodil SPAR University eist betere arbeidsvoorwaarden." In *De Pipet*, Webzine, 09-09-2017.
12. The article also hints at the simulacrum process: "look at what they did to me. It's not what a crocodile is supposed to look like!"
13. Derrida, *The animal that therefore I am*, p.80.
14. Lippit, *Electric Animal: Toward a Rhetoric of Wildlife*, p.1.
15. Berger, "Why Look at Animals?", p.7.
16. *Ibid.*, p.4.
17. *Ibid.*, p.5–6.
18. *Ibid.*, p.7.
19. *Ibid.*
20. Cf. supra, p.8.
21. Uexküll, *A Foray into the Worlds of Animals and Humans,* 2010, p.120.
22. Savage-Rumbaugh, E. Sue and Kelly McDonald. "Deception and Social Manipulation in Symbol-Using Apes." In Byrne, Richard W. and Whiten, Andrew (Eds.) *Machiavellian Intelligence: Social Expertise and the Evolution of Intellect in Monkeys, Apes, and Humans.* New York: Clarendon Press/ Oxford University Press (224-237), 1988, p.232.

23. Moe, *Zoopoetics*, 2014, p.19–20; see also supra, p.146 and p.202.
24. Pryor, Karen. "Creating Creative Animals." In Kaufman, J. and Kaufman, A. (Eds.), *Animal Creativity and Innovation*. London: Elsevier, 2005.
25. See supra, p.79, 157, and 202.
26. This is why sovereigns and beasts are paradoxically so deeply akin. Political sovereigns take advantage of portraying themselves as wild or even monstrous animals, although the sovereign ego was built by proclaiming itself radically different from animals. These ideas are central to Derrida's *The Beast and The Sovereign* (Paris: Galilée, 2 volumes, 2008–2009).
27. See Sartre, *L'idiot de la famille. Gustave Flaubert de 1821 à 1857*, Paris, Gallimard, 1971,; and Young, Iris Marion, "Throwing like a Girl," In *Human Studies*, 3 (1), 1980.
28. Shelly R. Scott also points to this link, in "The Racehorse as Protagonist: Agency, independence, and Improvisation," in the framework of a more specific comparison between two forms of oppression: "Here I demonstrate how thoroughbreds improvise in human-engineered races, revealing their capacity for agency. Agency by these animals is exercised in ways similar to that of colonized peoples, as it must be exerted within domains that do not belong to them. Both oppressed animals and people deal with limitations imposed on their capacity for agency by rebelliously or subversively exerting their own wills." *Animals and Agency. An Interdisciplinary exploration*. Sarah E. McFarland, Ryan Hediger (Eds.) Leiden, Boston: Brill, 2009, p.47).
29. See supra, p.157.
30. See M. Homans' illuminating analyses in *Bearing the Word*, 1986.
31. von Hofmannsthal, Hugo. "Ein Brief," In *Der Tag*. Berlin, Nr. 489, Oct 18 1902 (Teil 1); Nr. 491, Oct 19 1902 (Teil 2). Trans. by J. Rotenberg, *The Lord Chandos Letter and Other Writings*, New York: New York Review Books, 2005.
32. Cauvin, Jacques. *Naissance des divinités, naissance de l'agriculture: la révolution des symboles au Néolithique*. Paris: CNRS Editions, 1994. Trans. Trévor Watkins. *The Birth of the Gods and the Origin of Agriculture*. Cambridge: Cambridge University Press, 2000
33. See for instance Schmidt, Klaus. "Göbekli Tepe – the Stone Age Sanctuaries. New results of ongoing excavations with a special focus on sculptures and high reliefs." In *Documenta Praehistorica*, XXXVII, 2010, p.253 and the special issue of the journal *Paléorient*, 2011, Vol. 37, No. 1: "Néolithisations: nouvelles données, nouvelles interprétations. À propos du modèle théorique de Jacques Cauvin."
34. Aurenche, Olivier. "Jacques Cauvin et la religion néolithique. Genèse d'une théorie". In *Paléorient*, Vol. 37, No. 1 (15–27), 2011, p.16.
35. Cauvin, "The Birth of the Gods and the Origin of Agriculture," p.19–20.
36. *Ibid.*, p.33 and p.68.
37. *Ibid.*, p.125.
38. *Ibid.*, p.128–32.
39. *Ibid.*, p.128.
40. *Ibid.*, p.125–32.
41. *Ibid.*, p.71.
42. *Ibid.*, p.29 and 71.
43. *Ibid.*, p.32.
44. *Ibid.*, p.123–4.
45. *Ibid.*, p.125.
46. *Ibid.*, p.123.
47. The term "universal" corresponds to a consistent pattern of understanding in Cauvin's approach, see for instance "The Birth of the Gods and the Origin of Agriculture," p.132.

48. Thomas, Joël. "Le mythe du Taureau et les racines de la tauromachie De Dyonisos au duende." In *Du taureau et de la tauromachie. Hier et aujourd'hui* (H. Boyer Ed.), Perpignan, Presses Universitaires, 2012, §24.

49. Moschus, "Europa," in *The Greek Bucolic Poets*, Trans. A. S. F. Gow, Cambridge: Cambridge University Press, 1953, 2.89.

50. See also Louise Westling's illuminating analyses of *The Bacchae*, in *The Logos of the Living World*, 2014, p.56–60. Cauvin focuses too exclusively on a structuralist reading of the complementary values of the Goddess and the Bull. But, in fact, the latter is not unequivocally masculine. As noted by Pascal Darcque and René Treuil in "Un 'bucrane' néolithique à Dikili Tash (Macédoine orientale): parallèles et perspectives d'interprétation" (In: Bulletin de correspondance hellénique. Volume 122, livraison 1, 1998, p.9) archeologists too often hastily call "bullhead" what could be the head of an ox or a cow, thus projecting a certain symbolic reading on a complex material. Marija Gambutas has also demonstrated the role played by bucrania in many representations of the goddess and in the framework of fertility rites (*The Living Goddess*. Berkeley: University of California Press, 1999).

51. Cauvin, "The Birth of the Gods and the Origin of Agriculture", p.32.

52. See for instance Darcque and Treuil, 1998, p.25 and Guilaine, Jean. "La "Révolution des symboles" de Jacques Cauvin revisitée". In *Paléorient*, vol. 37, no. 1, 2011, p.178.

53. In "Merleau-Ponty and the Eco-Literary Imaginary" (Handbook of Ecocriticism and Cultural Ecology, Hubert Zapf Ed. Berlin: De Gruyter, (65–83), 2016), Louise Westling defines what she calls an eco-literary sedimentation process, which layers lines of meaning that are inherited from the history of animal species and from our cultural history. She borrows the concept of sedimentation from Husserl and Merleau-Ponty as well as from literary studies (texts as palimpsests). The most thought-provoking aspect of this theory resides in the way Westling establishes a continuity–in the sense of a dialogue justified by a deep essential kinship–between sedimentation in the DNA and sedimentation in cultural works. Westling finds indeed striking arguments that back up her claim in Hoffmeyer's approach to biosemiotics and in contemporary new takes on genetics (see also our analyses about pseudogenes and epigenetics, supra, §25).

54. Merleau-Ponty, Maurice. "La philosophie aujourd'hui", In *Notes de Cours 1959–1961*, Paris: Gallimard, 1996, p.127.

55. See §31 about Merleau-Ponty's concept of a symbolism that aggregates and disaggregates through accidents and derailments.

Bibliography

Abram, David. *Becoming Animal. An Earthly Cosmology.* New York: Vintage books, 2011

Adams, Carol. *The Sexual Politics of Meat.* New York: Bloomsbury Academic, 1990

Adorno, Theodor W., *Jargon der Eigentlichkeit. Zur deutschen Ideologie ist ein ideologiekritisches Werk,* Suhrkamp Verlag, 1964. Trans. by Knut Tarnowski and and Frederic Will Evanston: Northwestern University Press, 1973

Agamben, Giorgio. *L'aperto. L'uomo e l'animale.* Torino: Bollati Boringhieri, 2002. Trans. Kevin Attell. *The Open: Man and Animal.* Stanford: Stanford University Press, 2004

Alatalo, Rauno V. and J. Mappes. "Tracking the Evolution of Warning Signals." In *Nature,* Vol. 382 (708–709), 1996

Aristotle. *Parts of Animals.* Trans A.L. Peck. Cambridge: Harvard University Press, 1937

————. *On the Soul.* Trans. J.A. Smith. In *The Works of Aristotle,* Vol. 3, Oxford: Clarendon, 1930

————. *Metaphysics.* Trans. William David Ross. Oxford: Clarendon Press, 1924

Artmann, Stefan. "Computing Codes versus Interpreting Life. Two Alternative Ways of Synthesizing Biological Knowledge through Semiotics." In Marcello Barbieri (Ed.), *Introduction to Biosemiotics. The New Biological Synthesis* (209–233). Dordrecht: Springer, 2007

Ash, Mitchell G. *Gestalt Psychology in German Culture, 1890–1967. Holism and the quest for objectivity.* Cambridge: Cambridge University Press, 1995

Aurenche, Olivier. "Jacques Cauvin et la religion néolithique. Genèse d'une théorie". In *Paléorient,* Vol. 37, No. 1 (15–27), 2011

Bachelard, Gaston. *Lautréamont,* Paris: José Corti, 1939. Trans. James Hillman and Robert S. Dupree. Dallas: Dallas Institute, 1986

————. *L'eau et les rêves. Essai sur l'imagination de la matière.* Paris, José Corti, 1942. Trans. Edith R. Farrell. Dallas: Dallas Institute of Humanities and Culture, 1999

————. *L'air et les songes. Essai sur l'imagination du mouvement.* Paris: José Corti, 1943. Trans. Edith R. Farrell. Dallas: Dallas Institute of Humanities and Culture, 1988

Bailly, Jean-Christophe. *Le versant animal,* Paris: Bayard, 2007

————. *Le Dépaysement,* Paris: Seuil, 2011, reprint: Points, 2012

Barbieri, Marcello (Ed.). *Introduction to Biosemiotics. The New Biological Synthesis,* Dordrecht: Springer, 2007

Baracchi, David, I. Petrocelli, L. Chittka, G. Ricciardi, and S. Turillazzil. "Speed and Accuracy in Nest-Mate Recognition: A Hover Wasp Prioritizes Face Recognition Over Colony Odour Cues to Minimize Intrusion by Outsiders." In *Proceedings of the Royal Society of London B. Biological Sciences*, Vol. 282, No. 1802, 2015

Barrett, Paul, Peter J. Gautrey, Sandra Herbert, David Kohn, and Sydney Smith (Eds), *Charles Darwin's Notebooks, 1836-1844: Geology, Transmutation of Species, Metaphysical Enquiries*, London: British Museum (Natural History) and Cambridge University Press, 1987

Basil the Great. *Homily on Psalm, 28*. In *Exegetic Homilies*. (204–5). Trans. Sister Agnes Clare Way. Washington DC: Catholic University of America Press, 1963

Bates, Elizabeth and Frederic Dick. "Beyond Phrenology: Brain and Language in the Next Millennium." In *Brain and Language*, Vol. 71 (18–21), 2000

Bateson, Gregory. *Steps to an Ecology of Mind*. Northvale NJ: Jason Aronson, 1972

Bateson, Melissa, Suzanne Desire, Sarah E. Gartside, and Geraldine A. Wright. "Agitated Honeybees Exhibit Pessimistic Cognitive Biases". In *Current Biology* Jun 21, Vol. 21, No. 12 (1070–3), 2011

Baudrillard, Jean. *Simulacre et Simulation*. Paris: Galilée, 1981

Beer, Gillian. *Darwin's Plots. Evolutionary Narrative in Darwin, George Eliot and Nineteenth-Century Fiction*. Cambridge: Cambridge University Press, 1983, reprint 2004

Bekoff, Mark and Colin Allen. "The Evolution of Social Play: Interdisciplinary Analyses of Cognitive Processes." In *The Cognitive Animal*, M. Bekoff, C. Allen and Gordon M. Burghardt (Eds.). Cambridge: The MIT Press, 2002

Berger, John. "Why Look at Animals?". In *About Looking*, London: Writers & Readers, 1980

Beran, Michael. J. and David Smith. "The Uncertainty Response in Animal-Metacognition Researchers." In *Journal of Comparative Psychology*, May; Vol. 128, No. 2 (55–9), 2014

Bergson, Henri. *Matière et mémoire*. Paris: Alcan, 1896. Reprint Chicoutimi: Les classiques des sciences sociales, 2003

Berwick, Robert C., Okanoya, K., Beckers, G.J.L. and Bolhuis, J.J. "Songs to Syntax: The Linguistics of Birdsong." In *Trends in Cognitive Sciences*, Vol. 15, No. 3 (113–21), 2011

Bimbenet, Etienne. "L'homme ne peut jamais être un animal". In *Bulletin d'analyse phénoménologique* VI 2, 2010

_____. *L'animal que je ne suis plus*. Paris: Gallimard, 2011

Bird-Davis, Nurit. "Beyond 'the Original Affluent Society': A Culturalist Reformulation." In *Current Anthropology*, Vol. 33, 1992

Boehm, Gottfried. *Was ist ein Bild?* München: Fink, 1994

Boesch, Christophe. "Teaching Among wild chimpanzees," *Animal Behaviour*, Vol. 41, 1991

Boesch, Christophe, Hedwige Boesch,. "Tool Use and Tool Making in Wild Chimpanzees". In *Folia Primatologica*, Vol. 54, 1990

Botigué, Laura R., Shiya Song, Amelie Scheu, Shyamalika Gopalan, Amanda L. Pendleton, Matthew Oetjens, Angela M. Taravella, Timo Seregély, Andrea Zeeb-Lanz, Rose-Marie Arbogast, Dean Bobo, Kevin Daly, Martina Unterländer, Joachim Burger, Jeffrey M. Kidd, Krishna R. Veeramah. "Ancient European Dog Genomes Reveal Continuity since the Early Neolithic", in *Nature Communications*, Vol. 8, 2017

Bourgeois, Arthur. P. (Ed.). *Ojibwa narratives of Charles and Charlotte Kawbawgam and Jacques LePique, 1893–1895* (recorded with notes by H. H. Kidder). Detroit: Wayne State University Press, 1994

Bovet, Dalila and Jacques Vauclair, "Picture Recognition in Animals and Humans." In *Behavioural Brain Research*, Vol. 109, 2000

Bredekamp, Horst. *Darwin's Korallen: Die frühen Evolutionsdiagramme und die Tradition der Naturgeschichte.* Berlin: Klaus Wagenbach Verlag, 2005

Buchanan, Brett. *Onto-Ethologies: The Animal Environments of Uexkull, Heidegger, Merleau-Ponty, and Deleuze*, New York: Suny Press, 2008

Burgat, Florence. *L'animal dans les pratiques de la consommation*, Paris: PUF, coll. "Que sais-je?", 1995

_____. *Liberté et inquiétude de la vie animale.* Paris: Kimé, 2006

_____. *L'humanité carnivore.* Paris: Seuil, 2017

Burgat, Florence and Christian Ciocan (Eds.), *Phénoménologie de la vie animale*, Bucarest: Zeta books, 2016

Burkhardt, Frederick and James A. Secord (Eds.) *The Correspondence of Charles Darwin*, Volume 6, 1856 to 1857. Cambridge: Cambridge University Press, 1990

Burt, Jonathan. *Animals in Film*, London: Reaktion Books, 2002

Buytendijk, Frederik. *Wege zum Verständnis der Tiere*. Zürich, Leipzig: Max Niehans Verlag, 1938. New Edition in French (Trans. A. Frank-Duquesne), *Traité de Psychologie Animale*, Paris: Vrin, 1952

_____. "Versuche über die Steuerung der Bewegungen," *Archives néerlandaises de physiologie de l'homme et des animaux*, XVII, 1, 1932

Buytendijk, Frederik and Helmut Plessner. *Die Deutung des mimischen Ausdrucks: ein Beitrag zur Lehre vom Bewusstsein des anderen Ichs.* Bonn: Cohen, 1925

Byrne, Richard W. "The Evolution of Intelligence". In *Behaviour and Evolution* (P.J.B. Slater and T. R. Halliday. Eds.). Cambridge: Cambridge University Press, 1994

Byrne, Richard W. and Jennifer Byrne, M.E. "Complex Leaf-Gathering Skills of Mountain Gorillas (Gorilla g. beringei): Variability and Standardization." In *American Journal of Primatology*, 31, 1993

Byrne, Richard W. and Andrew Whiten (Eds.) *Machiavellian Intelligence: Social Expertise and the Evolution of Intellect in Monkeys, Apes, and Humans.* New York: Clarendon Press/Oxford University Press, 1988

Caillois, Roger. *Le mythe et l'homme.* Paris: Gallimard, 1938

Caro, Tim. M. and Sheila Girling. *Antipredator Defenses in Birds and Mammals*, Chicago: University of Chicago Press, 2005

Caro, Tim, Amanda Izzo, Robert C Reiner Jr, Hannah Walker, and Theodore Stankowich. "The Function of Zebra Stripes." In *Nature Communications* Vol. 5, No. 3535, 2014

Carroll, Lewis. *Alice's Adventures in Wonderland.* London: Macmillan and Co, 1865

Carruthers P. and J.B. Ritchie. "The Emergence of Metacognition: Affect and Uncertainty in Animals." In Beran, M., Brandl, J., Perner, J., Proust, J. (Eds). *Foundations of Metacognition.* Oxford University Press, 2012

Cauvin, Jacques. *Naissance des divinités, naissance de l'agriculture: la révolution des symboles au Néolithique.* Paris: CNRS Editions, 1994. Trans. Trévor Watkins. *The Birth of the Gods and the Origin of Agriculture.* Cambridge: Cambridge University Press, 2000

246   Bibliography

Ceccarelli, Fadia Sara. "Behavioral Mimicry in Myrmarachne Species (Araneae, Salticidae) From North Queensland, Australia," *Journal of Arachnology*, Vol. 36, No. 2 (344–51), 2008

Clayton, Nicola Susan, T.J. Bussey, and A. Dickinson. "Can Animals Recall the Past and Plan for the Future?" In *Nature Reviews Neuroscience*, Vol. 4 (685–91), 2003

Coetzee, John Maxwell. *The Lives of Animals*. Princeton, N.J.: Princeton University Press, 1999

Cook, Alison J., Clifford J Woolf, Patrick D. Wall, and Stephen B McMahon, "Dynamic Receptive Field Plasticity in Rat Spinal Cord Dorsal Horn Following C-Primary Afferent Input." In *Nature*, Vol 325, 8, January 1987

Comte de Lautréamont (Isidore Ducasse). *Les Chants de Maldoror*, Paris: L. Genonceaux, 1890, 47. Trans. Guy Wernham, New York: New Directions, 1965

Dalziell, Anastasia. H. and J.A. Welbergen. "Mimicry for All Modalities." In *Ecology Letters*, Vol. 19, No. 6, 2016

Darcque, Pascal and René Treuil. "Un 'bucrane' néolithique à Dikili Tash (Macédoine orientale): parallèles et perspectives d'interprétation." In: *Bulletin de correspondance hellénique*. Vol. 122, livraison 1, 1998

Darwin, Charles. *On the Origin of Species*. Oxford: Oxford University Press, 1996

Darwin, Francis and A.C. Seward (Eds.). *More Letters of Charles Darwin: A Record of His Work in a Series of Hitherto Unpublished Letters*, 2 vols. London: John Murray, 1903

Davidson, Donald. "Rational Animals." In *Dialectica*, Vol. 36, No. 4, 1982

Dawkins, Richard. *The Selfish Gene*. Oxford: Oxford University Press, 1976

Deacon, Terrence W. *The Symbolic Species: The Co-Evolution of Language and the Brain*. New York/London: W. W. Norton, 1997

Deane, C. Douglas. "The Broken-Wing Behavior of the Killdeer." *The Auk*, Vol., 61, 1944

Deleuze, Gilles. *Empirisme et subjectivité*. Paris: PUF, 1953

_____. *Proust et les Signes*, Paris: PUF, 1964, Reprint Quadrige, 1998

_____. *Différence et répétition*, Paris: PUF, 1968. Trans. Paul Patton. New York: Columbia University Press, 1994

_____. *Cinéma 1 L'image-mouvement*. Paris: Minuit, 1983. Trans. Hugh Tomlinson and Barbara Habberjam. Minneapolis: University of Minnesota Press, 1986

_____. *Cinéma 2 L'image-temps*. Paris: Minuit, 1985

_____. *Deux régimes de fou. Textes et entretiens, 1975-1995*. Paris: Minuit, 2003. Trans. Ames Hodges and Mike Taormina, New York: Semiotext(e), 2007

Deleuze, Gilles and Félix Guattari. *Kafka. Pour une littérature mineure*. Paris: Minuit, 1975. Trans. Dana Polan. *Kafka. Toward a Minor Literature*. Minneapolis: University of Minnesota Press, 1986

_____. *Mille plateaux*. Paris: Minuit, 1980. Trans. B. Massumi. *A Thousand Plateaus*. London: Athlone, 1988

_____. *Qu'est-ce que la philosophie?* Paris: Éditions de Minuit, Paris, 1991. Trans. H. Tomlinson and G. Burchell. New York: Columbia University Press, 1994

Deleuze, Gilles and Claire Parnet. *Dialogues*. Paris: Flammarion, 1977. Trans. Hugh Tomlinson and Barbara Habberjam. New York: Columbia University Press, 1987

Dennett, Daniel. "Do Animals Have Beliefs?" In Herbert Roitblat (Ed.), *Comparative Approaches to Cognitive Sciences*, London: MIT Press, 1995

Derrida, Jacques. *De la grammatologie*, Paris: Les Editions de Minuit, 1967, *Of Grammatology*. Trans. Gayatri Spivak, Baltimore: Johns Hopkins University Press, 1976

_____. "Che cos'è la poesia?" in *Poesia*, I, 11, Nov 1988

_____. *Spectres de Marx*. Paris: Galilée, 1993

_____. *L'animal que donc je suis*. Paris: Galilée, 2006. Trans. David Wills. *The Animal That Therefore I Am*. New York: Fordham University Press, 2008

_____. *La bête et le souverain*. Paris: Galilée, 2 volumes, 2008–2009

Depraz, Natalie, Francisco J. Varela, and Pierre Vermersch (Eds.). *On Becoming Aware. A Pragmatics of Experiencing*. Amsterdam: J. Benjamins, 2003

Despret, Vinciane. "The Body We Care For: Figures of Anthropo-zoo-genesis," In M. Akrich et M. Berg, (Eds.) *Body and Society*. Special Issue on "Bodies on Trial" Vol. 10, No. 2–3, 2004

Devor, Marshall and Patrick D. Wall, "Plasticity in the Spinal Cord Sensory Map Following Peripheral Nerve Injury in Rats." In *The Journal of Neuroscience*, Vol. 1, No. 7, 1981

De Waal, Frans. *Chimpanzee Politics: Power and Sex among Apes*. London: Jonathan Cape Ltd., 1982

_____. *Primates and Philosophers. How Morality Evolved*. Princeton: Princeton University Press, 2006

_____. *Are We Smart Enough to Know How Smart Animals Are?* London: Granta Books, 2016

Diaconu, Madalina and Christian Ciocan. *Phenomenology of Animality. Studia Phaenomenologica 17*. Bucarest: Zeta books, 2017

Dias, Brian G. and Kerry Ressler. "Parental Olfactory Experience Influences Behavior and Neural Structure in Subsequent Generations." In *Nature Neuroscience*, Vol. 17, No. 1, December 2013

Driscoll, Kári and Eva Hoffmann (Eds.). *What is Zoopoetics? Texts, Bodies, Entanglement*. Cham: Palgrave MacMillan, 2018

Driver, Peter M. and David Andrew Humphries. *Protean Behaviour: The Biology of Unpredictability*. Oxford: Clarendon Press, 1988

Dufourcq, Annabelle. *La dimension imaginaire du réel dans la philosophie de Husserl*. Dordrecht: Springer, Phaenomenologica, 2010

_____. *Merleau-Ponty: une ontologie de l'imaginaire*. Dordrecht: Springer, Phaenomenologica, 2012

_____. "Is a World without Animals Possible?", Trans. Ramon Fonkoué. In *Environmental Philosophy*, Vol. 11, No. 1, 2014

_____. "De la chair à la révolte: l'activité passive dans *L'idiot de la famille*," In *Horizon. Studies in Phenomenology*, Vol. 3, No. 2, 2014

_____. "Sous les masques il n'y a pas de visages: l'éthique merleau-pontyenne entre problème de l'altérité radicale, foi et institution" in *Chiasmi International*, Vol. 17, 2015

_____. "The Fundamental Imaginary Dimension of the Real in Merleau-Ponty's Philosophy," in *Research in Phenomenology*, Brill, Vol. 45, No. 1, 2015

_____. "Vies et morts de l'imagination: La puissance des actes fantômes". In *Bulletin D'Analyse Phénoménologique*, Vol. 13, No. 2, 2017

_____. "Who/What is Bête? From an Uncanny Word to an Interanimal Ethics", in *Environmental Philosophy*, Vol. 16, No. 1, 2019

Eason, Cassandra. *Fabulous Creature, Mythical Monsters, and Animal Power Symbol. A Handbook*. Wesport: Greenwood Press, 2008

El-Hani, Charbel Niño. "Between the Cross and the Sword: The Crisis of the Gene Concept." In *Genetics and Molecular Biology*, Vol. 30, No. 2, 2007

Empson, William. "Alice in Wonderland – The Child as Swain." In Robert Phillips (Ed.), *Aspects of Alice: Lewis Carroll's Dreamchild As Seen Through the Critics' Looking Glasses*, New York: The Vanguard Press, 1971

Endicott, Kirk. *Batek Negrito Religion. The World-View and Rituals of a Hunting and Gathering People of Peninsular Malaysia.* Oxford: Clarendon Press, 1979

Evreinov, Nikolai. *The Theatre in Life.* Trans. Alexander Nazaroff, New York: Brentano's, 1927

Feddersen-Petersen, Dorit Urd. "Vocalization of European Wolves (Canis lupus lupus L.) and Various Dog Breeds (Canis lupus f. fam.)." In *Archives Animal Breeding, Archiv Tierzucht*, 43, 2000

Fedigan, Linda. "Social and Solitary Play in a Colony of Vervet Monkeys (Cercopithecus aethiops)." In *Primates*, 13(4), 1972

Feit, Harvey. "The Ethnoecology of the Waswanipi Cree: Or How Hunters Can Manage Their Resources." In *Cultural Ecology: Readings on the Canadian Indians and Eskimos*, Cox, B. (Ed.). Toronto: McClelland and Stewart, 1973

Ferrari, Pier F., Elisabetta Visalberghi, Annika Paukner, L. Fogassi, and A. Ruggiero. "Neonatal Imitation in Rhesus Macaques." *PLoS Biology,* Vol. 4, No. 9, 2006

Fischer, Alice. "Why the Unicorn has Become the Emblem for Our Times," *The Guardian*, Sun 15 Oct 2017

Fontenay, Élisabeth de. *Le Silence des bêtes*, Paris: Fayard, 1999

Fouts, Roger S. "Communication with Chimpanzees." In *Hominisation and Behavior*, G. Kurth and I. Eibl-Eibesfeldt (Eds.), (137–58), Stuttgart: Gustav Fischer, 1975

Freeman, Margaret B. *The Unicorn Tapestries*, New York: The Metropolitan Museum of art, 1976

Freud, Sigmund, *The Interpretation of Dream.* In *The Complete Psychological Works*, New York: Norton & Company, 1976

———. *Fragment of an Analysis of a Case of Hysteria,* In *The Complete Psychological Works*, New York: Norton & Company, 1976

Frisch, Stefan. "How Cognitive Neuroscience could be More Biological—and What it might Learn from Clinical Neuropsychology." In *Frontiers in Human Neuroscience*, Vol. 8, 2014

Gambutas, Marija. *The Living Goddess.* Berkeley: University of California Press, 1999

Gibson, Gabriella. "Do Tsetse Flies 'See' Zebras? A Field Study of the Visual Response of Tsetse to Striped Targets." In *Physiological Entomology*, Vol. 17, No. 2 (141–7), June 1992

Glăveanu, Vlad Petre. "Commentary on Chapter 4: Proto-c Creativity?", In Kaufman, Allison B. and Kaufman, James C. (Eds). *Animal Creativity and Innovation.* Elsevier, 2015

Goodall, Jane. *The Chimpanzees of Gombe: Patterns of Behavior.* Cambridge: Harvard University Press, 1986

Gould, Stephen Jay and Richard C. Lewontin. "The Spandrels of San Marco and the Panglossian Paradigm: A Critique of the Adaptationist Programme." In *Proceedings of the Royal Society of London, Series B*, Vol. 205, No. 1161, 1979

Grene, Marjorie. "Beyond Darwinism: Portmann's Thought," in *Commentary XL*, 1965

Griffin, Donald. *Animal Thinking.* Cambridge: Harvard University Press, 1984

Gross, Aaron and Vallely, Anne (Eds). *Animals and the Human Imagination. A Companion to Animal Studies.* New York: Columbia University Press, 2012

Gudeman, Stephen. *Economics as Culture: Models and Metaphors of Livelihood*. London: Routledge & Kegan Paul, 1986

Guilaine, Jean. "La "Révolution des symboles" de Jacques Cauvin revisitée". In *Paléorient*, Vol. 37, No. 1 (177–85), 2011

Guillaume, Paul. *La Formation des habitudes*, Paris: Alcan, 1936

Haldane, John Burdon Sanderson. "Rituel humain et communication animale." In *Diogene* 4, 1953

Hallowell, A. Irving. "The Role of Dreams in Ojibwa Culture." In *Contributions to Anthropology: Selected Papers of A. Irving Hallowell*, R.D. Fogelson, F. Eggan, M.E. Spiro, G.W. Stocking, A.F.C. Wallace and W.E. Washburn (Eds.). Chicago: University of Chicago Press, 1976

Haraway, Donna. "A Cyborg Manifesto. Science, Technology, and Socialist-Feminism in the Late Twentieth Century." In *Simians, Cyborgs and Women: The Reinvention of Nature*. New York: Routledge, 1991

‒‒‒‒‒‒‒. "Situated Knowledges: The Science Question in Feminism and the Privilege of Partial Perspective." In *Feminist Studies*, Vol. 14, No. 3, 1988

‒‒‒‒‒‒‒. Modest_Witness@Second_Millennium. FemaleMan_Meets_OncoMouse: Feminism and Techno-Science. New York: Routledge, 1997

‒‒‒‒‒‒‒. *The Companion Species Manifesto: Dogs, People, and Significant Otherness*, Chicago: Prickly Paradigm Press, 2003

‒‒‒‒‒‒‒. "The Promises of Monsters: A Regenerative Politics for Inappropriate/d Others," in *The Haraway Reader*, London, New York: Routledge, 2004

‒‒‒‒‒‒‒. *When Species meet*. Minneapolis, London: University of Minnesota Press, 2008

Harman, Graham. *Guerrilla Metaphysics: Phenomenology and the Carpentry of Things*. Chicago: Open Court, 2005

Hearne, Vicki. *Adam's Task: Calling Animals by Name*, New York: Knopf, 1987

Heidegger, Martin. *Die Grundbegriffe der Metaphysik. Welt—Endlichkeit—Einsamkeit*. Frankfurt: Klostermann, 1983. Trans. William McNeill and Nicholas Walker. *The Fundamental Concepts of Metaphysics: World, Finitude, Solitude*. Bloomington: Indiana University Press, 1995.

Hekman, Jessica Perry. "The Epigenetics of Fear," 2013, in the Blog "Dog Zombie", http://dogzombie.blogspot.com/2013/12/the-epigenetics-of-fear.html

Hess, Elizabeth. *Nim Chimpsky: The Chimp Who Would Be Human*, New York: Bantam Books, 2009

Heyes, Cecilia. M. and Galef, Bennett. G., Jr. (Eds.). *Social Learning in Animals: The Roots of Culture*. San Diego: Academic Press, 1996

Hofmannsthal, Hugo von. "Ein Brief," in *Der Tag*. Berlin, Nr. 489, Oct 18 1902 (Teil 1); Nr. 491, Oct 19 1902 (Teil 2). *Letter of Lord Chandos*, in *The Lord Chandos Letter and Other Writings*, trans. J. Rotenberg. New York: New York Review Books, 2005

Hoffmeyer, Jesper. "Biosemiotics: Towards a New Synthesis in Biology." In *European Journal for Semiotic Studies*, Vol. 9, No. 2 (355–76), 1997

‒‒‒‒‒‒‒. "Semiosis and Biohistory: A Reply. Semiotics in the Biosphere: Reviews and a Rejoinder." In *Semiotica* (Special Issue) Vol. 120, No. 3/4, 1998.

‒‒‒‒‒‒‒. *Biosemiotics: An Examination into the Signs of Life and the Life of Signs*. Scranton: University of Scranton Press, 2008.

‒‒‒‒‒‒‒. "The Semiome: From Genetic to Semiotic Scaffolding." In *Semiotica*, Vol. 198, 2014

Hoffmeyer, J. and C. Emmeche. "Code-Duality and the Semiotics of Nature." In M. Anderson and F. Merrell (Eds). *On Semiotic Modelling*, Berlin & New York: Mouton & De Gruyter, 1991

Holmes, Nicholas P. and Charles Spence. "The Body Schema and the Multisensory Representation(s) of Peripersonal Space." In *Cognitive Processing*, Jun; Vol. 5, No. 2, 2004

Homans, Margaret. *Bearing the Word. Language and Female Experience in Nineteenth-Century Women's Writing*, Chicago: University of Chicago Press, 1986

Howse, Philip. "Lepidopteran Wing Patterns and the Evolution of Satyric Mimicry." In *Biological Journal of the Linnean Society*, Vol. 109, No. 1, 2013

Howse, Philip E., Allen, J.A. "Satyric Mimicry: The Evolution of Apparent Imperfection." *Proceedings of the Royal Society of London Series B: Biological Sciences*, Vol. 257, No. 1349, 1994.

Huber, L., Range, F., Voelkl, B., Szucsich, A., Viranyi, Z. and Miklosi, A. "The Evolution of Imitation: What do the Capacities of Non-Human Animals Tell us about the Mechanisms of Imitation?" In *Philosophical Transactions of the Royal Society of London. B. Biological Sciences*, Vol. 364, 2009

Huffman, Michael A. "Acquisition of Innovative Cultural Behaviors in Nonhuman Primates: A Case Study of Stone Handling, a Socially Transmitted Behavior in Japanese Macaques," in *Social Learning in Animals. The Roots of Culture* (Heyes and Galef Eds.). San Diego: Academic Press, 1996.

Humphrey, George. *The Nature of Learning: In Its Relation to the Living System*, London: Kegan Paul, Trench, Trubner & Co, 1933

Husserl, Edmund. *Logische Untersuchungen, Zweiter Teil, Untersuchungen zur Phänomenologie und Theorie der Erkenntnis*, Tübingen: Max Niemeyer, 1901. Trans. J.N. Findlay. *Logical Investigations*. Volume I. New York: Routledge, 2001

_____. "Letter to Hofmannsthal." January 12 1907, in *Husserliana. Briefwechsel*, Vol. 3.7, *Wissenschaftlerkorrespondenz*. Dordrecht: Kluwer, 1994

_____. *Ideen zu einer reinen Phänomenologie und phänomenologischen Philosophie, Erstes Buch: Allgemeine Einführung in die reine Phänomenologie*. The Hague: M. Nijhoff, 1950. Trans. F. Kersten. *Ideas pertaining to a pure phenomenology and to a phenomenological philosophy. First Book. General Introduction to a Pure Phenomenology*, The Hague, Nijhoff, 1982

_____. *Ideen zu einer reinen Phänomenologie und phänomenologischen Philosophie, Zweites Buch: Phänomenologische Untersuchungen zur Konstitution (Husserliana IV) (1912-1917)*. The Hague: M. Nijhoff, 1952. Trans. Rojcewicz, Richard and Schuwer André. *Ideas pertaining to a pure phenomenology and to a phenomenological philosophy. Second Book. Studies in the Phenomenology of Constitution*. Dordrecht: Kluwer, 1993

_____. *Cartesianische Meditationen und Pariser Vorträge (1929-1931)*. The Hague: M. Nijhoff, 1950. Trans. D. Cairns, Dordrecht:Springer, 1960

_____. *Die Krisis der europäischen Wissenschaften und die transzendentale Phänomenologie. Eine Einleitung in die phänomenologische Philosophie*. The Hague: M. Nijhoff, 1954. Trans. David Carr. *The Crisis of the European Sciences and Transcendental Phenomenology*. Evanston: Northwestern, 1970

_____. "Addendum XXIII of *The Crisis of European Sciences and Transcendental Phenomenology*: Edmund Husserl". Trans. Niall Keane, In *Journal of the British Society for Phenomenology*, Vol. 44, No. 1, January, 2013

_____. *Erste Philosophie (1923-1924), Zweiter Teil: Theorie der phänomenologis-chen Reduktion*. The Hague: M. Nijhoff, 1959

_____. *Zur Phänomenologie der Intersubjektivität. Texte aus dem Nachlaß, Zweiter Teil: 1921-1928*. The Hague: M. Nijoff, 1973

_____. *Phantasie, Bildbewußtsein, Erinnerung. Zur Phänomenologie der anschau-lichen Vergegenwärtigungen, Texte aus dem Nachlaß (1898-1925)*, Den Haag: M. Nijhoff, 1980. Trans. J.B. Brough. *Phantasy, Image Consciousness, and Memory*. Dordrecht: Springer, 2005

_____. *Husserliana. Briefwechsel*, Vol. 3.7, *Wissenschaftlerkorrespondenz*, Dordrecht: Kluwer, 1994

Huxley, Julian. "A Discussion on Ritualization of Behaviour in Animals and Man." In *Philosophical Transactions of the Royal Society of London, Series B, Biological Sciences*, Vol. 251, No. 772, 1966

Ingold, Tim (Ed.). *Key Debates in Anthropology*, London & New York: Routledge, 1996

_____. *The Perception of the Environment: Essays in Livelihood, Dwelling, and Skill*. London and New York: Routledge, 2000

Jackson, Michael. "Thinking Through the Body: An Essay on Understanding Metaphor." In *Social Analysis*, Vol. 14, 1983

Jonas, Hans. *The Phenomenon of Life. Toward a Philosophical Biology*. New York: A Delta Book, 1966

Jung, Carl Gustav. *The Archetypes and the Collective Unconscious*, Trans. R.F.C. Hull, Princeton: Princeton Univ. Press, 1968

Kabadayi, Can and Mathias Osvath,. "Ravens Parallel Great Apes in Flexible Planning for Tool-Use and Bartering." In *Science*, Vol. 357, No. 6347, 2017

Kaminski, Juliane, Sebastian Tempelmann, Josep Call, and Michael Tomasello. "Domestic Dogs Comprehend Human Communication with Iconic Signs." In *Developmental Science*, Vol. 12, No. 6, 2009

Kaminski, Juliane, Jennifer Hynds, Paul Morris, and Bridget M. Waller. "Human Attention Affects Facial Expressions in Domestic Dogs," in *Scientific Reports*, Vol. 7, No. 1, 2017

Kant, Immanuel. *Kritik der reinen Vernunft*, in *Gesammelte Schriften*, herausgegeben von der Königlich Preussischen Akademie der Wissenschaften, vols. 3 and 4 (Berlin, 1903/04), translated by P. Guyer and W. W. Wood, Vol. 3, Cambridge: Cambridge University Press, 1998

Kaplan, Gisela. *Bird Minds. Cognition and Behaviour of Australian Native Birds*. Melbourne: CSIRO Publishing, 2016

Katz, Joel, Anthony L. Vaccarlno, Terence J. Coderre, and Ronald Melzack, "Injury Prior to Neurectomy Alters the Pattern of Autotomy in Rats. Behavioral Evidence of Central Neural Plasticity." In *Anesthesiology*, Vol. 75, No. 5, 1991

Kaufman, Allison B. and James C. Kaufman (Eds). *Animal Creativity and Innovation*. London: Elsevier, 2015

Kaufman Matthew, Mark M Churchland, Stephen I Ryu, and Krishna V Shenoy, "Vacillation, Indecision and Hesitation in Moment-By-Moment Decoding of Monkey Motor Cortex." In *eLifeSciences*, May 5, 2015

Kendrick, Keith, Ana P. da Costa, Andrea E. Leigh, Michael R. Hinton, and Jon W. Peirce. "Sheep Don't Forget a Face," In *Nature*, Vol. 414 (165–6), 2001

Kennedy, George A. "A Hoot in the Dark: The Evolution of General Rhetoric." In *Philosophy and Rhetoric*, Vol. 25, No. 1, 1992

Koffka, Kurt. *Principles of Gestalt Psychology*. New York: Routledge, 1935

Köhler, Wolfgang. *Die physischen Gestalten in Ruhe und im stationären Zustand. Eine naturphilosophische Untersuchung*, Braunschweig: Friedrich Vieweg und Sohn, 1920
_____. *Intelligenzprüfungen an Menschenaffen*, 2nd ed., Berlin: Springer, 1921. Trans. E. Winters. *The Mentality of Apes*. London: Routledge & Kegan Paul, 1925

Khvatov, Ivan A., Alexey Sokolov, Alexandr Kharitonov, and Kseniya Kulichenkova. "Body Scheme in Rats Rattus norvegicus." In *Experimental Psychology*, Vol. 9, No. 1, 2016

Kistner, D.H. "Revision of the African species of the Termitophilous tribe Corotocini (Coeloptera: Stapylinidae). I. A New Genus and Species from Ovamboland and its Zoogeographic Significance." In *Journal of the New York Entomological Society*, Vol. 76, 1968

Kjernsmo, Karin. *Anti-predator Adaptations in Aquatic Environments*, Turku: Åbo Akademi University Press, 2014

Kleisner, Karel. "The Semantic Morphology of Adolf Portmann: A Starting Point for the Biosemiotics of Organic Form?" In *Biosemiotics*, Vol. 1 (207–19), 2008

Kleisner, Kleisner Karel, and Anton Markoš. "Semetic Rings: Towards the New Concept of Mimetic Resemblances," In *Theory in Biosciences*, Vol. 123 (209–22), 2005

Knolle, Franziska, Rita P. Goncalves, and A. Jennifer Morton. "Sheep Recognize Familiar and Unfamiliar Human Faces from Two-Dimensional Images." In *Royal Society Open Science*, Vol. 4, No. 11, 8 Nov 2017

Kornell, Nate. "Where is the "Meta" in Animal Metacognition?" In *Journal of Comparative Psychology*, Vol. 128 (143–9), 2014

Kull, Kalevi. "Ecosystems are Made of Semiosic Bonds: Consortia, Umwelten, Biophony and Ecological Codes." In *Biosemiotics*, Vol. 3, No. 3 (347–57), 2010

Lacan, Jacques. "Fonction et champ de la parole et du langage en psychanalyse", in *Ecrits*, Paris: Seuil, 1966
_____. *Le Séminaire. Les écrits techniques de Freud* (1953-1954), tome 1, Paris: Seuil, 1975. Trans. J. Forrester, New York: W.W. Norton, 1988
_____. *Le séminaire, Livre II, Le moi dans la théorie de Freud et dans la technique de la psychanalyse*, Paris: Seuil, 1977. Trans. J. Forrester, Cambridge: Cambridge University Press, 1988

Langerholc, John. "Facial Mimicry in the Animal Kingdom." In *Bolletino di zoologia*, 58.3 (185–204), 1991

Larison, Brenda, Ryan J. Harrigan, Henri A. Thomassen, Daniel I. Rubenstein, Alec M. Chan-Golston, Elizabeth Li, and Thomas B. Smith. "How the Zebra got its Stripes: A Problem with Too Many Solutions." In *Royal Society Open Science*, Vol. 2, No. 1, January 2015

Latour, Bruno. *Nous n'avons jamais été modernes*. Paris: La Découverte, 1991. Trans. by Catherine Porter, Harvard University Press, 1993
_____. *Politiques de la nature. Comment faire entrer les sciences en démocratie*. Paris: La Découverte, "Armillaire", 1999. Trans. Catherine Porter. Cambridge: Harvard University Press, 2004

Langmore, Naomi E., Golo Maurer Greg J. Adcock, and Rebecca M. Kilner. "Socially Acquired Host-Specific Mimicry and the Evolution of Host Races in Horsfield's Bronze-Cuckoo Chalcites basalis." In *Evolution*, Vol. 62 (1689–99), 2008

Lanzing, W.J.R. and Bower, C.C., "Development of Colour Patterns in Relation to Behaviour in Tilapia mossambica (Peters)." In *Journal of Fish Biology*, Vol. 6, No. 1 (29–41), 1974

Lestel, Dominique, Jeffrey Bussolini, and Matthew Chrulew,. "The Phenomenology of Animal Life." In *Environmental Humanities*, Vol. 5 (125–48), 2014

Levinas, Emmanuel. *La théorie de l'intuition dans la phénoménologie de Husserl*. Paris: Vrin, 1970

Levins, Richard and Lewontin, Richard. *The Dialectical Biologist*. Cambridge: Harvard University Press, 1985.

Lévi-Strauss, Claude. *La pensée sauvage*, Paris: Plon, 1962

Lewontin, Richard. *The Triple Helix*, Cambridge: Harvard University Press, 2000.

Li, Jianguo, Minfan Wu, Min Zhuo, and Zao C Xu "Alteration of Neuronal Activity After Digit Amputation in Rat Anterior Cingulate Cortex." In *International Journal of Physiology, Pathophysiology and Pharmacology*, Vol. 5, No. 1, (43–51), 2013

Lieberman, Philip. *The Biology and Evolution of Language*, Cambridge: Harvard University Press. 1984

Liu, Rebekah Yi. *The Background and Meaning of the Image of the Beast in Rev 13:14, 15*. PhD diss., Andrews University, 2016

Lippit, Akira. *Electric Animal: Toward a Rhetoric of Wildlife*. Minneapolis: University of Minnesota Press, 2000

Lipps, Theodor. "Einfühlung, innere Nachahmung und Organempfindungen," *Archiv für die gesamte Psychologie*, Vol. I, 1903

Lorenz, Konrad. "Der Kumpan in der Umwelt des Vogels. Der Artgenosse als auslösendes Moment sozialer Verhaltungsweisen," Journal für Ornithologie. Beiblatt, Vol. 83, 1935

Lovell-Smith, Rose. "The Animals of Wonderland: Tenniel as Carroll's Reader". In *Criticism*, 45.4 (383–415), 2003

Lyytinen, Anne, Paul M. Brakefield, and Johanna Mappes. "Significance of Butterfly Eyespots as an Anti-Predator Device in Ground-Based and Aerial Attacks," in *Oikos*, Vol. 100 (373–9), 2003

Maran, Timo. "Are Ecological Codes Archetypal Structures?" In T. Maran, K. Lindström, R. Magnus, and M. Tønnessen(Eds.), *Semiotics in the Wild: Essays in Honour of Kalevi Kull on the Occasion of his 60th Birthday*. Tartu: University of Tartu Press, 2012

————. *Mimicry and Meaning: Structure and Semiotics of Biological Mimicry*, Dordrecht: Springer, 2017

Maravita, Angelo and Atsushi Iriki. "Tools for the Body (Schema)." In *Trends in Cognitive Sciences* Vol. 8, No. 2 (79–86), 2004

Margulis, Lynn and Dorion Sagan. *Acquiring Genomes: A Theory of the Origins of Species*, New York, Basic Books, 2002

Masuda, Kazuo. "La dette symbolique de la Phénoménologie de la perception", in François Heidsieck (Ed.), *Merleau-Ponty. Le philosophe et son langage*. Paris: Vrin, 1993

Matsumiya, Lynn C., Robert E. Sorge, Susana G. Sotocinal, John M. Tabaka, Jeffrey S. Wieskopf, Austin Zaloum, Oliver D. King, and Jeffrey S. Mogil. "Using the Mouse Grimace Scale to Reevaluate the Efficacy of Postoperative Analgesics in Laboratory Mice." In *Journal of the American Association for Laboratory Animal Science*, Vol. 51, No. 1, (42–9), 2012

Matyjasiak, Piotr. "Birds Associate Species-Specific Acoustic and Visual Cues: Recognition of Heterospecific Rivals by Male Blackcaps." In *Behavioral Ecology*, Vol. 16, No. 2, 1 (467–71), March 2005

May-Collado, Laura J. "Changes in Whistle Structure of Two Dolphin Species During Interspecific Associations", in *Ethology*, Vol. 116, No. 11, 1065–74, November 2010

McFarland, David. *Animal Behaviour* (Third Edition), Edinburgh: Pearson Education, 1999

Melville, Herman. *Moby-Dick; or, The Whale*. New York: Harper & Brothers publishers, 1851

Merleau-Ponty, Maurice. *La structure du comportement*. Paris: PUF. 1942. Reprint Quadrige 1990. Trans. Alden Fisher. *The Structure of Behavior*. Boston: Beacon Press, 1960

————. *Phénoménologie de la perception*. Paris: Gallimard. 1945. Trans. D. Landes. *Phenomenology of Perception*. New York: Routledge, 2012

————. *Sens et non-sens*, Paris: Nagel, 1948. Reed. Paris: Gallimard, 1996. Trans. Hubert L. Dreyfus and Patricia Allen Dreyfus, *Sense and Nonsense.*, Evanston: Northwestern University Press, 1964

————. *Signes*, Paris: Gallimard, 1960. Trans. Richard McCleary, Evanston: Northwestern University Press, 1964

————. *L'Œil et l'Esprit*. Paris: Gallimard. 1961. Trans. Michael B. Smith. *Eye and Mind*. In *The Merleau-Ponty Aesthetics Reader*, Galen A. Johnson, Michael B. Smith (Eds.). Evanston: Northwestern University Press, 1993

————. *Le visible et l'invisible*. Paris: Gallimard, Tel, 1964. Trans. A. Lingis, Evanston: Northwestern University Press, 1968

————. *La Prose du monde*, Paris, Gallimard, 1969

————. *La Nature. Notes de cours au Collège de France* (1957–1960). Paris: Seuil (collection Traces écrites), 1995. Trans. Robert Vallier. Evanston: Northwestern University Press, 2003

————. "La philosophie aujourd'hui" In *Notes de Cours 1959-1961*, Paris: Gallimard, 1996

————. *Notes de cours sur L'origine de la géométrie de Husserl*, Paris, PUF, 1998

————. *Psychologie et pédagogie de l'enfant, Cours de Sorbonne 1949-1952*. Lagrasse: Verdier, 2001

————. *Causeries 1948*. Paris: Seuil, "Traces écrites," 2002. Trans. O. Davis, *The World of Perception*. New York: Routledge, 2004

————. *L'institution. La passivité. Notes de cours au Collège de France* (1954–1955). Paris: Belin, 2003. Institution and Passivity. Trans. L. Lawlor and H. Massey. Evanston: Northwestern University Press, 2010

————. "Preface to *Signs*" in *The Merleau-Ponty Reader*, Toadvine and Lawlor (Eds.), Evanston: Northwestern University Press, 2007

Milgram, Stanley. "Behavioral Study of Obedience". *Journal of Abnormal and Social Psychology*. Vol. 67, No. 4, 1963

Miller, Amy L. and Matthew C. Leach,. "The Mouse Grimace Scale: A Clinically Useful Tool?", in PLOS One. Vol. 10, No. 9, 2015

Mitchell, William J. Thomas. *Iconology: Image, Text, Ideology*. Chicago: University of Chicago Press, 1987

————. "The Pictorial Turn," *Picture Theory: Essays on Verbal and Visual Representation*. Chicago: University of Chicago Press, 1994

————. *What Do Pictures Want? The Lives and Loves of Images*. Chicago: University of Chicago Press, 2005

Moe, Aaron. *Zoopoetics: Animals and the Making of Poetry*. Lanham: Lexington Books, 2014

Molnár, Csaba, Frédéric Kaplan, Pierre Roy, François Pachet, Péter Pongrácz, Antal Dóka, and Ádám Miklósi. "Classification of Dog Barks: A Machine Learning Approach." In *Animal Cognition*, Vol. 11, 2008

Montaigne, Michel de. "Of the Force of Imagination." In *Essays*. Trans. Charles Cotton. Project Gutenberg ebook, 2006

Morin, Olivier. *How Traditions Live and Die*. Oxford: Oxford University Press, 2016

Moore, Bruce. R. "Avian Movement Imitation and a New Form of Mimicry: Tracing the Evolution of a Complex Form of Learning." *Behaviour*, 122, 1992

Moschus, "Europa," in *The Greek Bucolic Poets*, Trans. A.S.F. Gow, Cambridge: Cambridge University Press, 1953

Mullally, Sinéad L. and Eleanor A. Maguire, "Memory, Imagination, and Predicting the Future: A Common Brain Mechanism?" In *The Neuroscientist*, Vol. 20, No. 3, 2013

Nagel, Thomas. "What Is It Like to Be a Bat?". *The Philosophical Review*, Vol. 83, No. 4, 1974

Nadasdy, Paul. "The Gift in the Animal: The Ontology of Hunting and Human-Animal Sociality." In *American Ethnologist*, Vol. 34, No. 1, (25–43), February 2007

Nelson, Ximena J., Daniel T. Garnett, and Christopher S. Evans, "Receiver Psychology and the Design of the Deceptive Caudal Luring Signal of the Death Adder Ximena." In *Animal Behaviour*, Vol. 79, No. 3, (555–61), 2010

Newport, Cait, Guy Wallis, Yarema Reshitnyk, and Ulrike E. Siebeck. "Discrimination of Human Faces by Archerfish (Toxotes chatareus)." In *Scientific Reports*. Vol. 6, No. 27523, 2016

Nigel J.T. Thomas. "Are Theories of Imagery Theories of Imagination? An Active Perception Approach to Conscious Mental Content." In *Cognitive Science*, Vol. 23, 1999

Nimmo, Richie. "Animal Cultures, Subjectivity, and Knowledge: Symmetrical Reflections beyond the Great Divide." In *Society & Animals*, Vol. 20, No. 2 (173–92), 2012

Norman, Mark D., Finn, J., and Tregenza, T. "Dynamic Mimicry in an Indo-Malayan Octopus." In *Proceedings of the Royal Society of London B*, Vol. 268, No. 1478, (1755–8), 2001

Norris, Margot. *Beasts of the Modern Imagination. Darwin, Nietzsche, Kafka, Ernst, & Lawrence*. Baltimore and London: The Johns Hopkins University Press, 1985

Nüsslein-Volhard, Christiane. *Coming to Life: How Genes Drive Development*. Carlsbad: Kales Press, 2006

Oliveira, Paulo S. "Ant-Mimicry in Some Brazilian Salticid and Clubionid Spiders (Araneae: Salticidae, Clubionidae)," in *Biological Journal of the Linnean Society*, Vol. 33, (1–15), 1988

Oliver, Kelly. "Stopping the Anthropological Machine: Agamben with Heidegger and Merleau-Ponty," *PhaenEx*, Vol. 2 (1–23), 2007

————. *Animal Lessons: How They Teach Us to Be Human*. New York: Columbia University Press, 2009

Olsen, Stanley. *Origins of The Domestic Dog: The Fossil Record*, Tucson: University of Arizona Press, 1985

Ouattara, Karim, Alban Lemasson, and Klaus Zuberbuhler. "Generating Meaning with Finite Means in Campbell's Monkeys." In *Proceedings of the National Academy of Sciences*, Vol. 106, No. 48, December 7, 2009

Oyama, Susan. *The Ontogeny of Information: Developmental Systems and Evolution*. Cambridge: Cambridge University Press, 1985

Painter, Corinne and Lotz, C. (Eds). *Phenomenology and the Non-Human Animal: At the Limits of Experience*, Dordrecht: Springer, 2007

Payne, Mark. *The Animal Part: Human and Other Animals in the Poetic Imagination*. Chicago: The University of Chicago Press, 2010

Peirce, Charles. *Collected Papers*, 8 vols. Charles Hartshorne, Paul Weiss, and Arthur Burks (Eds.). Cambridge: Belknap Press of Harvard University Press, 1931-1958

Pelkwijk, J.J. Ter and N. Tinbergen,. "Eine reizbiologische Analyse einiger Verhaltensweisen von Gasterosteus aculeatus L." In *Zeitschrift für Tierpsychologie*, Vol. 1 (193–200), 1937

Pellis, Sergio M. "Keeping in Touch: Play Fighting and Social Knowledge," in Marc Bekoff, Colin Allen, and Gordon M. Burghardt (Eds.) *The Cognitive Animal. Empirical and Theoretical Perspectives on Animal Cognition*, Cambridge: The MIT Press, 2002

Pereira, Carlos. *Parler aux Chevaux Autrement - approche sémiotique de l'équitation*, Paris, Amphora, 2009

Petermann, Bruno. *The Gestalt Theory and the Problem of Configuration*. London: Routledge and Kegan Paul, 1932, Reprint 1999

Petitot, Jean, Varela, F.J., Pachoud, B., Roy, J.-M. (Eds.). *Naturalizing Phenomenology: Issues in Contemporary Phenomenology and Cognitive Science*, Stanford CA: Stanford University Press, 1999

Pharies, David A. *Charles S. Peirce and The Linguistic Sign*. Amsterdam & Philadelphia: John Benjamins, 1985

Plato. *The Republic*. Trans. Alan Bloom. New York: HarperCollins, 1968

Pongrácz, Peter, Csaba Molnár, Ádam Miklósi, and Vilmos Csányi. "Human Listeners Are Able to Classify Dog (Canis familiaris) Barks Recorded in Different Situations." In *Journal of Comparative Psycholog*, Vol. 19, No. 2 (136–44), 2005

Portmann, Adolf. *Das Tier als Soziales Wesen*. Zürich: Rhein-Verlag, 1953. Trans. Oliver Coburn. *Animals as Social Beings*. New York and Evanston: Harper Torchbooks, 1964

———. *Tarnung im Tierreich*. Berlin-Göttingen-Heidelberg: Springer-Verlag, 1956. Trans A. J. Pomerans. *Animal Camouflage*. Ann Arbor: The University of Michigan Press, 1959

———. *Biologie und Geist*, Zürich: Rhein Verlag, 1956. Reprint Frankfurt am Main, Suhrkamp Taschenbuch, 1973

———. "Selbstdarstellung als Motiv der lebendigen Formbildung," in *Geist und Werk. Aus der Werkstatt unserer Autoren. Zum 75. Geburtstag von Dr. Daniel Brody*, Zurich: Rhein Verlag, 1958. French Translation by Jacques Dewitte, "L'autoprésentation, motif de l'élaboration des formes vivantes" in *Etudes phénoménologiques*, No. 23–24, 1996

———. *Die Tiergestalt*. Freiburg/Basel/Wien: Herder, 1965. Trans. Hella Czech. *Animal Forms and Patterns*. New York: Schocken Books, 1967

Portmann, Adolf and Richard Carter. *Essays in Philosophical Zoology by Adolf Portmann. The living Form and the Seeing Eye*, Lewiston: The Edwin Mellen Press, 1990

Preuss, Mattias. "Spinning Theory: Three Figures of Arachnopoetics." In *What is Zoopoetics? Texts, Bodies, Entanglement* (Kári Driscoll, Eva Hoffmann Eds.). Cham: Palgrave MacMillan, 2018

Pryor, Karen. *Reaching the Animal Mind: Clicker Training and What it Teaches Us about All Animals*. New York, NY: Scribner, 2009

———. "Creating Creative Animals." In J. Kaufman and A. Kaufman (Eds.), *Animal Creativity and Innovation*. London: Elsevier, 2005

Pryor, Karen, Haag, R., and O'Reilly, J. "The Creative Porpoise: Training for Novel Behavior." In *Journal of the Experimental Analysis of Behavior*, Vol. 12 (653–61), 1969

Pryor, Karen. and K.R. Ramirez. "Modern Animal Training: A Transformative Technology." In F. McSweeney, and E. Murphy (Eds.), *A Handbook of Operant and Classical Conditioning*. New York: Wiley and Blackwell, 2014

Renier, Laurent, A.G. De Volder, and J.P. Rauschecker, "Cortical Plasticity and Preserved Function in Early Blindness." In *Neuroscience & Biobehavioral Reviews*, Vol. 14, April 2014

Rees, Amanda. "Reflections on the Field: Primatology, Popular Science and the Politics of Personhood." In *Social Studies of Science*, Vol. 37, No. 6, (881–907), 2007

Rilke, Rainer Maria. *Duineser Elegien*, Insel Verlag Leipzig 1923, Translated by A.S. Kline, in *The Poetry of Rainer Maria Rilke*, Poetry in translation, CreateSpace Independent Publishing Platform

Rimbaud, Arthur. *Lettre à Izambard du 13 mai 1871*, in *Œuvres Complètes*, Paris: Gallimard, Pléiade, 2009

Roberts, William A. "Mental Time Travel: Animals Anticipate the Future," in *Current Biology*, Vol. 17, No. 11, 2007

Roper, T.J. and S.E. Cook. "Responses of Chicks to Brightly Coloured Insect Prey." *Behaviour*, Vol. 110, (276–93), 1989

Roth, Lina and Olle Lind. "The Impact of Domestication on the Chicken Optical Apparatus." In *PLOS One*, 2013

Rottman, Joshua. "Evolution, Development, and the Emergence of Disgust," in *Evolutionary Psychology*, Vol. 12, No. 2, (417–33) 2014

Rousseau, Jean-Jacques. *A Discourse on the Origin of Inequality*, Trans. G.D.H. Cole, London: Cosmo Classic, 2005

Rowland, Hannah, Johanna Mappes, Graeme D. Ruxton, and Michael P. Speed. "Mimicry Between Unequally Defended Prey can be Parasitic: Evidence for Quasi-Batesian Mimicry," in *Ecology Letters*, Vol. 13 (1494–1502), 2010

Rozin, Paul, Jonathan Haidt and Clark R. McCauley. "Disgust." In M. Lewis, J. Haviland, and L. F. Barrett (Eds.). *Handbook of Emotions*. New York: Guilford Press, 2008

Ruonakoski, Erika. "Phenomenology and the Study of Animal Behavior," in *Phenomenology and the Non-Human Animal: At the Limits of Experience*, Corinne Painter and Christian Lotz(Eds.), (75–84). Dordrecht: Springer, 2007

Sallis, John. "Hovering: Imagination and the Spacing of Truth," in *Spacings: Of Reason and Imagination in Texts of Kant, Fichte, Hegel*. Chicago: University of Chicago Press, 1987

Sartre, Jean-Paul. *The Transcendence of the Ego. An Existentialist Theory of Consciousness*. Trans. Forrest Williams and Robert Kirkpatrick. New York: Hill and Wang, 1957

————. *L'imaginaire. Psychologie phénoménologique de l'imagination*. Paris: Gallimard, 1940. Folio essai. Trans. Jonathan Webber. *The Imaginary*. London: Routledge, 2004

————. *L'idiot de la famille. Gustave Flaubert de 1821 à 1857*, Paris, Gallimard, 1971, nouvelle édition revue et complétée, Paris, Gallimard, 1988. Trans. Carol Cosman, Chicago: University of Chicago Press, 1981-1993

Savage-Rumbaugh, Sue. *Ape Language: From Conditioned Response to Symbol*. New York: Columbia University Press, 1986

————. "Communication, Symbolic Communication, and Language: Reply to Seidenberg and Petitto". In Journal of Experimental Psychology, Vol. 116, No. 3, 1987

Savage-Rumbaugh, Sue and Roger Lewin, *Kanzi: The Ape at the Brink of the Human Mind*, New York: Wiley, 1994

Savage-Rumbaugh, Sue, Duane M. Rumbaugh and Sarah Boysen. "Do Apes Use Language? One research group considers the evidence for representational ability in apes". In *American Scientist*, Vol. 68, No. 1 (49–61), January-February 1980

Savage-Rumbaugh, Sue, Jeannine Murphy, Rose A. Sevcik, Karen E. Brakke, Shelly L. Williams, Duane M. Rumbaugh, and Elizabeth Bates. "Language Comprehension in Ape and Child." In *Monographs of the Society for Research in Child Development*, Vol. 58, 1993

Savage-Rumbaugh, Sue and Kelly McDonald. "Deception and Social Manipulation in Symbol-Using Apes". In Byrne, Richard W., and Andrew Whiten (Eds.) *Machiavellian Intelligence: Social Expertise and the Evolution of Intellect in Monkeys, Apes, and Humans*. New York: Clarendon Press/Oxford University Press (224–37), 1988

Scheler, Max. *Wesen und Formen der Sympathie*. Bern: Friedrich Cohen Verlag, 1923. *The Nature of Sympathy*. Trans. Peter Heath, London: Transaction, 2008

Schelling, Friedrich Wilhelm Joseph. *Philosophical Investigations into the Essence of Human Freedom*. Trans. Jeff Love and Johannes Schmidt. New York: State University of New York Press, 2006

Schilder, Paul. *The Image and Appearance of the Human Body: Studies in the Constructive Energies of the Psyche*, Londres, K. Paul, Trench, Trubner, 1935. Reprint, London: Routlegde, 1950

Schleidt, Wolfgang M. and Michael D. Shalter, "Co-Evolution of Human and Canids: An Alternative View of Dog Domestication: Homo homini lupus?" in *Evolution and cognition*, Vol. 9, No. 1, 2003

Schlenker, Philippe, Emmanuel Chemla, Kate Arnold, Alban Lemasson, Karim Ouattara, Sumir Keenan, Claudia Stephan, Robin Ryder, and Klaus Zuberbühler,. "Monkey Semantics: Two 'Dialects' of Campbell's Monkey Alarm Calls." In *Linguistics and Philosophy*, Vol. 37, No. 6, 2014

Schmidt, Klaus. "Göbekli Tepe – the Stone Age Sanctuaries. New Results of Ongoing Excavations with a Special Focus on Sculptures and High Reliefs." In *Documenta Praehistorica*, XXXVII, 2010

Scott, Shelly R. "The Racehorse as Protagonist: Agency, Independence, and Improvisation." In *Animals and Agency. An Interdisciplinary Exploration*. Sarah E. McFarland, Ryan Hediger(Eds.). Leiden, Boston: Brill, 2009

Seeman, Alex (Ed.). *Joint Attention. New Developments in Psychology, Philosophy of Mind, and Social Science*, Ed., (43–72). Cambridge: MIT Press, 2011

Shakespeare, William. *Macbeth*. Cambridge: Cambridge University Press, 1997

Sheehan, Michael J. and Elizabeth A. Tibbetts. "Evolution of Identity Signals: Frequency-Dependent Benefits of Distinctive Phenotypes used for Individual Recognition." In *Evolution*, Vol. 63, No. 12, 2009

Shepherd, Stephen V. "Following Gaze: Gaze-Following Behavior as a Window into Social Cognition", *Frontiers in Integrative Neuroscience*, Vol. 4, 2010

Simondon, Gilbert. *L'individu et sa genèse physico-biologique*, Paris: PUF, 1964

Smuts, Barbara. "Gestural Communication in Olive Baboons and Domestic Dogs." In Marc Bekoff, Colin Allen, and Gordon M. Burghardt (Eds.) *The Cognitive Animal Empirical and Theoretical Perspectives on Animal Cognition*, Cambridge: MIT Press, 2002

Speed, Michael. P. "Muellerian Mimicry and the Psychology of Predation." *Animal Behavior*, Vol. 45, 1993

Spence, Kenneth Wartinbee. "Experimental Studies of Learning and Higher Mental Processes in Infrahuman Primates." *Psychological Bulletin*, Vol. 34, (806–50), 1937

Steinbock, Anthony J. *Home and Beyond: Generative Phenomenology After Husserl.* Evanston: Northwestern University Press, 1995

Stephan, Claudia, Anna Wilkinson, andLudwig Huber. "Have We Met Before? Pigeons Recognise Familiar Human Faces." In *Avian Biology Research*, Vol. 5, No. 2, (75–80), 2012

Stoinsky, Tara S., Joanna L. Wrate, Nicky Ure, and Andrew Whiten. "Imitative Learning by Captive Western Lowland Gorillas (Gorilla Gorilla) in a Simulated Food-Processing Task." *Journal of Comparative Psychology*, Vol. 115, 2001

Stone, Sherril M. "Human Facial Discrimination in Horses: Can They Tell Us Apart?" In *Animal Cognition*, Vol. 13, No. 1, (51–61), July 2009

Souriau, Etienne. *Le sens artistique des animaux.* Paris: Hachette, 1965

Suda, Chikako and Josep Call. "What does an Intermediate Success Rate Mean? An Analysis of a Piagetian Liquid Conservation Task in the Great Apes." In *Cognition*, Vol. 99, (53–71), 2006

Talairach-Vilmas, Laurence. *Fairy Tales, Natural History and Victorian Culture.* Basingstoke: Palgrave Macmillan, 2014

Tanner, Adrian. *Bringing Home Animals: Religious Ideology and Mode of Production of the Mistassini Cree Hunters.* New York: St Martin's Press, 1979

Tate, Andrew J., Hanno Fischer, Andrea E Leigh, and Keith M Kendrick. "Behavioural and Neurophysiological Evidence for Face Identity and Face Emotion Processing in Animals." In *Philosophical Transactions of the Royal Society*, Vol. 361, (2155–72), 2006

Tayler, C. and G. Saayman. "Imitative Behaviour by Indian Ocean Bottlenose Dolphins (Tursiops aduncus) in Captivity." *Behaviour*, Vol. 44, No. 3-4, 286–98, 1973

Tedore, Cynthia and Sönke Johnsen,. "Pheromones Exert Top-Down Effects on Visual Recognition in the Jumping Spider Lyssomanes viridis." In *Journal of Experimental Biology*, Vol. 216, (1744–56), 2013

Tejada, Ricardo. "Deleuze face à la phénoménologie." In *Papiers du Collège International de Philosophie*, Vol. 41, 1998

Thierry, Bernard. "Social Transmission, Tradition and Culture in Primates: From the Epiphenomenon to the Phenomenon." In *Techniques & Cultures*, Vol. 23, No. 24, 1994

Thomas, Joël. "Le mythe du Taureau et les racines de la tauromachie De Dyonisos au duende." In *Du taureau et de la tauromachie. Hier et aujourd'hui* (Boyer, H. Ed.), Perpignan, Presses Universitaires (coll. Etudes), 2012

Thompson, Evan. *Mind in Life. Biology, Phenomenology, and the Sciences of Mind.* Cambridge: Harvard University Press, 2007

Thorpe, William Homan. *Learning and Instinct in Animals.* London: Methuen, 1956

Toadvine, Ted. "How Not to be a Jellyfish: Human Exceptionalism and the Ontology of Reflection". In *Phenomenology and the Non-Human Animal: At the Limits of Experience* (Lotz and Painter Eds.). Dordrecht: Springer, 2007

―――. *Merleau-Ponty's Philosophy of Nature.* Evanston: Northwestern, 2009

Toadvine, Ted and Leonard Lawlor (Eds.). *The Merleau-Ponty Reader.* Evanston: Northwestern University Press, 2007

Tomasello, Michael. "Do Apes Ape?" In *Social Learning in Animals: The Roots of Culture* (Heyes, C.M. and Galef, B.G. Jr, eds), (319–46). San Diego: Academic Press, 1996

_____. "Emulation Learning and Cultural Learning." In *Behavioral and Brain Sciences*, Vol. 21, (703–4), 1998

_____. *The Cultural Origins of Human Cognition*. Cambridge: Harvard University Press, 1999

Tomasello, Michael, Ann C. Kruger, and Hilary H. Ratner. "Cultural Learning." In *Behavioral and Brain Sciences*, Vol. 16, (495–552), 1993

Tomasello, Michael, Savage-Rumbaugh, Sue and Kruger, Ann Cale. "Imitative Learning of Actions on Objects by Children, Chimpanzees, and Enculturated Chimpanzees." In Child Development, Vol. 64, No. 6, (1688–1705), 1993

Toronchuk, Judith A. and George F. R. Ellis. "Disgust: Sensory Affect or Primary Emotional System?", in *Cognition and Emotion*, Vol. 21, No. 8

Truppa, Valentina, Giovanna Spinozzi, Tommaso Stegagno, and Joël Fagot. "Picture Processing in Tufted Capuchin Monkeys (Cebus apella)." In *Behavioural Processes*, Vol. 82, (140–52), 2009

Uexküll, Jakob von. *Theoretische Biologie*. Berlin: J. Springer Verlag, 1920. Trans. D. L. Mackinnon *Theoretical Biology*. New York: Harcourt, Brace, 1926

_____. *Streiftzüge durch die Umwelten von Tieren und Menschen Ein Bilderbuch unsichtbarer Welten*. Berlin: Springer, 1934. Trans. Joseph D. O'Neil, *A Foray into the Worlds of Animals and Humans. With A Theory of Meaning*. Minneapolis, London: University of Minnesota Press, 2010

Van Leeuwen, Edwin J.C., Katherine A. Cronin, and Daniel B.M. Haun. "A Group-Specific Arbitrary Tradition in Chimpanzees (Pan Troglodytes)," in *Animal Cognition*, Vol. 17, No. 6, (1421–25), November 2014

Van Zandt Brower, Jane. "Experimental Studies of Mimicry in some North American Butterflies," In *Foundations of Animal Behavior: Classic Papers with Commentaries*, Chicago: University of Chicago Press, 1996

Varela, Francisco J. "Organism: A Meshwork of Selfless Selves," in *Organism and the Origin of Self*, Tauber, A. (Ed.). Dordrecht: Kluwer, 1991

Vatan, Florence. "L'obscur attrait des formes: Wolfgang Köhler et la catégorie de Gestalt". *Revue d'Histoire des Sciences Humaines*, Vol. 5, (95–116), 2001

Vercammen, Maurits. "Roze krokodil SPAR University eist betere arbeidsvoorwaarden". In *De Pipet*, 09-09-2017

Vialatte, Alexandre. *Bestiaire*, Paris: Arléa, 2002

Voelkl, Bernhard and Ludwig Huber. "True Imitation in Marmosets." In *Animal Behaviour*, Vol. 60, (195–202), 2000

_____. "Imitation as Faithful Copying of a Novel Technique in Marmoset Monkeys." PLOS One, Vol. 2, No. 7, 2007

Voss, Julia. *Darwin's Pictures: Views of Evolutionary Theory, 1834–1874*. New Haven and London: Yale University Press, 2010

Walsh, D.M. *Organisms, Agency, and Evolution*, Cambridge: Cambridge University Press, 2015

Wathan Jennifer, L. Proops, K. Grounds, and K. McComb. "Horses Discriminate Between Facial Expressions of Conspecifics," in *Scientific Reports*, Vol. 6, 2016

Wemelsfelder, Françoise. "The Scientific Validity of Subjective Concepts in Models of Animal Welfare." In *Applied Animal Behaviour Science*, Vol. 53, 1997

Westling, Louise. *The Logos of the Living World*. New York, Fordham University Press, 2014

———. "Merleau-Ponty and the Eco-literary imaginary." In *Handbook of Ecocriticism and Cultural Ecology* (Hubert, Zapf Ed.). Berlin: De Gruyter, 2016

———. "Deep History, Interspecies Coevolution, and the Eco-Imaginary", in M. Calarco and Dominik Ohrem (Eds.) *Exploring Animal Encounters: Philosophical, Cultural, and Historical Perspectives*, London: PalgraveMacmillan, 2018

Wheeler, Wendy. *The Whole Creature: Complexity, Biosemiotics and the Evolution of Culture*, London: Lawrence And Wishart Ltd, 2006

Whiten, A., & R. Ham. "On the Nature and Evolution of Imitation in the Animal Kingdom: Reappraisal of a Century of Research." In *Advances in the Study of Behavior*, Vol. 21, (239–83), (P.J.B. Slater, J.S. Rosenblatt, C. Beer, and M. Milinski. Eds.). New York: Academic Press, 1992

Whiten, Andrew and Custance, Deborah. "Studies of Imitation in Chimpanzees and Children." In *Social Learning in Animals: The Roots of Culture* (Heyes, C.M. and Jr. B.G., Galef Eds.). San Diego: Academic Press, 1996

Whiten, Andrew, McGuigan, Nicola, Marshall-Pescini, Sarah and Hopper, Lydia M. "Emulation, Imitation, Over-Imitation and the Scope of Culture for Child and Chimpanzee." In *Philosophical Transactions of the Royal Society B: Biological Sciences*, Vol. 364, No. 1528, 2009

Young, Iris Marion. "Throwing like a Girl." In *Human Studies*, Vol. 3, No. 1, (137–56), 1980

Young, Rosamund. *The Secret Life of Cows*, London: Faber & Faber, 2017

Zahavi, Dan. *Self and Other: Exploring Subjectivity, Empathy, and Shame*. Oxford: Oxford University Press, 2014

Zentall, Thomas R. "An Analysis of Imitative Learning in animals." In *Social Learning in Animals: The Roots of Culture* (Heyes, C.M. and Jr, B.G. Galef Eds),. San Diego: Academic Press, (221–43), 1996

———. "Animals Represent the Past and the Future." In *Evolutionary Psychology*, Vol. 11, No. 3, 2013

Zhuangzi, *Wandering on the Way. Early Taoist Tales and Parables of Chuang Tzu*, Trans. Victor H. Mair, Honolulu: University of Hawai'i Press, 1994

Index

Printed in the United States
by Baker & Taylor Publisher Services